ENERGY TECHNOLOGY INNOVATION

Energy technology innovation – improving how we produce and use energy – is critical for a transition toward sustainability. This book presents a rich set of twenty historical case studies of energy technology innovation embedded within a unifying conceptual framework. It provides insights into why some innovation efforts have been more successful than others, and it draws important policy conclusions. The case studies cover a wide range of energy technologies, ranging from energy supply to energy end use; from successes to failures; and from industrialized, emerging, and developing economies. The case studies are presented by an international group of eminent scholars under the auspices of the Global Energy Assessment (GEA), whose main volume was published in 2012 by Cambridge University Press. *Energy Technology Innovation* presents new data, new concepts, and novel analytical and policy perspectives. It will prove invaluable for researchers, policy makers, economists, industrial innovators, and entrepreneurs in the field of energy technology.

Arnulf Grubler is a world-leading scholar on the history of energy systems and on technological change and innovation policy. He is acting program leader of the Transitions to New Technologies Program at the International Institute for Applied Systems Analysis (IIASA), Laxenburg, Austria, and professor in the field of energy and technology at the School of Forestry and Environmental Studies at Yale University. He has served as lead and contributing author and review editor for the Intergovernmental Panel on Climate Change (IPCC) since 1996. He has authored or edited several books, including *Technology and Global Change* (1998, Cambridge University Press) and *Technological Change and the Environment* (with N. Nakicenovic and W. D. Nordhaus, 2002). He is also a convening lead author of three chapters in the *Global Energy Assessment* (2012, Cambridge University Press).

Charlie Wilson is a researcher with the Tyndall Centre for Climate Change Research and a lecturer in the School of Environmental Sciences at the University of East Anglia, UK. He is a scholar on innovation studies and on the history of technological change in energy systems. His current research focuses on historical and future technology diffusion dynamics, and on the adoption of energy-efficient and smart home technologies. Previously he held positions with the London School of Economics and the IIASA. He is also a lead author of two chapters in the *Global Energy Assessment* (2012, Cambridge University Press).

Energy Technology Innovation

Learning from Historical Successes and Failures

Edited by

Arnulf Grubler

International Institute for Applied Systems Analysis, Laxenburg, Austria, and School of Forestry and Environmental Studies, Yale University, United States

Charlie Wilson

Tyndall Centre for Climate Change Research, University of East Anglia, United Kingdom

CAMBRIDGE
UNIVERSITY PRESS

CAMBRIDGE
UNIVERSITY PRESS

32 Avenue of the Americas, New York, NY 10013-2473, USA

Cambridge University Press is part of the University of Cambridge.

It furthers the University's mission by disseminating knowledge in the pursuit of education, learning, and research at the highest international levels of excellence.

www.cambridge.org
Information on this title: www.cambridge.org/9781107023222

First published 2014

Printed in the United States of America

A catalog record for this publication is available from the British Library.

Library of Congress Cataloging in Publication Data
Energy technology innovation : learning from historical successes and failures / edited by Arnulf Grubler, International Institute for Applied Systems Analysis, Laxenburg, Austria, Charlie Wilson, University of East Anglia, UK.
pages cm
Includes bibliographical references.
ISBN 978-1-107-02322-2 (hardback)
1. Power resources. I. Grubler, Arnulf, 1955– II. Wilson, Charlie, 1972–
TJ163.2.E4933 2014
621.042–dc23 2013027352

ISBN 978-1-107-02322-2 Hardback

Contents

Author Bios

Laura Díaz Anadón is assistant professor of public policy at the Harvard Kennedy School of Government, co–faculty chair of the Energy Technology Innovation Policy research group, and associate director of the Science Technology and Public Policy program of the Belfer Center for Science and International Affairs also at Harvard. Her research focuses on energy- and environment-oriented technical change and seeks to quantify the complex factors that contribute to it, identify its benefits, and develop tools for policy makers interested in promoting it.

Per Dannemand Andersen is a professor in the Management Engineering Department at the Technical University of Denmark (DTU). His recent research has focused on innovation and research strategies for energy technologies, technology foresight, and strategy and strategy processes in industrial sectors.

Evandro Luíz Dall'Oglio is a professor in the chemistry department of the Federal University Mato Grosso in Brazil, where he also serves as coordinator of postgraduate studies in chemistry.

Sabine Fuss is head of the Resources and International Trade Group at the Mercator Research Institute on Global Commons and Climate Change (MCC). Previously she worked on the development of economic methods with emphasis on uncertainty and risk in the Ecosystems Services and Management (ESM) Program at the International Institute for Applied Systems Analysis (IIASA). Her research interests are in economic modeling, especially with respect to decision making under uncertainty, and with particular focus on energy planning and climate change.

Kelly Sims Gallagher is an associate professor of energy and environmental policy at the Fletcher School, Tufts University. She directs the Energy, Climate, and Innovation (ECI) research program in the Center for International Environment and Resource Policy. She is also senior associate and a member of the board of directors of the Belfer Center for Science and International Affairs at Harvard University, where she previously directed the Energy Technology Innovation Policy (ETIP) research group. Broadly, she focuses on energy and climate policy in both the United States and China.

Arnulf Grubler is a senior research scholar at the IIASA in Laxenburg, Austria, and a part-time resident faculty member at the School of Forestry and Environmental Studies at Yale University. His research and teaching interests focus on the interplay between energy and technology systems and their implications on the environment, in particular on climate change.

Arne Jacobson is director of the Schatz Energy Research Center and an associate professor of Environmental Resources Engineering at Humboldt State University in California. His areas of research interest include renewable energy technology, energy and climate change mitigation policy, and energy access for low-income people in developing countries. He holds a PhD from the University of California, Berkeley.

Martin Jakob is executive partner at TEP Energy. He is a longtime researcher and project manager at the engineering firm Paul Widmer, the Paul Scherrer Institute (PSI), and the Centre for Energy Policy and Economics (CEPE) at the ETH Zurich, where he also received his PhD. His research focuses on energy demand modeling, quantitative estimates of energy efficiency potentials, and economic and policy studies and concepts in the fields of household energy service and electricity demands.

Kejun Jiang is director of the Energy Research Institute of the National Development and Reform Commission in China. He is a lead author for the Intergovernmental Panel on Climate Change and for the Global Environmental Outlook (GEO-4). He began research on climate change related to energy policy analysis in 1993, focusing on energy technology policy assessment, energy supply policy assessment, renewable energy development, and energy conservation.

Daniel M. Kammen is the Class of 1935 Distinguished Professor of Energy at the University of California, Berkeley; founding director of the Renewable and Appropriate Energy Laboratory; co-director of the Berkeley Institute of the Environment; and director of the Transportation Sustainability Research Center. He is also a climate advisor to the Obama administration and a coordinating lead author for the Intergovernmental Panel on Climate Change. During 2010–2011, he served as the World Bank Group's Chief Technical Specialist for Renewable Energy and Energy Efficiency.

Ruud Kempener is an analyst in the Innovation and Technology Center at the International Renewable Energy Agency (IRENA). Previously he worked as a postdoctoral research Fellow in the Energy Research, Development, Demonstration & Deployment (ERD3) Policy project at the Belfer Center for Science and International Affairs, Harvard University.

Osamu Kimura is a research scientist at the Socio-economic Research Center of the Central Research Institute of Electric Power Industry (CRIEPI) in Japan.

Bernadett Kiss is a research associate in the International Institute for Industrial Environmental Economics (IIIEE) at Lund University, where she received her PhD. Her research interest lies in the fields of energy efficiency in buildings and policy promoting technological change.

Volker Krey is Deputy Program Leader of the Energy Program at the IIASA, Austria, where he has been working since 2007. He received his PhD in mechanical engineering in 2006 from the Ruhr-University of Bochum (Germany). He was a lead author of the recently published *IPCC Special Report on Renewable Energy Sources and Climate Change Mitigation*, a lead author of the *Global Energy Assessment*, and a lead author of the IPCC *Fifth Assessment Report*. Since 2010 he has also been an Associate Deputy Editor of the journal *Climatic Change*. His scientific research focuses on the integrated assessment of climate change and major global energy challenges, including energy security and energy access. His work focuses on the development and application of integrated assessment models with different regional focuses (national to global scale) and time horizons.

David McCollum is a Research Scholar with the Energy Program at the IIASA, Austria, having joined the group in February 2011. He received his doctorate in transportation technology and policy from the Institute of Transportation Studies at the University of California, Davis (USA), following completion of an MSc in agricultural and resource economics from the same institution. He is a contributing author of the *Fifth Assessment Report* of the Intergovernmental Panel on Climate Change and was previously a lead analyst on the *Global Energy Assessment*. His main fields of scientific interest include techno-economic analysis of advanced energy and transport technologies and the development and application of energy-economic and integrated assessment models.

Dustin Meyer is a graduate of the School of Forestry and Environmental Studies at Yale University.

Lynn Mytelka is a professorial Fellow with the Knowledge and Industrial Dynamics group at the UNU Maastricht Economic and Social Research Institute on Innovation and Technology (UNU-MERIT). Her research covers a broad range of issues in development, science, and technology, including innovation systems, clustering, technological upgrading in traditional industries, North-South cooperation, biotechnology, strategic partnerships, multinational corporations, and competitiveness.

Lena Neij is professor and director at the IIIEE at Lund University, Sweden. Her research focuses on the analysis of governance and policies for more sustainable development, including the dynamics of energy systems in terms of technical change.

Gregory F. Nemet is a professor at the University of Wisconsin–Madison in the La Follette School of Public Affairs and the Nelson Institute's Center for Sustainability and the Global Environment. His research and teaching focus on improving analysis of the global energy system and, more generally, on understanding how to expand access to energy services while reducing environmental impacts.

Anastasia O'Rourke is an experienced consultant and researcher who has worked in Sweden, France, the United States, and Australia, most recently authoring several pioneering reports on venture capital and cleantech. In 2008 she cofounded the VC-backed start-up company Big Room Inc., which runs www.ecolabelling.org and is building a large database of certified green products. Her research interests include cleantech entrepreneurship, the emergence of innovation systems, environmental finance, and the growth of green markets.

Rich Press is a graduate of the School of Forestry and Environmental Studies at Yale University.

Keywan Riahi leads the Energy Program at the IIASA, Austria. He also holds a part-time position as Visiting Professor in the field of energy systems analysis at the

Graz University of Technology, Austria. He is a member of the Scientific Steering Committee of the Integrated Assessment Modeling Consortium, and since 1998 he has served as a lead author and review editor to various international Assessments including the *Global Energy Assessment* and the Intergovernmental Panel on Climate Change. His main research interests are the long-term patterns of technological change and economic development and, in particular, the evolution of the energy system.

Paulo Teixeira de Sousa Jr. is a professor at the Federal University Mato Grosso in Brazil. He holds degrees in chemistry from the University of Brasilia and the University of Sao Paulo, and a PhD from the University of East Anglia. His research focuses on biodiesel production processes and on maximizing associated by-products value chains.

Charlie Wilson is a lecturer in energy and climate change research at the Tyndall Centre at the University of East Anglia. His research interests lie at the intersection of innovation, behavior, and policy in the field of energy and climate change mitigation.

Acronyms

ACEA	Association des Constructeurs Européens d'Automobiles
ARPA-E	U.S. Advanced Research Projects Agency – Energy
a-Si	amorphous silicon
BEV	battery electric vehicle
BOS	balance of system
BRIMCS	Brazil, Russia, India, Mexico, China, South Africa
CAFC	corporate average fuel consumption
CAFE	corporate average fuel economy
CCS	carbon capture and storage
CEA	French state nuclear R&D organization
CONACYT	Mexican National Council on Science and Technology
COP	coefficient of performance
CRTF	Central Receiver Test Facility
CTC	Copersucar Technology Center
DM	dual mode
DOE	U.S. Department of Energy
EAA	Energy Alternatives Africa
EC	European Commission
ÉDF	Électricité de France
EMBRAPA	Brazilian Agricultural Research Corporation
EPA	Environmental Protection Agency
EPR	European Pressurized Water Reactor
ESDA	Energy for Sustainable Development, Africa
ESTIF	European Solar Thermal Industry Federation
EV	electric vehicle
EVA	ethylene vinyl acetate
FAW	First Auto Works
FCAI	Australian Federal Chamber of Automotive Industries
FFVs	flex-fuel vehicles
FSU	former Soviet Union
GSHPs	ground source heat pumps
HEVs	hybrid electric vehicles
IAA	Institute for Sugar and Alcohol

IAC	Campinas Institute of Agronomy
ICT	information and communication technology
IEA	International Energy Agency
IEC	International Electrotechnical Commission
IGCC	integrated gasification combined cycle
JAMA	Japanese Automobile Manufacturers Association
KAMA	Korean Automobile Manufacturers Association
KBS	Kenya Bureau of Standards
KEREA	Kenya Renewable Energy Association
LNG	liquefied natural gas
M&A	mergers and acquisitions
METI	Japanese Ministry of Economy, Trade and Industry
MOST	Chinese Ministry of Science and Technology
mpg	miles per gallon
NAPS	Neste Advanced Power Systems
NHTSA	National Highway Traffic Safety Administration
NiMH	nickel metal hydride
NUTEK	Swedish Agency for Economic and Regional Growth
O&M	operation and maintenance
PE	private equity
PNGV	Partnership for a New Generation of Vehicles
ProAlcool	Brazil's First Ethanol Program
PURPA	Public Utility Regulatory Policy Act
PV	solar photovoltaic
PWR	Pressurized Water Reactor
R&D	research and development
RD&D	research, development, and demonstration
SANERI	South African National Energy Research Institute
SEFI	Sustainable Energy Finance Initiative
SEGS	Solar Energy Generating Station
SFC	Synthetic Fuels Corporation
SOEs	state-owned enterprises
SRES	Special Report on Emissions Scenarios
SRREN	Special Report on Renewable Energy Sources and Climate Change Mitigation
STE	solar thermal electricity
SWH	solar water heaters
T&D	transport and distribution
VC	venture capital

Acknowledgements

Many of the chapters in this book were researched, discussed, drafted, and compiled under the auspices of the Global Energy Assessment (GEA). The GEA provided an opportunity and support for a diverse and international team of researchers to meet periodically over a four-year period to work on energy technology innovation, a process that ultimately led to this book as a companion volume to a more condensed chapter in the GEA, which also covers energy technology innovation (Grubler et al., 2012). We gratefully acknowledge the many fruitful interactions and ideas with the full team of lead and contributing authors to this GEA chapter, and to the GEA support staff for facilitating the process, in particular Eri Nagai for her help with compiling, checking, and referencing the chapter manuscripts.

ENERGY TECHNOLOGY INNOVATION

PART I

INTRODUCTION

1 Energy Technology Innovation

Charlie Wilson and Arnulf Grubler

1 Introduction

1.1 What is Energy Technology Innovation?

Asked to "vision" their ideal future, schoolchildren agree almost invariably on one common feature of the year 2050: solar-powered flying skateboards.[1] What better example of *energy technology innovation*? Materials and knowledge combined in a novel application containing an energy conversion chain in miniature. An energy supply technology (solar panel) converts an energy resource (solar radiation) into an energy carrier (electricity). An end-use technology (electric motor and fan) converts the energy carrier into a useful service (aerial mobility). A novel combination of technologies; a novel energy service. And a very satisfied end user (schoolchildren).

The useful services provided by energy underpin life. Mobility – particularly of the flying skateboard kind – is a prominent example. Others include heating, cooling, cooking, cleaning, entertaining, communicating, information processing, industrial processing, manufacturing – the list of energy services goes on. Energy technologies are used to extract, capture, process, convert, transport, and distribute energy in its myriad forms through the transformation chains that provide us these services.

Energy technology innovations range from radical new inventions to marginal performance improvements, and encompass social and behavioural changes alongside more visible material changes in technological hardware. Innovation is most simply conceived of as novelty. Unlike natural resources, innovation originates from human endeavour or inspiration, so is an entirely human-made resource. Despite popular conceptions of the visionary genius, innovation is more sweat and application than eureka and ease. Innovation success typically means widespread diffusion and commercial uptake. But this outcome is the culmination of an often lengthy process that runs from research and development through demonstration and trials to early market formation and then diffusion. There are countless pitfalls along the way. The majority of innovation journeys end in failure, some abject, others marginal (Edgerton, 2011). Innovation is neither costless nor certain.

[1] From the author's experience in Canadian classrooms with eight to ten year olds as part of a climate change–related lesson. The aerodynamically questionable assumption of onboard electric motors is the author's interpretation.

1.2 Global Energy Challenges and the Rationale for This Book

First and foremost, this book is about what we have learnt about energy technology innovation based on successes and failures recorded in the pages of history. The book is developed around a varied set of twenty case studies of energy technology innovation. At the risk of disillusioning readers at such an early stage, what follows is not a lengthy treatment of solar-powered skateboards nor other technologies from the realms of science fiction. Our interest is rather more earthbound, but still far from prosaic. We are concerned with energy technology innovations ranging from solar photovoltaics in Kenya to appliances in Japan to wind power in Europe to hybrid cars in China. What we do share with solar-powered skateboards is an interest in alternatives: technological innovations largely at the fringes of our current energy system, but hoped and heralded to play an ever-greater part in our future.

Why is technology innovation important? It has played and continues to play a central role in economic growth and development. Likewise, energy technology innovation has been key in historical transformations of energy systems and services and is central to future sustainability. The context for this book is the enormous challenges facing the global energy system in mitigating climate change, in providing universal access to modern energy carriers like electricity, and in ensuring the supply and distribution of energy is secure and resilient. The *Global Energy Assessment* sets out these challenges as well as the possible pathways describing how they can be surmounted (GEA, 2012). This book is a companion volume to a chapter in the *Global Energy Assessment* that covers energy technology innovation (Grubler et al., 2012). The *Global Energy Assessment* develops the theory and practice of energy technology innovation, and explains the integral role that energy technology innovation can and will play in addressing global energy challenges. This book enriches and deepens this assessment, and offers detailed empirical support for the analysis and arguments of the *Global Energy Assessment.*

1.3 Aims and Intended Audiences

The aim of this book is twofold. The first is to develop and validate a comprehensive, integrated framework for thinking about and for analysing energy technology innovation. The second is to identify critical elements of successful innovation efforts as a basis for informing and supporting the wide range of innovation activities and policies in the energy technology innovation system.

These two aims – analysing energy technology innovation from a systemic perspective and developing insights to support successful innovation outcomes – makes this book of potential interest to a broad audience: researchers and students of innovation processes; social scientists and engineers working on energy technologies; policy makers in the domains of innovation and technology policy as well as within the energy sector; scenario developers and system planners concerned with energy transitions and technological change; financiers in the clean technology and "green" energy space; and concerned citizens curious with what we know and don't know about the role energy technologies and innovation may play in the context of future sustainability.

The historical case studies around which the book is developed require a familiarity with energy technology and an understanding of innovation concepts and terms. While the book is pitched at the informed reader, these introductory sections should provide the necessary background to understand what follows.

The policy community is one of the key constituencies for our findings. Policy can and does strongly shape the various interrelated stages of the innovation process, particularly in the case of energy technologies given their environmental and social implications. The concluding sections of the book apply the analytical framework developed to abstract generalisable policy guidelines that should work in favour of innovation success and guard against innovation failure.

2 Energy Technology Innovation: An Overview

2.1 Energy Technology Innovation Historically

Technological change has long been recognised as integral to economic growth and development (Freeman and Perez, 1988; Solow, 1957). Moreover, technological change and social change are inseparable, caught in lockstep by a web of mutual interdependencies. These change processes have played a driving role in past energy transitions from preindustrial, traditional practices, through the coal and steam era, and to the infrastructure of today (Grubler, 1998; Smil, 1994). Technological change also plays a central role in future scenarios of climate change mitigation (Halsnæs et al., 2007; Nakicenovic et al., 2000). One of the case studies in this book synthesises the "grand patterns" of technological change in the energy system, drawing parallels between historical evidence and future expectation. Four of these patterns are summarised here as they provide important context for what follows.

Firstly, the history of energy transitions is predominantly characterised by changing types and amounts of energy end-use services. The characteristic large-scale features of the energy supply – mines, rigs, pipelines, grids, power plants, refineries, dams, and turbines – are ultimately driven and shaped by growth and changes in the demand for energy services. End-use applications and their contribution to the proliferating array of consumer goods and services are the most important markets for new energy technologies.

Secondly, many technology innovations, including in the energy field, tend initially to be crude, imperfect, and expensive (Rosenberg, 1994). But they gain a market foothold by offering a novel energy service attractive to a distinctive set of users in a particular market niche. Thus shielded from the winnowing force of full market competition, technologies undergo an iterative process of testing, improvement, adaptation, even redesign, until the costs of providing the particular energy service fall to the point at which they become attractive to the wider market. Such technologies may then enjoy widespread commercialisation and diffusion: the ultimate outcome of innovation success. But attractiveness beats cheap, at least initially.

Thirdly, technological change from innovation through to widespread diffusion is generally slow, lasting as a rule many decades. The innovation process itself is constantly iterating back and forth. Moreover, a technology is not adopted and used in isolation. Dependent on related technologies and infrastructures, but also business models and wider market and social institutions, an energy technology's diffusion is contingent and necessarily gradual (Hughes, 1983). But these same interdependencies, which take considerable time to develop, also give rise to a self-reinforcing process by which successful technologies reach a position of dominance and so themselves become entrenched, a condition known as "lock-in" (Cowan and Hulten, 1996; Unruh, 2000).

Fourthly, the transformative potential of energy technologies arises through clustering and spillovers (Grubler, 1998). Clustering means combinations of interrelated technologies. Spillovers mean applications of technologies beyond their initial designation or use. In short, technologies operate more effectively as "gangs" than as individuals. This again implies slower potential rates of change, exacerbated by the capital intensiveness and long-lived nature of many energy technologies and their associated infrastructures.

These four "grand patterns" observed historically are reflected clearly in the treatment of energy technology innovation developed in this book. Indeed, as we will argue, the importance to historical energy transitions of end-use services, of performance and cost advantages, of interdependencies and inertia, and of clustering and spillovers, underscores the need for a *systemic perspective* on energy technology innovation.

2.2 Energy Technology Innovation: Concepts and Terms

Like any field of study, energy technology innovation has its own particular concepts and terms. While we have tried to ensure the writing is accessible to those with a general grounding in energy and innovation studies, these introductory chapters go over the key ideas and define the key terms needed to engage with the case study chapters that follow (see also Table 1.1).

Figure 1.1 starts by introducing and organising the key elements of energy technology innovation. These are: (i) the stages of the innovation process; (ii) the flows and feedbacks between these stages; (iii) the influence of both technology-push and market-pull drivers; (iv) the relevance of both energy supply and energy end-use technologies.

Technologies move sequentially through a "life cycle" from birth (invention, innovation), to adolescence (growth), maturity (saturation), and ultimate senescence (decline driven by competition from newer and more attractive innovations). Models of innovation describe the drivers and mechanisms behind this technology life cycle. Their intellectual history goes back to the nineteenth century and Marx's conceptualisation of technological innovation in his economic theories. In the early and mid twentieth century, Schumpeter (1942) emphasised the importance of radical, disruptive, or "breakthrough" technologies driven by entrepreneurship and competition, while scholars such as Usher (1929) pointed to the compounded effects of numerous, small, "incremental" innovations.

Early models of the innovation process, such as those formulated in the influential U.S. report, "Science the Endless Frontier" (Bush, 1945), emphasised the role of basic, publicly funded science in a linear innovation process from basic research to applied development, then demonstration in a commercial setting, concluding with widespread diffusion.

These innovation stages are shown in the centre of Figure 1.1. The innovation processes are the linkages from stage to sequential stage. But as can be seen, the

Table 1.1. Definition of key terms used to describe energy technology innovation.

	Key term	Definition as used in this book
Innovation Processes & Stages	invention	origination of an idea as a technological solution to a perceived problem or need
	innovation	putting ideas into practice through a (iterative) process of design, testing, application, and improvement
	research & development (R&D)	knowledge generation by directed activities (e.g., evaluation, screening, research) aimed at developing new or improving on existing technological knowledge
	demonstration	construction of prototypes or pilots for testing and demonstrating technological feasibility and/or commercial viability
	research, development & demonstration (RD&D)	a commonly used grouping of the main precommercial stages of the innovation cycle
	niche markets	application of a technology in a limited market setting (or niche) based on a specific relative performance advantage (or on public policy incentives) and typically protected in some way from full market competition
	market formation	activities designed to create, enhance, or exploit niche markets and the early commercialisation of technologies in wider markets
	diffusion [also: deployment]	widespread uptake of an energy technology throughout the market of potential adopters
	innovation or technology life cycle	the sequence of processes and stages of an innovation's journey from invention right through to senescence or obsolescence
Types of Innovation	radical innovation [also: breakthrough, disruptive]	a novel technology that strongly deviates from prevailing norms and so often entails a disruptive change over existing commercial technologies and associated institutions
	incremental innovation [also: continuous]	an improvement in performance, cost, reliability, design, etc. to an existing commercial technology without any fundamental novelty in end-use service provision
Drivers of Innovation	technology-push [also: supply-push]	forces driving the generation of innovations (e.g., by reducing innovation costs)
	market-pull [also: demand-pull]	forces driving the market provision of innovations (e.g., by increasing innovation payoffs)
Types of Energy Technology	energy supply technologies	technologies used to extract, harness, transport primary energy resources (e.g., coal, uranium, sunlight) and convert them into secondary and final energy (e.g., petrol, electricity)
	energy end-use technologies	technologies that convert final energy into a useful service to end users (e.g., heating, mobility, entertainment)

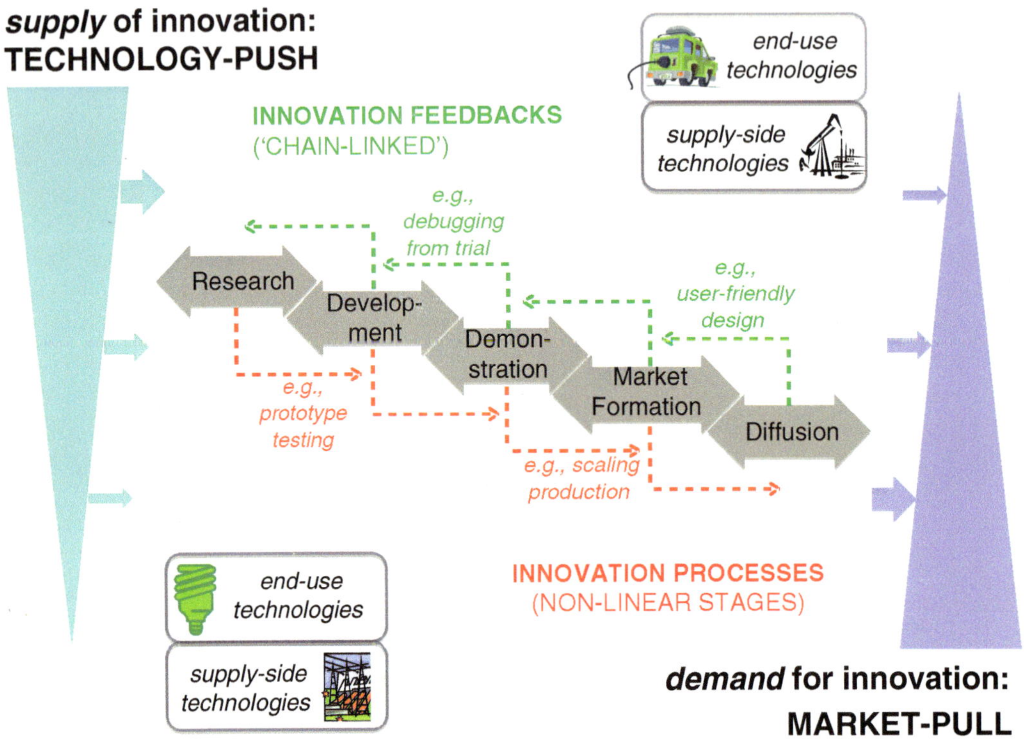

Figure 1.1. Key elements of energy technology innovation.

innovation life cycle is no longer considered linear nor unidirectional (Freeman, 1994; Mowery and Rosenberg, 1979). This has given rise to the term *chain-linked model* (Brooks, 1995; Kline and Rosenberg, 1986) shown in the centre of Figure 1.1 with the stages of innovation linked in both directions by innovation processes and innovation feedbacks. So, as an example, knowledge does not just flow from basic scientific research to technology development and commercialisation; technological applications can also enable breakthroughs or discoveries in basic scientific research. The sequence of stages is also not prescriptive. Some technologies are successful without having proceeded through each stage in the innovation process (Grubler, 1998).

Related to these bidirectional flows and feedbacks is a recognition that the innovation process is driven by forces of both supply and demand, reducing the costs of innovation on one hand, and increasing the payoffs from innovation on the other (Nemet, 2009). These are shown on the left and right sides of Figure 1.1. Supply-side forces push technologies through the development process to commercialisation, hence also "technology-push." The supply of innovation emphasises the interests and roles of scientists, researchers, prototypers, product developers, engineers and designers, and venture capitalists seeking to promote and provide new goods and services. In contrast, demand-side forces signal a market appetite that pulls technologies through the development process, hence also "market-pull." The demand for innovation emphasises the interests and roles of consumers, end users, and public institutions expressing their needs or desires for new goods and services. As we will see, these "technology-push" and "market-pull" drivers do not work in isolation; not only are they complementary, they are also both necessary.

Figure 1.1 also includes an additional stage of "market formation" that separates widespread diffusion from the key innovation stages of research, development, and demonstration (RD&D). The specific inclusion of market formation recognises the major hurdle faced by technologies as they start to compete with their incumbent rivals in a market environment. Apparently successful innovations may fall in this "valley of death" if they are too expensive, offer too indistinct performance advantages, are too difficult to scale up, or lack a clear perceived market demand. Market formation activities support new technologies through this early competition. In some cases, natural market niches are formed if a particular group of end users is less price sensitive and accords particular value to the relative advantages of the new technology (Kemp, Schot, and Hoogma, 1998).

Finally, Figure 1.1 highlights that this simple model of the innovation process in an energy context applies both to energy supply technologies and to end-use technologies. Energy supply technologies are used to extract, process, transport, and convert energy resources into a form useful to end users. Examples include solar panels, oil refineries, natural gas pipelines, and nuclear power plants. End-use technologies are used to convert final energy into useful energy and then to services like heating, mobility, or communication. Examples include boilers, cars, and mobile phones. The relevance of this energy supply and end-use distinction is discussed further in the next chapter.

2.3 Energy Technology Innovation in a Nutshell

So what is energy technology innovation? The simple definition with which the chapter opened captures its essence: materials and knowledge combined in some novel application involving energy conversion and the provision of a useful energy service. More formally, energy technology innovation can be regarded as the embodied result of institutionalised research, development, and collective learning processes involving both suppliers and users of energy technologies operating within specific incentive structures and adoption environments. The "embodied result" is most easily visualised in physical terms (capital stock, hardware, gadgets), but can also be in the form of new social arrangements or patterns of behaviour associated with end-use services. As a general rule, technological change involves changes in hardware as well as in social or institutional settings. The "innovation environment" thus ranges from research laboratories and testing facilities for the suppliers of technologies to households or consumer goods markets and social networks.

This working definition of energy technology innovation points to the importance of the actors and institutions involved in the development and commercialisation of an energy technology innovation. This is one of the additional dimensions to the innovation processes and feedbacks shown in Figure 1.1 that we set out in the next chapter.

REFERENCES

Brooks, H., 1995. What We Know and Do Not Know about Technology Transfer: Linking Knowledge to Action. *Marshaling Technology for Development.* Washington, DC: National Academy Press.

Bush, V., 1945. Science the Endless Frontier. *A Report to the President* [Online].

Cowan, R. & Hulten, S., 1996. Escaping Lock-In: The Case of the Electric Vehicle. *Technological Forecasting and Social Change*, 53(1): 61–79.
Edgerton, D., 2011. In Praise of Luddism. *Nature*, 471(7336): 27–29.
Freeman, C., 1994. The Economics of Technical Change. *Cambridge Journal of Economics*, 18(5): 463–514.
Freeman, C. & Perez, C., 1988. Structural Crises of Adjustment, Business Cycles and Investment Behaviour. In Dosi, G., Freeman, C., Nelson, R., Silverberg, G., & Soete, L. (eds.) *Technical Change and Economic Theory*. London: Pinter Publishers.
GEA, 2012. *Global Energy Assessment – Toward a Sustainable Future*, Cambridge, UK and New York, NY, USA and the International Institute for Applied Systems Analysis, Laxenburg, Austria, Cambridge University Press.
Grubler, A., 1998. *Technology and Global Change*, Cambridge, UK, Cambridge University Press.
Grubler, A., Aguayo, F., Gallagher, K., Hekkert, M., Jiang, K., Mytelka, L., Neij, L., Nemet, G., & Wilson, C., 2012. Chapter 24 – Policies for the Energy Technology Innovation System (ETIS). *Global Energy Assessment – Toward a Sustainable Future*. Cambridge University Press, Cambridge, UK and New York, NY, USA and the International Institute for Applied Systems Analysis, Laxenburg, Austria.
Halsnæs, K., Shukla, P., Ahuja, D., Akumu, G., Beale, R., Edmonds, J., Gollier, C., Grubler, A., Ha Duong, M., Markandya, A., McFarland, M., Nikitina, E., Sugiyama, T., Villavicencio, A., & Zou, J., 2007. Framing issues. In Metz, B., Davidson, O., Bosch, P., Dave, R., & Meyer, L. (eds.) *Climate Change 2007: Mitigation*. Cambridge, UK: Contribution of Working Group III to the Fourth Assessment Report of the Intergovernmental Panel on Climate Change, Cambridge University Press.
Hughes, T. P., 1983. *Networks of Power: Electrification in Western Society, 1880–1930*, Baltimore, MD and London: Johns Hopkins University Press.
Kemp, R., Schot, J., & Hoogma, R., 1998. Regime Shifts to Sustainability through Processes of Niche Formation: The Approach of Strategic Niche Management. *Technology Analysis & Strategic Management*, 10(2): 175–98.
Kline, S. J. & Rosenberg, N., 1986. An Overview of Innovation. In Landau, R. & Rosenberg, N. (eds.) *The Positive Sum Strategy: Harnessing Technology for Economic Growth*. Washington, DC: National Academy Press.
Mowery, D. & Rosenberg, N., 1979. The Influence of Market Demand upon Innovation: A Critical Review of Some Recent Empirical Studies. *Research Policy*, 8(2): 102–53.
Nakicenovic, N., Alcamo, J., Davis, G., de Vries, B., Fenhann, J., Gaffin, S., Gregory, K., Grubler, A., Tae, Y. J., Kram, T., La Rovere, E. L., Michaelis, L., Mori, S., Morita, T., Pepper, W., Pitcher, H., Price, L., Riahi, K., Roehrl, A., Rogner, H.-H., Sankovski, A., Schlesinger, M., Shukla, P., Smith, S., Swart, R., Van Rooijen, S., Victor, N., & Zhou, D., 2000. *Special Report on Emissions Scenarios*, Cambridge, UK, IPCC and Cambridge University Press.
Nemet, G. F., 2009. Demand-Pull, Technology-Push, and Government-Led Incentives for Non-Incremental Technical Change. *Research Policy*, 38(5): 700–709.
Rosenberg, N., 1994. *Exploring the Black Box: Technology, Economics, and History*, Cambridge, UK, Cambridge University Press.
Schumpeter, J. A., 1942. *Capitalism, Socialism and Democracy*, New York: Harper.
Smil, V., 1994. *Energy in World History*, Boulder, CO: Westview Press.
Solow, R. M., 1957. Technical Change and the Aggregate Production Function. *The Review of Economics and Statistics*, 39(3): 312–20.
Unruh, G. C., 2000. Understanding Carbon Lock-In. *Energy Policy*, 28(12): 817–30.
Usher, A. P., 1929. *A History of Mechanical Invention*, New York: McGraw-Hill.

2 The Energy Technology Innovation System

Charlie Wilson and Arnulf Grubler

1 A Systemic Perspective on Energy Technology Innovation

Throughout this book, we argue the case for a systemic perspective on energy technology innovation. We set out an analytical framework with all the key elements of the energy technology innovation system. We show how a wide range of historical innovation case studies all validate this innovation systems approach. And we use this systemic perspective to generalise policy guidelines to support energy technology innovation. But why is this all needed?

Firstly, a systemic perspective provides the necessary context for understanding the innovation stages and processes outlined in the previous chapter, and for explaining why some innovations succeed while others fail. Effectively functioning innovation systems are characterised by a web of interrelationships and dependencies (Hekkert et al., 2007). This certainly applies to energy technology innovation in response to environmental and sustainability challenges. Innovation systems are equally relevant to the study of incremental improvements to existing technologies and an emphasis on diffusion (Pacala and Socolow, 2004; Socolow, 2011), and to radical breakthroughs of new technologies and an emphasis on basic and applied research (Hoffert, 2010; Hoffert et al., 2002).

Secondly, the literature on energy technology innovation reviewed for the *Global Energy Assessment*, and the case studies researched and compiled for this book, make it abundantly clear that a piecemeal, partial approach to energy technology innovation is wholly insufficient. The challenges facing the global energy system are immense: mitigating climate change; bringing modern energy services to 2 billion currently disenfranchised people; ensuring a secure and low-polluting energy supply. The magnitude of these multiple energy challenges set out in the *Global Energy Assessment* means moving beyond a narrow attention on particular technologies, particular stages of the innovation life cycle, or particular innovation drivers or processes.

In short, a systemic perspective on energy technology innovation is necessary to understand how innovation works, why innovations succeed or fail, and how innovation efforts can be geared up to meet pressing sustainability challenges. We return to these advantages of a systemic perspective on energy technology innovation at the end of this chapter. But first, we set out exactly what we mean by an energy technology innovation system.

2 Overview of the Energy Technology Innovation System

2.1 Four Dimensions of the Energy Technology Innovation System

Figure 2.1 sets out the basic stages, processes, and push/pull drivers of energy technology innovation. What does a systemic perspective add? The energy technology innovation system consists of four important dimensions:

- *knowledge*;
- *actors and institutions*;
- *resources*;
- *adoption and use*.

Firstly, the energy technology innovation system makes salient the role of knowledge, the essence of innovation. Knowledge generation, exchange, and learning are key processes in which innovation actors engage.

Secondly, the energy technology innovation system considers the constellation of actors and institutions that bring innovation processes to life and shape a technology's life cycle from invention through to senescence. From a systemic perspective, innovation is understood as an interactive process involving a network of firms and other economic actors who bring new products, processes, and forms of organisation into economic use in an institutional context that influences their innovative behaviour and performance (Freeman and Perez, 1988; Lundvall, 1992; Nelson and Winter, 1982).

Thirdly, the energy technology innovation system explicitly recognises the importance of financial, human, and other resources that are mobilised as key inputs to support and sustain innovation processes. Public policy is an important influence on this resource mobilisation, and warrants special attention.

Fourthly, the energy technology innovation system necessarily accommodates the adoption and use of an innovation in both niche and widespread market ("real-world") settings. The relative advantage an innovation offers end users relative to the incumbent technology is a key determinant of its likely success (Rogers, 2003). For energy technologies, the two most important attributes of this relative advantage are performance (e.g., flexibility, reliability, versatility, cleanliness, usefulness), and the cost with which a useful service can be provided. These attributes are in turn related to and determined by specific characteristics of a technology.

From the outset, we should emphasise that these four dimensions are strongly interlinked and not substitutable for each other. More knowledge generation does not obviate the need for effective actors and institutions; improved technology performance does not mean fewer resources need to be mobilised. We return to these interlinkages throughout the case studies and in the concluding chapters.

These four dimensions – *knowledge, actors and institutions, resources, and adoption and use* – are the systemic framework within which a technology's life cycle plays out. Figure 2.1 depicts these four dimensions as wedges and associated processes as arrows. In the centre of the diagram are the innovation stages and processes described in the previous chapter. The wedges in Figure 2.1 represent stocks and the arrows represent flows. The tapering of the wedges away from the centre represents how stocks are gradually accumulated through the processes depicted by the incoming arrows; the tapering of the wedges towards the centre represents how stocks can be depreciated. Innovation processes are both accretive and depletive.

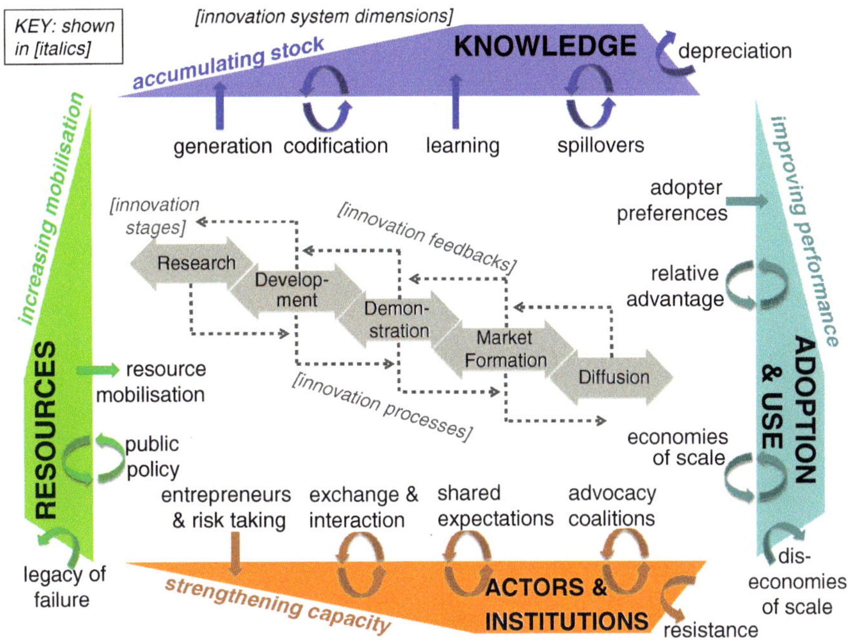

Figure 2.1. The energy technology innovation system.

To take one example, knowledge (the wedge at the top of Figure 2.1) is accumulated by processes relating to knowledge generation and learning. These are represented as inward arrows. Depreciation, represented as an outward arrow, is a process by which accumulated knowledge is lost (and the dedicated subject of one of our case studies). The accumulation of knowledge is further supported by spillovers and codification (e.g., formal documentation). These two processes are expressed as cyclical arrows, both inward and outward, as they are potentially self-reinforcing: spillovers can generate knowledge that can result in more spillovers, and so on. However, the cycle is not necessarily virtuous. Ultimately, the effect of a particular process is contingent on other processes in both the same and other dimensions of the energy technology innovation system. Thus knowledge generation and spillovers are themselves influenced by resource mobilisation (resources dimension) as well as exchanges and interactions between actors (actors and institutions dimension).

More broadly, we emphasise the interdependence between the dimensions and their associated processes. Their neat representation in Figure 2.1 is inevitably stylised. The various interactions and relationships are explained more fully later in this chapter. For now, the important point is that our framework representing the energy technology innovation system contains three key elements:

- the stages, processes, and feedbacks of energy technology innovation – represented as block and dotted arrows in the centre of Figure 2.1;
- the four dimensions of the innovation system that are of critical importance to energy technologies – represented as wedges around the sides of Figure 2.1;
- the principal innovation processes associated with these four dimensions – represented by arrows flowing into and out of the wedges around the sides of Figure 2.1.

Table 2.1. Definition of key terms used to describe the energy technology innovation system.

Key term	Definition as used in this book
energy technology innovation system	an analytical framework applied to energy technologies that places innovation stages, processes, and drivers within the context of innovation systems described by the four dimensions of knowledge, actors and institutions, resources, and technology adoption and use
learning	improved (technological) knowledge derived from production experience (learning-by-doing) and/or user experience (learning-by-using) leading to performance improvements, including cost reductions
knowledge spillovers	knowledge transfer between different innovation actors and fields of technology application through mechanisms such as imitation, trade, licensing, foreign direct investment, and movement of personnel
actors	the individuals, end users, consumers, producers, firms, and public bodies involved in the activities of an innovation system
networks	the form, strength, interactions, and reciprocity of relationships between actors in an innovation system
institutions	organisations and formal structures like rules and regulations, but also habits, practices, routines, and norms of the various actors in an innovation system

As is evident by now, this energy technology innovation systems framework has its own particular terms and concepts, which we have summarised in Table 2.1. These are in addition to the more commonly used terms describing innovation that were explained in the previous chapter.

2.2 Energy Technology Innovation Systems in Context

What is not shown in Figure 2.1 is that the broader context of energy technology innovation systems also matters. National, geographical, infrastructural, and other factors affect the relative importance of the roles and relationships between elements of the innovation system. The relevance of "national systems of innovation" provides one example of why this broader context is important (Lundvall, 2009; Nelson, 1993). Differences in the broader context also help explain why innovation systems for specific energy technologies may vary in their composition and functioning, involving different sets of actors, interacting in different ways, incentivised for solving different problems, and acting at different spatial scales (Hekkert et al., 2007; Jacobsson and Lauber, 2006). We address these issues of context through our comparative assessment of a wide range of energy technology innovation case studies, drawn from different countries, geographies, and innovation environments.

It is also important to note that, just as it takes time for a technology to move iteratively through the stages of its life cycle (shown in the centre of Figure 2.1), so too does it take time for knowledge to be built up, for performance and cost attributes to improve, for innovation actors to form and develop networks, for institutions to be established and strengthened, for resources to be mobilised and applied. Innovation systems have a temporal as well as a geographic context. In functional terms, the life cycle of a particular technology develops in tandem with its corresponding innovation system (Jacobsson and Johnson, 2000). Our case studies therefore tell the unfolding histories of particular energy technology innovations, but also the developments over time in their corresponding innovation systems.

2.3 Where Does the Energy Technology Innovation System Come From?

The systemic perspective on energy technology innovation originates principally in the comparative analysis of the case studies of innovation successes and failures contained in this book. These were commissioned from an array of authors involved in the *Global Energy Assessment* who themselves represent different traditions of energy technology and innovation research. No structure or theoretical framework was imposed ex ante. The emphasis was on applied insights from historical evidence. In compiling and assessing these insights, it became clear that common elements repeatedly emerged. Moreover, these elements were interrelated. Some related to innovation stages and processes and the respective roles of technology-push and market-pull activities. Others related to the actors and institutions in innovation systems and to their roles in generating knowledge and mobilising resources. Still others related to the performance and cost attributes of technologies, and how these affected the attractiveness of innovations to particular niches of end users or to the mass market.

So we constructed a framework of the energy technology innovation system to describe, organise, and interrelate the many insights that the case studies generated. As such, this systemic perspective should be treated as empirically founded, inductive, and interpreted ex post from available evidence. It should not be treated as normative, nor as a coherent theory. Rather it encapsulates what we have learnt about successful energy technology innovation, and what may be missing in cases of failure. It is intended as an analytical framework whose development has clearly been influenced by both long-standing and more recent theoretical contributions in the energy, technology, and innovation literatures spanning economics, management, and sociological research traditions. This literature review is set out in full in the chapter on innovation in the *Global Energy Assessment* (Grubler et al., 2012), with key elements summarised in the following sections.

The energy technology innovation systems approach has many advantages. It allows the back-and-forth relationships between different stages of the innovation process to be characterised and their importance to successful innovation outcomes identified. It allows technology-push and market-pull efforts to be made commensurate and compared. It supports a move away from policies aimed at isolated components of the system to comprehensive strategies to support effective functioning of the innovation system as a whole. It uncovers the roles and relationships between innovation actors and institutions, and relates innovation system inputs to outputs. It provides for the synthesis and analysis of data covering energy-supply and end-use technologies in both developed and developing countries from different points in history and with different degrees of success. For all these reasons, the systemic perspective offers new insights that complement and improve upon conventional analysis and resulting policies.

3 Dimensions and Processes of the Energy Technology Innovation System

Figure 2.1 depicts the main dimensions and processes of the energy technology innovation system. The sections that follow expand on the conceptual basis for this framework, drawing on a range of theoretical and empirical work. These sections are dry, and as brief as possible. Their main purpose is to provide the reader with sufficient knowledge of concepts and terms to understand and interpret the innovation case studies that follow. Readers already familiar with the terms and concepts shown in Figure 2.1 can skip the next sections and return to them as a reference guide should details in the case study chapters not be clear.

3.1 Dimensions and Processes (I) – Knowledge

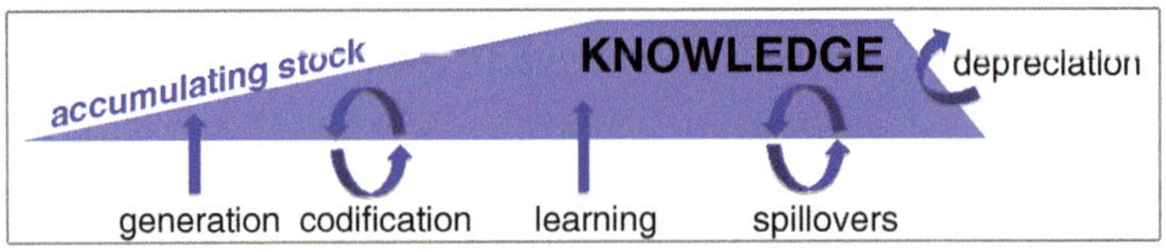

3.1.1 Introduction – Knowledge

"*The most fundamental resource in the modern economy is knowledge and, accordingly, the most important process is learning*" (Lundvall, 1998, emphasis added). Knowledge is a ubiquitous and powerful driver of technological change. It can be basic ("know-why") or applied ("know-how"). It can be publicly available (e.g., codified through publication in scientific journals), embodied in hardware or artefacts, or entirely tacit or disembodied (e.g., in the accumulated experience of a production engineer). Understanding the mechanisms by which knowledge is generated, reproduced, exchanged, and lost is critical to understanding energy technology innovation systems.

Knowledge has some distinctive characteristics. It is non-rival, which means its use by someone does not preclude its use by someone else. It is non-exhaustible, which means its use does not result in its depletion. It is also cumulative and combinatorial, which means it builds on preexisting versions of itself as well as accumulating from a range of different sources (Foray, 2004).

3.1.2 Knowledge Generation, Flows, and Spillovers

Science is a strong source of technical innovation, and the disciplines that support it are numerous (Ausubel and Marchetti, 1997). However, technical change in energy systems has tended to be based more in engineering practices and disciplines, with major innovations originating from outside classical basic science. The best historical example is the development and application of steam engines, which predated (and stimulated) the discovery of the Laws of Thermodynamics (Grubler, 1998). Applied research and development (R&D) relating to energy supply technologies has been oriented by four main goals (or problems to be solved), also termed "focusing devices" (Rosenberg, 1982):

- Harnessing new energy sources and carriers with qualities that confer relative advantages (e.g., energy density, abundance, transportability, flexibility, modularity, etc.);
- Increasing efficiency of conversion and use (both energetic and economic);
- Improving controllability, security, and stability of energy conversion and delivery infrastructures;
- Reducing adverse social and environmental impacts.

Whereas the first three goals are endogenous or inherent economic drivers of change in the energy system, regulation has historically driven improvements in energy technologies' social and environmental performance.

Knowledge is also generated and exchanged through producer-user collaborations in the development phase of technologies, and through producer-producer collaborations in the manufacturing phase (Fridlund, 2000; Lundvall, 1992; Von Hippel, 1988). These forms of knowledge generation and exchange cannot be generated by scientific research nor predicted from first principles. Indeed, much applied knowledge often flows routinely between actors within an industry and across sectors and countries.

Public entities and institutions can play a direct role in knowledge flows. But the majority of knowledge flows relating to energy technology innovation take place through private means. Internationally, for example, knowledge flows through licensing and technology transfer agreements; through trade in final, intermediate, and capital goods; and through foreign direct investment or intrafirm technology transfers in internationalised companies.

Less formal, or even unintended, knowledge flows through nonmarket mechanisms are known as knowledge "spillovers." Knowledge spillovers are enabled by a wide range of mechanisms including: formal and informal networks of scientists, engineers, and technicians; training and movement of personnel; universities, R&D labs, and international research collaborations; scientific publications and patents; and reverse engineering. Knowledge spillovers are difficult to measure (Mohnen, 1997; Van Pottelsberghe de la Potterie, 1997) but have been found to be important for energy technologies. Examples include the cases of photovoltaics (Lako, 2004; Watanabe et al., 2002) and wind energy (Kamp et al., 2004; Lako, 2004).

3.1.3 Knowledge Codification

Knowledge spillovers are not the simple transfer of technological information. Knowledge implies the cognitive capability to interpret, process, and articulate information. By analogy, knowledge spillovers imply the capacity to absorb knowledge produced elsewhere. The capacity of a firm to assimilate knowledge spillovers requires an already existing knowledge base derived from its own R&D, production experience, or training (Cohen and Levinthal, 1990).

In general, codified and embodied knowledge is easier to capture through spillovers (Cowan and Foray, 1997). Codified knowledge means reproducible, transparent, accessible knowledge documented or enshrined in blueprints, manuals, or sets of instructions. Much knowledge remains tacit or uncodified in the form of personal or institutional knowledge and skills. In general, new knowledge generated in the early phases of the innovation life cycle is less likely to be codified. This in turn makes it more difficult to reproduce, memorise, recombine, and learn, and so more costly to exchange or transfer in a usable form (Von Hippel, 1994). In this case, knowledge generation implies the local practising and accumulating of experience (Nelson and Winter, 1982). As experience accumulates, practices and procedures become increasingly codified as hardware is developed and adapted.

3.1.4 Learning

Learning is a particular form of knowledge generation relating to the experience of producing and/or using a technology. Insights about learning originated from observations that workers in manufacturing plants became more efficient as they produced more units (Alchian, 1963; Rapping, 1965; Wright, 1936). This became formalised in a model explaining technical change as a function of "learning-by-doing" through the accumulation of production experience (Arrow, 1962). Originally referring to changes in labour productivity, learning has since been generalised to include changes in capital or organisational productivity. The concept has also been extended to "learning-by-using," which expresses the same effect of experience but from the perspective of technology users such as plant operators or consumers.

Learning results in improved and standardised production processes and products, which in turn can result in cost reductions. The relationship between production experience and cost reductions is commonly expressed as a learning curve that describes the unit cost of a technology as a function of cumulative production. (The experience curve is a related concept more commonly used to describe reductions in aggregate manufacturing costs over whole industries or generations of technologies (Dutton and Thomas, 1984), although no standard terminology is used in the literature.) Derived from learning curves, learning rates describe the cost reduction associated with a doubling of production volumes of a particular technology.

Learning is far from an automatic, deterministic outcome of production experience, and a number of misconceptions exist regarding how to interpret and use learning curves (Neij, 2004; Nemet, 2006). One misconception is that policies can "buy down a technology's costs" or "drive a technology down its learning curve" by using market subsidies without due regard for the broader elements of the innovation system. Learning is increasingly recognised as a complement and not as a substitute for R&D and other sources of knowledge generation (Halsnæs et al., 2007). Another misconception is that learning rates can be reliably forecast for particular technologies (Van der Zwaan and Seebregts, 2004). Future technology costs may be influenced not only by learning effects, but by changes in input prices, quality improvements, changes in competitive environment and market structure, growth in demand, and so on. The use and misuse of learning curves is a recurring theme in our innovation case studies.

3.1.5 Knowledge Depreciation

Just as knowledge can be generated, exchanged, transferred, and learned, so too can it be lost or unlearned. Knowledge stocks need continuous replenishment and can depreciate rapidly in instances of high personnel turnover and retirement, technological obsolescence, institutional inertia, or "stop-go" production schedules. Knowledge depreciation particularly affects settings in which knowledge remains largely tacit or uncodified, residing in specialised personnel or managers, for example. This particular vulnerability of tacit knowledge implies that knowledge accumulated through learning-by-doing is especially prone to depreciation because production experience is less likely to be codified than research and development activities. Depreciation can also occur as knowledge becomes obsolete such that it is no longer relevant for updated processes or techniques, and inadequately recharged by new learning. Knowledge depreciation is an underexplored field, so we have dedicated one case study to reviewing what we know about it from innovation histories.

3.2 Dimensions and Processes (II) – Actors and Institutions

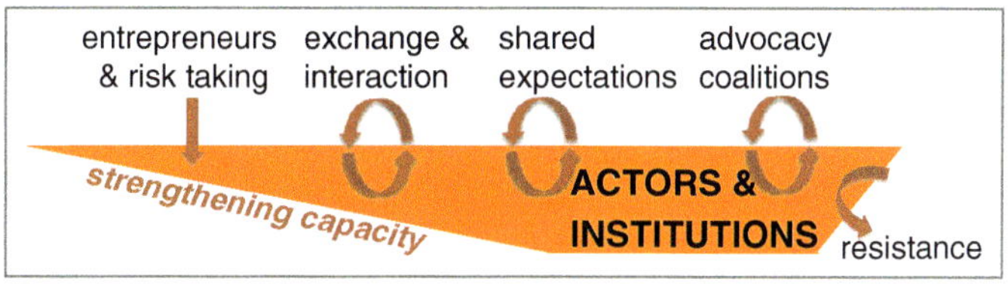

3.2.1 Introduction – Actors and Institutions

Actors and institutions is the second dimension of the energy technology innovation system and is represented by the wedge-shaped block at the bottom of Figure 2.1. By actors, we mean the users, producers, firms, and public entities involved in innovation systems (see also Table 2.1 for definitions). Interaction between actors takes place in networks. Actor networks describe the form, strength, interactions, and reciprocity of relationships between actors in an innovation system. The systemic perspective on energy technology innovation emphasises that innovation is a collective activity involving many actors in specific institutional settings with corresponding incentive structures (Edquist and Johnson, 1997; Lundvall, 2007). By institutions, we mean organisations and formal structures like rules and regulations, but also the practices, routines, and norms of the various actors in an innovation system.

Many of the different actors and institutions involved in energy technology innovation systems have been mentioned already. Actors are commonly differentiated by their involvement in particular innovation stages or processes (e.g., R&D or market adoption) or by their status as public or private entities (e.g., governments, firms, industry associations, entrepreneurs, consumers). Actors are highly heterogeneous in terms of their technological knowledge and competence, their resources available for innovation, as well as their characteristics and preferences (e.g., risk aversion). The roles of particular actors in innovation systems can vary considerably: a firm can be a developer of one innovation and an adopter of another innovation; moreover both such roles may change over time (Edquist, 2001). As innovation systems become more globally integrated, so too have actors with a global reach grown in relevance (e.g., multinational firms, direct foreign investors, multilateral and nongovernmental institutions).

The distinction between public and private sector actors is of particular importance for energy technology innovation systems. Private sector actors dominate innovation processes relating to knowledge generation and technological improvements, and also in bringing innovations to market with associated requirements for manufacturing, marketing, and so on. A clear understanding of end users' needs from the outset of the innovation process is critical in distinguishing ultimate successes from failures (Freeman and Soete, 2000: 204–18). Typically, the relative importance of private actors increases over the course of an innovation system's development (Suurs and Hekkert, 2009b). However, public sector actors play an essential role in establishing appropriate incentives for innovation. The role of public policy as an institution is discussed further later in this chapter.

In addition to public policy, many other institutions play an important role. Innovation processes through which institutions evolve and strengthen have been termed "social techniques" (Ruttan, 1996). One such process involves exchange and interaction between actors. This exchange and interaction helps mediate the knowledge flows and spillovers discussed earlier (Carlsson and Stankiewicz, 1991).

Three additional innovation processes relating to actors and institutions are: entrepreneurs and risk taking, shared expectations, and advocacy coalitions. Along with exchange and interaction, these are represented by inward or circular arrows in Figure 2.1. One further innovation process – resistance – can lead to a weakening of the actors and institutions dimension of the energy technology innovation system. This is represented by an outward arrow in Figure 2.1.

3.2.2 Entrepreneurs and Risk Taking

There is no such thing as an innovation system without entrepreneurs (Carlsson and Stankiewicz, 1991). Entrepreneurial risk taking is essential to cope with the large and irreducible uncertainties surrounding new combinations of technological knowledge, applications, and markets (Meijer and Hekkert, 2007). The role of the entrepreneur is to harness the potential of innovations through concrete actions that both generate and take advantage of new commercial opportunities. Thus entrepreneurs take risk, create variety, and "field test" innovations to test the reactions to new applications of end users as well as competitors, develop business networks and markets, and in so doing, reduce uncertainties to pave the way for more risk-averse innovation actors. Feedback from end users can also be an important source of innovation in and of itself (Thomke and Hippel, 2002). Entrepreneurs can either be new entrants who see new market opportunities (e.g., spin-offs from university labs) or incumbents who diversify their business strategy to take advantage of new developments (e.g., large energy companies).

3.2.3 Shared Expectations

Innovation is always characterised by uncertainty. Shared or collective expectations are an important institution for reducing uncertainty and stimulating entrepreneurial activity (Borup et al., 2006; Van Lente and Rip, 1998). Shared expectations help "guide the search" of actors within the innovation system along more promising technological avenues. This may involve selecting alternatives from the variety of contrasting and competing designs created by knowledge generation activities in the early stages of the innovation life cycle (Hekkert et al., 2007). This initial "motor of innovation" can create a virtuous cycle as expectations build and stimulate further investment and activities in knowledge generation (Suurs and Hekkert, 2009a). In turn, this can attract entrepreneurs and risk-taking private sector investors whose involvement in knowledge flows and spillovers and whose integration into emerging networks of innovation actors helps establish bridges between the research and commercial communities.

Developing shared expectations also involves risk. If developed too soon, shared expectations may result in the premature selection and emphasis on a technology that ultimately proves inferior. If developed too fast or too exuberantly, shared expectations may lead to disappointment and a lasting legacy of failure. Various of our case studies describe how this can happen.

Many actors can be instrumental in shaping shared expectations. Policy makers can represent social preferences by defining policy objectives such as mitigating climate change. Firms can orient their R&D budgets and field-testing activities towards certain innovations. Technology road maps involving a range of both public and private innovation actors can enshrine a collective goal and strategy towards that goal. Strong political or civil society leadership can shape public opinion and acceptance, and so the ultimate payoffs for successful innovations in the mass market. These different processes rarely occur in isolation. In the case of vehicles, for example, fuel efficiency regulations and large-scale R&D programmes into low-emission alternatives to the internal combustion engine have jointly shaped the research direction of car manufacturers. Various of our case studies cover this innovation history.

3.2.4 Advocacy Coalitions and Resistance

New technologies face resistance from innovation actors whose commercial interests are vested in the incumbent system. This produces a natural inertia or resistance to innovations – particularly in the energy system with its characteristics of technological interdependence, infrastructure, and long-lived capital stock. Advocacy coalitions are an important institution to help counteract this resistance and inertia and to support the development of an innovation system around a promising innovation (Sabatier, 1987, 1988). The actors involved in advocacy coalitions are typically from NGOs and industry as political lobbying is an important advocacy strategy. Desired outcomes might be to implement or enhance market formation activities, or to align markets or other institutions with the needs and characteristics of a new technology. If successful, this collective activity can help overcome the resistance to innovations from incumbents. Public actors may also contribute to advocacy coalitions (Fligstein, 1997). An example would be planning authorities advising local or national governments to develop framework policies for emerging technologies to ensure their fair treatment relative to the incumbents. In general, innovation actors in advocacy coalitions try to convince other actors to take particular actions that they cannot take themselves.

3.3 Dimensions and Processes (III) – Adoption and Use

3.3.1 Introduction – Adoption and Use

The ultimate measure of success for a particular technology is its widespread adoption and use. Technology adoption and use is therefore the third dimension of the energy technology innovation system and is represented by the wedge-shaped block on the right of Figure 2.1.

End users or technology adopters in either specific market niches or in the wider mass market have preferences for technologies that provide them with useful services. These preferences interact with the relative advantage of an innovation over prevailing technologies and practices. This relative advantage, particularly in terms of performance and cost, strongly influences an innovation's prospects. The learning effects that were introduced as a process for knowledge generation clearly affect these attributes. So too do specific characteristics of a technology, particularly those relating to scale and so the potential for scaling up unit sizes or manufacturing runs. For energy technologies, economies (and diseconomies) of scale have proven important determinants of cost, as well as part of the shift towards standardisation and mass production as a precursor to widespread diffusion.

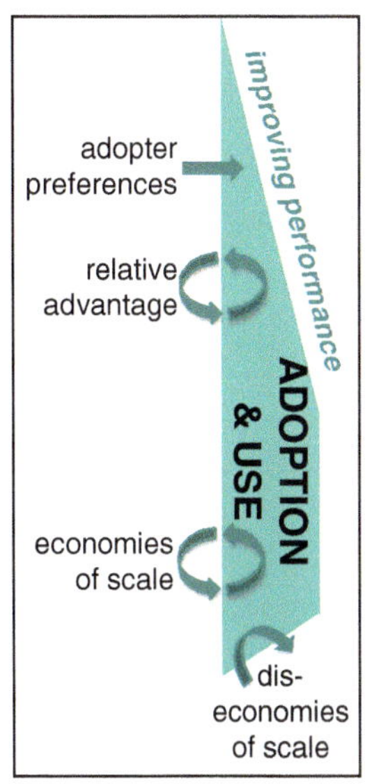

These various innovation processes influencing the adoption and use of an innovation are shown as arrows in Figure 2.1.

3.3.2 Adopter Preferences and Relative Advantage

Innovation systems research has typically paid less attention to the diffusion and use of a technology (Geels, 2004). Yet the needs and preferences of technology adopters distinguish innovation successes from failures.

The relative advantage of a technology and so its attractiveness to potential users is fundamentally related to the end-use service it provides. The two major energy transitions since the Industrial Revolution – from wood and horse power to coal and steam, then from coal and steam to oil and electricity – were not driven by scarce resources or price signals, but by innovations in end-use technologies and the services they provided to consumers (Grubler, 2008). The widespread diffusion of steam and gasoline engines and of electric motors and appliances then drove change in the energy supply and the use of energy resources (Wilson and Grubler, 2011). As well as being novel, these innovations were attractive to consumers as they dramatically altered the performance of end-use technologies and the cost of providing end-use services such as mobility, industrial power, or lighting.

New technologies when introduced are often crude, imperfect, and expensive (Rosenberg, 1994). The initial costs of steam engines in the mid eighteenth century amounted to a phenomenal US$12,000 per kW in an economy 130 times smaller than today with per capita incomes around US$1,500 (all values in 2003$) (Crafts, 2004). Yet despite their high cost, the performance benefits of steam engines in terms of increased power output and density ensured their attractiveness relative to the incumbent power providers, horses and water. These performance advantages are sought initially by end users in particular settings with particular needs: in other words, market niches. In the case of early steam engines, a key early niche was the dewatering of deep coal mines. End users in market niches are generally less sensitive to the relative cost of the energy service or have a higher willingness to pay for its performance advantages (Fouquet, 2010). Other performance characteristics relevant for energy technologies include versatility, reliability, storability, environmental impact, and so on.

Widespread diffusion is then initiated and sustained by increasing improvements in terms of cost (Fouquet, 2010). Costs tend to start to fall significantly only after an extended period of commercial testing, experimentation, learning, up-scaling, efficiency improvements, and other incremental refinements. Continuing reductions in the effective cost of providing an energy service allow the innovation to compete economically with the incumbent technology and so diffuse out of its niches and into the broader mass market. In the case of steam engines, costs started to come down during the mid nineteenth century, 100 years after their introduction. By the beginning of the twentieth century, costs had fallen by a factor of four.

3.3.3 Economies of Scale

Economies (and diseconomies) of scale have been influential processes in the historical evolution of energy industries such as electric power (Lee and Loftness, 1987) or petroleum refining (Enos, 1962). Economies of scale describe reductions in average unit costs as output or production increase over the long run, that is, assuming

Table 2.2. Economies of scale at different levels using wind power as an example.

Level of economy of scale[a]	Example of underlying driver of economy of scale	*Outcome* of economy of scale
unit level	taller towers capture stronger and more laminar winds *N.B. technical returns to scale: inputs proportional to blade length, outputs proportional to swept area (blade length squared)*	lower $/MWh for larger MW wind turbines *higher MWh/MW for larger MW wind turbines (no reference to costs)*
plant or facility level	fixed balance of system components spread over larger numbers of operating units	lower $/MW for larger MW wind farms
manufacturing or production level	fixed capital investments spread over higher output volumes; labour specialisation	lower $/turbine for larger manufacturing facilities

[a] Additional nontechnical or "pecuniary" economies of scale can result from larger firms or industries exerting political influence to secure increased subsidies (Rosegger, 1996). And external or inter-industry economies of scale can result from developments in enabling infrastructure such as transmission grids or in complementary industries such as equipment suppliers (Henriksen, Steen, and Ulltveit-Moe, 2001).

all factors of production (capital, labour, land, etc.) are variable (see chapter 4 of Rosegger, 1996) for underlying theory). Economies of scale are often conflated with technical returns to scale that describe a more than proportional increase in output for a given increase in inputs (with no reference to cost). Economies of scale can act at different levels. Cost reductions associated with up-scaling individual units of a technology (e.g., the capacity of a single wind turbine) or manufacturing plants (e.g., the number of turbine units produced in a single wind turbine factory) are both important processes influencing costs. Table 2.2 illustrates these scale effects using wind power as an example.

As well as affecting costs, economies of scale are associated with performance. The bywords of Fordist manufacturing include *specialisation* of machine tools, *routinisation* of labour tasks, *standardisation* of output, and *sequencing* of manufacturing along assembly lines (Raff, 1991). These technical, process, and organisational innovations affected labour productivity but also product quality. Standardisation in particular can lead to sustained improvements in the regularity and reliability of product performance.

3.3.4 Diseconomies of Scale

Diseconomies of scale describe *rising* average unit costs as output or production increase over the long run. These often arise because of the complexity of designing, building, and operating integrated technologies close to the unit scale frontier (Rosegger, 1996: 73). An example in the energy sector is the experience with large-scale power plants in the late 1970s. Their complexity meant delays during the construction phase, which added to borrowing costs (Koomey and Hultman, 2007). In the case of nuclear power plants (the subject of one of our case studies), this was exacerbated by further delays during the planning and permitting phases. Another consequence of complexity was an increase in the frequency and duration of unplanned plant shutdowns due to technical problems during the operating phase, which reduced revenues (Lovins et al., 2003; Ruttan, 2001). Increasingly stringent health, safety, and environmental regulations also impacted larger-scale plants more severely, illustrating that (dis)economies of scale are contingent on both technical and market factors.

3.4 Dimensions and Processes (IV) – Resources

3.4.1 Introduction – Resources

The fourth dimension to the energy technology innovation system is resources, represented by the wedge-shaped block on the left of Figure 2.1. Resources describe the key input into innovation systems, and mobilising resources is an essential innovation process. Resources can be financial, human, and intellectual property related, and can include the time and effort expended by innovation actors and nonfinancial investments in strengthening institutions and networks. Public policy plays a critical role, both as a direct source of resources (e.g., R&D financing and subsidies to support market formation), and in establishing an enabling framework for the mobilisation of private sector resources. The innovation processes relating to resource mobilisation and public policy shown as arrows in Figure 2.1 are therefore fundamentally connected to the other dimensions of the innovation system.

An additional innovation process shown as an outward arrow in Figure 2.1 concerns the legacy of failure. Resources mobilised that fail to bear fruit can negatively impact the availability of further resources in a particular innovation system, leading to stop-start progress or in the worst-case scenario, a failure of the innovation system to support a technology through its life cycle.

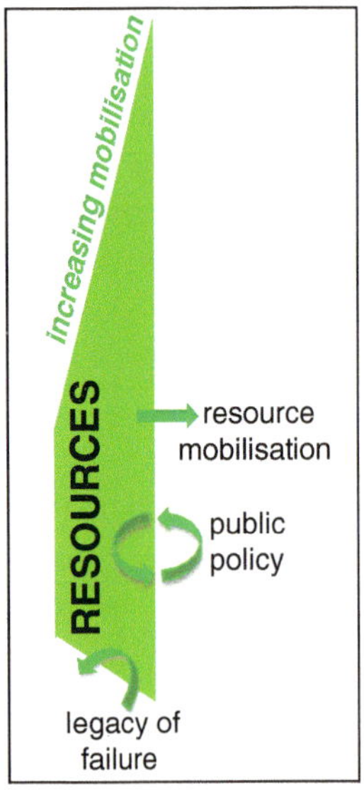

3.4.2 Resource Mobilisation

Resource mobilisation is an important indicator of the type and quantity of inputs into an energy technology innovation system. Many different types of resources are needed for innovation system development: RD&D activities, skills and competences, trained personnel, experienced engineers and financiers, codified knowledge, supportive institutions and markets, robust networks between innovation actors, and so on. Financial investments are often the most visible and quantifiable resource, and so are discussed in depth in a dedicated case study. Financial investments also serve as a useful indicator for firms and policy makers operating within budget constraints. As with other types of resource, financial investments can be classified according to innovation stage.

R&D investments, for example, are a well-defined expenditure category in macroeconomic and corporate accounts. Market formation investments are less clearly defined but typically relate either to risk-taking private finance such as venture capital, or to market incentives such as production subsidies. Diffusion investments dominate the costs of energy technology innovation systems as they involve the large-scale adoption of capital-intensive, embodied technological artefacts. Whereas the magnitude of energy supply investments are reasonably well understood, diffusion investments in energy end-use technologies go largely unrecorded (an omission that one of our case studies redresses).

The scale of resource mobilisation increases across these successive innovation stages from R&D to demonstration, to market formation and widespread diffusion. A first quantitative synthesis (described in detail in another of our case studies) suggests R&D investments of $50 billion, niche market investments of $150 billion, and diffusion investments of between $1,000 billion and $4,000 billion (estimates for 2005 and in 2005$). A wide range of public and private actors are involved in this resource mobilisation effort to support the energy technology innovation system, and in developing their skills and competences as a result (Carlsson and Stankiewicz, 1991).

3.4.3 Public Policy

There are good reasons for public policy to support innovation processes and the development of innovation systems. First and foremost, knowledge is largely a public good. Once produced and disclosed, it is difficult to control or restrict its use. So while expensive to generate through R&D, it may be cheap to reproduce. Although intellectual property rights are designed to reward private investors in knowledge generation, they may be unable to fully appropriate the benefits of their investments. The public good characteristics of knowledge lead to underinvestment, which is suboptimal from a social perspective (Nemet and Kammen, 2007). Policy makers thus have an important role to play at particular stages of the innovation life cycle, including supporting basic R&D with highly uncertain or distant commercial payoffs, or investing in R&D and testing infrastructure.

A second reason for public policy to support innovation is more specific to energy technologies. The many environmental costs and benefits associated with energy resources and technologies are treated as externalities by markets. Private investors in energy technologies with lower environmental impact cannot reap these benefits, again resulting in suboptimal investment from a social perspective. These two market failures relating to innovation and the environment have been termed the "double externality" problem that characterises much energy technology innovation (Jaffe, Newell, and Stavins, 2005). Policies to regulate or internalise within markets environmental externalities have historically focussed on local and regional air pollutants and now on greenhouse gas emissions. But in an energy context, security, reliability, land use, and resilience are all examples of broader social consequences of energy technology innovation to which private actors involved in resource mobilisation are incompletely or insufficiently exposed.

More generally, policy makers are concerned with energy technologies and the services they provide as these are integral to fundamental human needs and well-being. This is taken for granted and so less evident in developed countries. Yet as the *Global Energy Assessment* sets out in detail, expanding and improving access to modern forms of energy like electricity is of vital developmental importance in many developing countries, particularly in rural areas (GEA, 2012). The energy sector is also a major part of the economy whose fortunes are inextricably linked to economic growth, employment, manufacturing output, and so on. The potential impact of energy technology innovation on the broader economy is a further reason for policy makers' attention.

Innovation policies can directly target particular stages of the innovation process. Many examples have been noted already; the distinction between technology-push and market-pull policies and activities was also explained in the opening chapter. But policies can also support the development of the innovation system across all of its dimensions (see Figure 2.1). Examples include policies that support exchanges and interactions between innovation actors, policies that support testing and experimentation linking knowledge generation to adoption and use, or policies that require codification (e.g., performance disclosure) as a quid pro quo for public resource mobilisation (e.g., via subsidies). All these and more are discussed in detail in various of the case studies.

Finally, policies can unintentionally impact innovation systems while targeting an unrelated concern. Examples include policies on education, taxes and subsidies, and market regulation, which can all exert an important, yet indirect influence over both the supply of and demand for energy technology innovations. This points to the need for consistency between innovation policies and the broader regulatory and institutional environments comprising innovation systems. We return to this point throughout the case studies and again in the concluding chapters.

REFERENCES

Alchian, A., 1963. Reliability of Progress Curves in Airframe Production. *Econometrica: Journal of the Econometric Society*, 31(4): 679–93.

Arrow, K. J., 1962. The Economic Implications of Learning by Doing. *The Review of Economic Studies*, 29(3): 155–73.

Ausubel, J. H. & Marchetti, C., 1997. Elektron: Electrical Systems in Retrospect and Prospect. In Ausubel, J. H. & Langford, H. D. (eds.) *Technological Trajectories and the Human Environment*. Washington, DC: National Academy Press.

Borup, M., Brown, N., Konrad, K., & Van Lente, H., 2006. The Sociology of Expectations in Science and Technology. *Technology Analysis & Strategic Management*, 18(3–4): 285–98.

Carlsson, B. & Stankiewicz, R., 1991. On the Nature, Function and Composition of Technological Systems. *Journal of Evolutionary Economics*, 1(2): 93–119.

Cohen, W. M. & Levinthal, D. A., 1990. Absorptive Capacity: A New Perspective on Learning and Innovation. *Administrative Science Quarterly*, 35(1): 128–52.

Cowan, R. & Foray, D., 1997. The Economics of Codification and the Diffusion of Knowledge. *Industrial and Corporate Change*, 6(3): 595–622.

Crafts, N., 2004. Steam as a General Purpose Technology: A Growth Accounting Perspective. *Economic Journal*, 114(495): 338–51.

Dutton, J. M. & Thomas, A., 1984. Treating Progress Functions as a Managerial Opportunity. *Academy of Management. The Academy of Management Review*, 9(2): 235–48.

Edquist, C., 2001. Innovation Systems and Innovation Policy: The State of the Art. *DRUID's Nelson-Winter Conference*. Aarlborg.

Edquist, C. & Johnson, B., 1997. Institution and Organizations in Systems of Innovation. In Edquist, C. (ed.) *Systems of Innovation: Technologies, Institutions and Organization*. London: Pinter Publishers.

Enos, J. L., 1962. *Petroleum Progress and Profits: A History of Process Innovation*. Cambridge, MA: MIT Press.
Fligstein, N., 1997. Social Skill and Institutional Theory. *American Behavioral Scientist*, 40(4): 397–406.
Foray, D., 2004. *The Economics of Knowledge*. Cambridge, MA: MIT Press.
Fouquet, R., 2010. The Slow Search for Solutions: Lessons from Historical Energy Transitions by Sector and Service. *Energy Policy*, 38: 6586–96.
Freeman, C. & Perez, C., 1988. Structural Crises of Adjustment, Business Cycles and Investment Behaviour. In Dosi, G., Freeman, C., Nelson, R., Silverberg, G., & Soete, L. (eds.) *Technical Change and Economic Theory*. London: Pinter Publishers.
Freeman, C. & Soete, L., 2000. *The Economics of Industrial Innovation*. Cambridge, MA: MIT Press.
Fridlund, M., 2000. Procuring Products and Power, Developing International Competitiveness in Swedish Electrotechnology and Electric Power. In Edquist, C., Hommen, L., & Sipouri, L. (eds.) *Public Technology Procurement and Innovation*. Dordrecht: Kluwer Academic Publishers.
GEA, 2012. *Global Energy Assessment – Toward a Sustainable Future*, Cambridge, UK and New York, USA and the International Institute for Applied Systems Analysis, Laxenburg, Austria: Cambridge University Press.
Geels, F. W., 2004. From Sectoral Systems of Innovation to Socio-technical Systems: Insights about Dynamics and Change from Sociology and Institutional Theory. *Research Policy*, 33(6–7): 897–920.
Grubler, A., 1998. *Technology and Global Change*, Cambridge, UK: Cambridge University Press.
Grubler, A., 2008. Energy Transitions. In Cleveland, C. J. (ed.) *Encyclopedia of Earth*. Washington, DC: Environmental Information Coalition, National Council for Science and the Environment.
Grubler, A., Aguayo, F., Gallagher, K., Hekkert, M., Jiang, K., Mytelka, L., Neij, L., Nemet, G., & Wilson, C., 2012. Chapter 24 – Policies for the Energy Technology Innovation System (ETIS). *Global Energy Assessment – Toward a Sustainable Future*. Cambridge University Press, Cambridge, UK and New York, NY, USA and the International Institute for Applied Systems Analysis, Laxenburg, Austria: Cambridge University Press.
Halsnæs, K., Shukla, P., Ahuja, D., Akumu, G., Beale, R., Edmonds, J., Gollier, C., Grubler, A., Ha Duong, M., Markandya, A., McFarland, M., Nikitina, E., Sugiyama, T., Villavicencio, A., & Zou, J., 2007. Framing Issues. In Metz, B., Davidson, O., Bosch, P., Dave, R., & Meyer, L. (eds.) *Climate Change 2007: Mitigation*. Cambridge, UK: Contribution of Working Group III to the Fourth Assessment Report of the Intergovernmental Panel on Climate Change, Cambridge University Press.
Hekkert, M. P., Suurs, R. A. A., Negro, S. O., Kuhlmann, S., & Smits, R. E. H. M., 2007. Functions of Innovation Systems: A New Approach for Analysing Technological Change. *Technological Forecasting and Social Change*, 74(4): 413–32.
Henriksen, E., Steen, F., & Ulltveit-Moe, K., 2001. *Economies of Scale in European Manufacturing Revisited*. London: Centre for Economic Policy Research.
Hoffert, M. I., 2010. Farewell to Fossil Fuels? *Science*, 329(5997): 1292–94.
Hoffert, M. I., Caldeira, K., Benford, G., Criswell, D. R., Green, C., Herzog, H., Jain, A. K., Kheshgi, H. S., Lackner, K. S., Lewis, J. S., Lightfoot, H. D., Manheimer, W., Mankins, J. C., Mauel, M. E., Perkins, L. J., Schlesinger, M. E., Volk, T., & Wigley, T. M. L., 2002. Advanced Technology Paths to Global Climate Stability: Energy for a Greenhouse Planet. *Science*, 298(5595): 981–87.
Jacobsson, S. & Johnson, A., 2000. The Diffusion of Renewable Energy Technology: An Analytical Framework and Key Issues for Research. *Energy Policy*, 28(9): 625–40.
Jacobsson, S. & Lauber, V., 2006. The Politics and Policy of Energy System Transformation-Explaining the German Diffusion of Renewable Energy Technology. *Energy Policy*, 34(3): 256–76.
Jaffe, A. B., Newell, R. G., & Stavins, R. N., 2005. A Tale of Two Market Failures: Technology and Environmental Policy. *Ecological Economics*, 54(2–3): 164–74.

Kamp, L., Smits, R., Andriesse, R., 2004. Notions on Learning Applied to Wind Turbine Development in the Netherlands and Denmark. *Energy Policy*, 32(1625–37).

Koomey, J. & Hultman, N. E., 2007. A Reactor-Level Analysis of Busbar Costs for US Nuclear Plants, 1970–2005. *Energy Policy*, 35(11): 5630–42.

Lako, P., 2004. *Spillover Effects from Wind Power: Case Study in the Framework of the Project Spillovers of Climate Policy*. Petten, the Netherlands: Energy Research Centre of the Netherlands.

Lee, T. H. & Loftness, R. L., 1987. Managing Electrotechnology Innovation in the USA. *Working Papper WP-87–54*. Laxenburg, Austria.

Lovins, A., Kyle Datta, E., Feiler, T., Rabago, K. R., Swisher, J., Lehmann, A., & Wicker, K., 2003. *Small is Profitable: The Hidden Economic Benefits of Making Electrical Resources the Right Size*, Snowmass, CO: Rocky Mountain Institute.

Lundvall, B.-Å., 1998. Why Study National Systems and National Styles of Innovation? *Technology Analysis & Strategic Management*, 10(4): 407–22.

Lundvall, B.-Å. (ed.) 1992. *National Systems of Innovation: Towards a Theory of Innovation and Interactive Learning*, London: Pinter Publishers.

2007. National Innovation Systems-Analytical Concept and Development Tool. *Industry and Innovation*, 14(1): 95–119.

2009. Innovation as an Interactive Process: User-Producer Interaction to the National System of Innovation. *African Journal of Science, Technology, Innovation and Development*, 1(2&3): 10–34.

Meijer, I. & Hekkert, M. P., 2007. Managing Uncertainties in the Transition Towards Sustainability: Cases of Emerging Energy Technologies in The Netherlands. *Journal of Environmental Policy and Planning*, 9(3–4): 281–98.

Mohnen, P., 1997. Introduction: Input-Output Analysis of Interindustry R&D Spillovers. *Economic Systems Research*, 9(1): 3–9.

Neij, L., 2004. The Development of the Experience Curve Concept and its Application in Energy Policy Assessment. *International Journal of Energy Technology and Policy*, 2(1–2): 3–14.

Nelson, R. (ed.) 1993. *National Innovation Systems: A Comparative Analysis*. New York: Oxford University Press.

Nelson, R. & Winter, S., 1982. *An Evolutionary Theory of Economic Change*. Cambridge, MA: Harvard University Press.

Nemet, G. F., 2006. Beyond the Learning Curve: Factors Influencing Cost Reductions in Photovoltaics. *Energy Policy*, 34(17): 3218–32.

Nemet, G. F. & Kammen, D. M., 2007. U.S. Energy Research and Development: Declining Investment, Increasing Need, and the Feasibility of Expansion. *Energy Policy*, 35(1): 746–55.

Pacala, S. & Socolow, R., 2004. Stabilization Wedges: Solving the Climate Problem for the Next 50 Years with Current Technologies. *Science*, 305(5686): 968–72.

Raff, D. M. G., 1991. Making Cars and Making Money in the Interwar Automobile Industry: Economies of Scale and Scope and the Manufacturing behind the Marketing. *The Business History Review*, 65(4): 721–53.

Rapping, L., 1965. Learning and World War 2 Production Functions. *The Review of Economics and Statistics*, 47(1): 81–86.

Rogers, E. M., 2003. *Diffusion of Innovations*. New York: Free Press.

Rosegger, G., 1996. *The Economics of Production and Innovation: An Industrial Perspective*. Oxford, UK: Butterworth-Heinemann Ltd.

Rosenberg, N., 1982. *Inside the Black Box: Technology and Economics*. Cambridge, UK: Cambridge University Press.

1994. *Exploring the Black Box: Technology, Economics, and History*. Cambridge, UK: Cambridge University Press.

Ruttan, V., 2001. *Technology, Growth, and Development: An Induced Innovation Perspective*. Oxford, UK, Oxford University Press.

Ruttan, V. W., 1996. Induced Innovation and Path Dependence: A Reassessment with Respect to Agricultural Development and the Environment. *Technological Forecasting and Social Change*, 53(1): 41–59.
Sabatier, P. A., 1987. Knowledge, Policy-Oriented Learning, and Policy Change. *Science Communication*, 8(4): 649–92.
1988. An Advocacy Coalition Framework of Policy Change and the Role of Policy-Oriented Learning Therein. *Policy Sciences*, 21(2–4): 129–69.
Socolow, R., 2011. Wedges Reaffirmed. *Climate Central*, September 27, 2011.
Suurs, R. A. A. & Hekkert, M. P., 2009a. Competition between First and Second Generation Technologies: Lessons from the Formation of a Biofuels Innovation System in the Netherlands. *Energy*, 34(5): 669–79.
2009b. Cumulative Causation in the Formation of a Technological Innovation System: The Case of Biofuels in The Netherlands. *Technological Forecasting and Social Change*, 76(8): 1003–20.
Thomke, S. & Hippel, E. V., 2002. Customers as Innovators: A New Way to Create Value. *Harvard Business Review*, 80(4): 74–81.
Van der Zwaan, B. & Seebregts, A., 2004. Endogenous Learning in Climate-Energy-Economic Models – An Inventory of Key Uncertainties. *International Journal of Energy Technology and Policy*, 2(1–2): 130–41.
Van Lente, H. & Rip, A., 1998. The Rise of Membrane Technology: From Rhetorics to Social Reality. *Social Studies of Science*, 28(2): 221–54.
Van Pottelsberghe de la Potterie, B., 1997. Issues in Assessing the Effect of Interindustry R&D Spillovers. *Economic Systems Research*, 9(4): 331–57.
Von Hippel, E., 1988. *The Sources of Innovation*, Oxford, UK: Oxford University Press.
1994. "Sticky Information" and the Locus of Problem Solving: Implications for Innovation. *Management Science*, 40(4): 429–39.
Watanabe, C., Griffy-Brown, C., Zhu, B., & Nagamatsu, A., 2002. Inter-Firm Technology Spillover and the "Virtuous Cycle" of Photovoltaic Development in Japan. In Grubler, A., Nakicenovic, N., & Nordhaus, W. (eds.) *Technological Change and the Environment*. Washington, DC: Resources for the Future.
Wilson, C. & Grubler, A., 2011. Lessons from the History of Technological Change for Clean Energy Scenarios and Policies. *Natural Resources Forum*, 35: 165–84.
Wright, T. P., 1936. Factors Affecting the Cost of Airplanes. *Journal of Aeronautical Sciences*, 3: 122–28.

3 Historical Case Studies of Energy Technology Innovation

Arnulf Grubler and Charlie Wilson

1 Rationale for a Case Study Approach

Our evolving understanding of energy technology innovation and its systemic nature highlights the importance of contextual factors and the specific characteristics of technologies' relative advantages, of influential actors and institutions, and of the public policies that support the mobilisation of resources. These can be taken into account only by looking at specific innovation histories. For this reason, we have adopted a comparative case study approach for this book.

Although the growing body of quantitative analysis and models has greatly improved our understanding of innovation processes and energy technology innovation, they tend to emphasise particular elements of the energy technology innovation system that constrain their broader explanatory power. Reducing the complex dynamics of innovation to a few factors and relationships – as with learning curve models, for example – omits important characteristics of the broader innovation system and so offers overly mechanistic explanations of causality. The case study approach adopted here aims to mitigate potential oversimplifications or overgeneralisations.

Quantitative assessments are also limited by data availability. This may explain the lack of empirical work on energy end-use technologies relative to energy supply technologies. It may also limit cross-country comparisons if there are insufficient data or variables to control for between-country heterogeneity.

Most of the case studies included in this book are quantitatively founded as well as qualitatively rich. They allow an explicit assessment of the many and varied dimensions and processes of an innovation system. They allow for complexities, feedbacks, and interdependencies to be characterised and explored. These characteristics of comparative case study analysis are well suited to the development and application of the framework set out in the previous chapter.

But a case study approach also has inherent limitations. Generalising from case studies is limited by the specifics of contextual and technical characteristics. One consequence is that insights for energy technology and innovation policy that are extracted and discussed in the concluding chapters are principles and guidelines only, and not prescriptions nor universals. Selection bias is another issue with case study analysis. Technology diffusion scholars recognise the bias towards innovation successes rather than failures, with the conditions of failure often assumed to be the inverse of those for success (Rogers, 2003). We try to address some of these problems through our rationale for selecting case studies.

2 Selecting the Case Studies

The selection and commissioning of case studies for inclusion in this book were guided by a number of criteria.

Firstly, in terms of emphasis, we wanted studies focussed on individual energy technologies in specific contexts as well as comparative assessments of technologies across different contexts. But we also wanted meta-analytic or thematic studies of key innovation system processes or dimensions to provide a broader overview or to draw out particular points that the technology-focussed studies might not cover.

Secondly, in terms of technologies, we wanted studies of both end-use and energy supply technologies, of both developed and developing country contexts, and of both current and historical developments. This was to ensure we covered the heterogeneity of experience with energy technology innovation so we could generalise an analytical framework of the energy technology innovation system. This aim of broad coverage was, however, constrained by the available data, which are most sparse for end-use technologies in developing countries.

Thirdly, in terms of innovation outcomes, we wanted studies of both (relative) successes and failures, and of innovations expected to play an increasing role in sustainable energy transitions, as well as those with less clear social or environmental benefits. The success bias of innovation studies has already been noted, and we intentionally identified cases of relative failure or evident imperfections in energy technology innovation.

Fourthly, in terms of insights, we wanted studies that elucidated the particular role played by public policy. This was guided by our aim to generalise guidelines on energy technology innovation policy (which are synthesised in the concluding chapter).

3 Overview of the Case Studies

The twenty case studies selected are summarised in Table 3.1 with short descriptions of each. We have organised them in five groups as follows:

- *patterns and linkages*: case studies 1–4;
- *knowledge*: case studies 5–8;
- *adoption and use*: case studies 9–12;
- *actors and institutions*: case studies 13–16;
- *resources*: case studies 17–20.

The first group (case studies 1–4) corresponds to the energy technology innovation system as a whole. Case studies in this group explicitly adopt a systemic perspective and demonstrate the benefits of so doing. The other four groups correspond to the four dimensions of the energy technology innovation system: knowledge, adoption and use, actors and institutions, and resources. The case studies within each of these groups are particularly good illustrations or analyses of one or more innovation processes associated with that particular dimension. This does not mean they are

Table 3.1. Summary description of the historical case studies.

Short title		Authors	Summary description
SYSTEMIC PATTERNS			
1	Energy Transitions	Grubler	Review of patterns, drivers, and dynamics of energy systems, both historically and in future scenarios.
2	Technology Diffusion	Wilson	Comparison of rates and extents of growth in nine energy technologies historically.
3	Assessment Metrics	Wilson	Quantitative metrics and qualitative approaches for assessing innovation.
4	Technology Portfolios	Grubler, Fuss, McCollum, Krey, & Riahi	Tools to guide the selection of innovation portfolios under conditions of uncertainty.
KNOWLEDGE			
5	Solar Water Heaters	Nemet	Early success and later failure of the solar water heater industry, particularly in the United States.
6	Heat Pumps	Kiss, Neij, & Jakob	Different stages of heat pump diffusion in Sweden and Switzerland, emphasising the role of public policy.
7	Knowledge Depreciation	Grubler & Nemet	Loss or obsolescence of knowledge, with examples in the context of energy innovation.
8	Nuclear Power (France)	Grubler	Review of pressurised water reactor nuclear program in France, including cost escalation.
ADOPTION & USE			
9	Solar Thermal Electricity (United States)	Nemet	Early experience of solar thermal electricity in the United States, and spillovers to later stage production.
10	Vehicle Efficiency	Nemet	Standards for vehicle efficiency in the United States and elsewhere, and their influence on technological change.
11	Hybrid Cars	Gallagher	Development of hybrid electric vehicles in Japan, the United States, and China, emphasising the role of public policy.
12	Solar Photovoltaics	Nemet	Development of solar photovoltaics (PV) in different markets worldwide, focussing on drivers of cost reduction.
ACTORS & INSTITUTIONS			
13	Wind Power	Neij & Andersen	Evolution of innovation stages and strategies in different wind power markets worldwide.
14	End-Use Efficiency (Japan)	Kimura	The "Top Runner" program to improve end-use efficiencies in Japan, and the role of dynamic incentives.
15	Rural Solar (Kenya)	Kammen & Jacobson	Market dynamics in the solar photovoltaics market in Kenya, emphasising product quality issues.
16	Synfuels (US)	Anadón & Nemet	History of U.S. government investment in synthetic fuel production as oil substitute, and ultimate innovation "failure."

Short title		Authors	Summary description
RESOURCES			
17	Ethanol (Brazil)	Meyer, Mytelka, Press, Dall'Oglio, de Sousa, & Grubler	History of ethanol production and developments in automotive technologies in Brazil.
18	Global Financial Resources	Grubler, Anadón, Gallagher, Kempener, O'Rourke, & Wilson	Comparative assessment of financial resources mobilised globally at innovation, market formation and diffusion stages for energy technologies.
19	R,D&D Investments (Emerging Economies)	Kempener, Anadón, Gallagher, & Jiang	Review of public and private R,D&D investments in six major emerging economies, disaggregated by technology type.
20	Global End-Use Investments	Wilson & Grubler	Bottom-up estimation of current market investments in end-use technologies.

confined to that particular dimension only. In almost all cases, the case studies cover many different innovation system processes from across the four dimensions of the energy technology innovation system revealing its high degree of interdependence.

Additional characteristics of each case study are summarised in Table 3.2, which maps the case studies against the selection criteria set out earlier.

4 A Reader's Guide to the Case Studies

The introductory chapters should provide the reader with a sufficient overview of the energy technology innovation system framework, its dimensions and processes, and its key terms and concepts. The case studies add empirical detail and in-depth analysis to this framework (which ultimately was constructed on the basis of the case studies themselves). Each case study is designed as a standalone chapter. The reader may pick and choose according to the dimensions of the energy technology innovation system of interest (see Table 3.1) or according to the technology characteristics and contexts of interest (see Table 3.2). For the avid reader intending to read all case studies, this does not have to be done linearly.

Each case study begins with a succinct summary and a list of key points from the authors' perspective and we invite readers to read them all before examining a particular case study in more detail. In each case study, we also include an "editors' guide" box on the parts of the energy technology system innovation that the case study particularly serves to illustrate. These are intended to help the reader relate the case study material to the energy technology innovation system framework set out earlier. This framework should take clearer and clearer empirical shape as the reader progresses through the case studies. These "editors' guides" also provide a common voice or narrative throughout the case studies written by different authors

Table 3.2. Characteristics of the case studies.

		Criteria for case study selection	Thematic / meta-analytic	Supply technologies	End-Use technologies	Single context	Comparative context	Current	Historical	Developed country(s)	Developing country(s)	Influential public policy
Systemic Patterns	1	Energy Transitions	X	X	X		X		X	X	X	
	2	Technology Diffusion	X	X	X		X		X	X	X	
	3	Assessment Metrics	X									
	4	Technology Portfolios	X	X	X			X				
Knowledge	5	Solar Water Heaters			X	X			X	X		X
	6	Heat Pumps			X		X	X	X	X		X
	7	Knowledge Depreciation	X	X	X							
	8	Nuclear Power (France)		X		X			X	X		
Adoption & Use	9	Solar Thermal Electricity (United States)		X		X			X	X		X
	10	Vehicle Efficiency			X	X		X	X	X		X
	11	Hybrid Cars			X		X	X		X	X	X
	12	Solar Photovoltaics		X			X	X	X	X		X
Actors & Institutions	13	Wind Power		X			X	X	X	X	X	X
	14	End-Use Efficiency (Japan)			X	X		X	X	X		X
	15	Rural Solar (Kenya)		X		X		X	X		X	
	16	Synfuels (United States)		X		X			X	X		
Resources	17	Ethanol (Brazil)		X	X	X		X	X		X	X
	18	Global Financial Resources	X	X	X		X	X		X	X	X
	19	R,D&D Investments (Emerging Economies)	X	X	X		X	X			X	X
	20	Global End-Use Investments	X		X		X	X		X	X	

from different research traditions. In these editorial notes, we also preempt some of the conclusions by drawing out the key implications of the case study for energy technology innovation policy.

To conclude the book, we synthesise the findings from the case studies. In the first of two concluding chapters, we show how the case studies validate the energy technology innovation system framework. How this framework can be applied is then illustrated in the second concluding chapter, in which we generalise guidance for policy makers seeking to promote energy technology innovation and strengthen corresponding innovation systems.

REFERENCES

Rogers, E. M., 2003. *Diffusion of Innovations*. New York: Free Press.

PART II

PATTERNS AND LINKAGES IN THE ENERGY TECHNOLOGY INNOVATION SYSTEM

4 Grand Designs: Historical Patterns and Future Scenarios of Energy Technological Change

Arnulf Grubler

Author's Summary

The case study reviews patterns, drivers, and dynamics (rates of change) in energy systems from a historical as well as futures (scenario) perspective. From a historical perspective, two major energy transitions, each of which took up to a century to unfold, can be identified: the phase of growth in coal-fired steam power, and its subsequent displacement by oil and electricity-related end uses and technologies. Similar far-reaching future transitions are also described in the scenario literature as a function of alternative assumptions on rates and direction of innovative activities and performance and cost improvements of new energy technologies. These historical and future transitions share various characteristics. First, technological and social innovations have been core drivers of historical energy transitions and remain so in future scenarios. Second, energy history and futures are characterised by four "grand" patterns of technological change in energy systems: (i) clustering and spillover effects dominate over singular technologies; (ii) performance dominates over costs in the early phases of technology development; (iii) end-use applications dominate over energy supply; and (iv) time constants of change are substantial, spanning from many decades to a century. Finally, there is evidence that, contrary to popular conception, rates of system change have substantially slowed since the 1970s.

Editors' Guide

Relationship of Case Study to Energy Technology Innovation System Framework

This case study describing the grand patterns of technological change historically, and in the forward-looking scenario literature, captures many of the key interacting elements of the energy technology innovation system. Perhaps most striking are the long time constants of change attributed to the interplay between *knowledge generation* activities, the shifting *preferences of technology adopters* as the technologies mature, the *institutional capacity* needed to support broad-based industrial growth, and the magnitude of *resource mobilisation* this implies. These processes exemplify the four interacting dimensions of the energy technology innovation system.

The *relative advantage* of a technology is typically expressed as a function of better performance or lower cost. The long-term historical view, however, shows that novelty of energy service as a specific type of performance advantage is influential in the early phase of a technology's development. (This explains why technology *adoption and use* is included as one of the dimensions of the energy technology innovation system, even though on the surface it is more concerned with outputs and outcomes than inputs and processes). As the technology consolidates its position in *niche markets* protected by the price insensitivity of the early adopters, *learning* processes improve performance, reduce uncertainty, and reduce cost, building market demand among more risk-averse *potential adopters*. This greater clarity and confidence in expected market potential underwrites the necessary *resources needed to be mobilised* for industrial-scale production and deployment. This in turn relies on a strengthening constellation of *institutions* upon which the market builds, exemplified by the slow capturing of productivity benefits made available by electric drives in U.S. industry through the twentieth century. The synoptic historical view in this case study also illustrates clearly how energy technologies developed in one setting are applied in other settings. During the Industrial Revolution, stationary steam power initially developed to pump water out of coal mines was adapted for use in manufacturing, agricultural, and transport applications. These *knowledge spillovers* are pervasive.

The interactions between these processes – relating to *knowledge*, to *adoption and use*, to *resources*, and to *institutions* – helps explain the long time constants of change in the energy system. Also evident are the timing of particular innovation system processes over the innovation life cycle. The historical view shows clearly how niche markets play a critical role as a bridge across the "valley of death" between the R,D&D stages and the mass market. During the market formation stage, technology-push (supply-side) drivers to develop and improve performance most clearly interact with market-pull (demand-side) drivers that provide learning feedbacks for performance improvements and firm up expectations for market demand.

Insights for Energy Technology Innovation Policy

- End users and end-use services have historically been in the driving seat of technological change in the energy system. The case study also points out that forward-looking scenario analysis predominantly explores changes in the energy supply, given the difficulties of developing scenarios and modelling future changes in end-use services. There is a consequent skew (or bias) in the emphasis of innovation policy towards the supply side. End-use technologies dominate market investments and opportunities for efficiency improvements and so determine the size and composition of the energy supply. It is important to frame technology policy with the underlying purpose of the energy system, which is to provide useful services to satisfy human needs.

- One reason long-run change in the energy system is driven first and foremost by end-use technologies is that energy supply technologies are capital intensive. This magnifies the consequence of potential innovation failure and reduces the pool of actors (particularly investors) who can take risk and drive change. Low-carbon energy supply technologies and efficient energy end-use technologies share this capital intensiveness relative to the incumbent technologies. To the extent markets inadequately provide financing mechanisms to share upfront capital demands and associated risks among a wider set of actors (e.g., institutional investors) beyond the usual energy system suspects, public policy should play an important catalyst role.
- Technology studies and policies are often dominated by the view that reducing costs is the essential guarantor of future success. The historical view in this case study, represented empirically by the trajectory of cost decline for steam power, emphasises that cost reductions are secondary to performance advantages, at least initially, and that market-pull policies to support diffusion and so "buy-down" costs through learning processes may have to be sustained over very long periods with the large resource commitments this implies. An alternative emphasis on "performance-driven niche markets" warrants far greater attention.
- Climate change mitigation may require rapid change and turnover in the energy system, yet both historical analysis and the futures scenario literature emphasises the path dependence and inertia of complex technological systems. One implication is that "wait and see" policies are highly problematic. Near-term policy decisions open or close down the space for future technological diffusion. Another implication is the need to consider premature retirement of capital stock. This has occurred "naturally" with some information and telecommunication technologies as expanding services and improving performance advantages over successive generations of short-lived technologies overcome the relative cost advantages of amortised investments.

1 Introduction

Technological and congruent institutional and social change is widely recognised as the main driver of long-run economic growth ever since Solow (1957), and of development in general (Freeman and Perez, 1988). In terms of causality, technology and institutional/social settings *co-evolve*, mutually depending on and mutually cross-enhancing each other. Reality is a far cry away from simplistic, extreme notions of "technological determinism" (technology acting as a main agent of change) or "social construction" perspectives (in which the technological landscape is perceived as a mere outcome of the shaping of social forces or of class struggle). Energy technologies are no exception to this dichotomy of views, but no matter what the particular perspective, all scholars agree on the importance of technological change in past and future energy transitions (e.g., Grubler, 1998, 2008; Halsnæs et al., 2007; Nakicenovic et al., 2000; Smil, 1994).

The following sections synthesise our current understanding of the linkages of technological change and the evolution of energy systems from a historical as well as a futures (scenarios) perspective.

2 Grand Patterns of Change

Four "grand" patterns characterise technological change and their corresponding energy transitions.

First, no individual technology, as important as it may be, is able to transform whole energy systems that are large and complex. The importance of technology arises in particular through *clustering* effects (combinations of interrelated individual technologies) and *spillover* effects (applications outside the initial sector/use for which a technology was initially devised). In other words, technologies operate more effectively as families or as "gangs" and not as individuals. Because of clustering and spillovers, it is very difficult to dislodge a dominant technological "regime," a fact referred to in the technology literature as "path dependency" or "technology lock-in" (e.g., Arthur, 1989).

Second, any new technology introduced is initially crude, imperfect, and very expensive (Rosenberg, 1994). Performance (the ability to perform a particular task of delivering a novel energy service) initially dominates economics as a driver of technological change and diffusion. Only after an extended period of experimentation, learning, and improvements, and the establishment of a corresponding industrial base (in many cases profiting from standardisation, mass production, and scale economies of a growing industry), can new technologies begin to compete with existing ones on a pure cost basis. In other words: attractiveness beats cheap, at least initially.

Third, the history of past energy transitions highlights the critical importance of *end use* (i.e., consumers, energy demand) that dominates technology applications. Historically, energy supply has *followed* energy demand in technology applications, and energy end-use markets have been and remain the most important market outlets for new energy technologies. In other words: new energy technologies need to find consumers, and better many of them.

Finally, *fourth*, the process of technological change (from innovation to widespread diffusion) takes considerable time, as a rule many decades, and rates of change become slower the larger the energy system (components) affected (Grubler, 1996). These slow rates of change of energy technologies arise from four phenomena:

- the extended time required for experimentation, learning, and technology development from invention to innovation, to initial specialised niche market applications, and finally, in case of success, to pervasive adoption across many sectors, markets, and countries;
- the considerable time required for technology clustering and spillover effects to emerge;
- the capital intensiveness of many new energy technologies that *ceteris paribus* slows technology diffusion;
- the longevity of the capital stock of energy systems in many end-use applications (buildings), conversion technologies (refineries, power plants), and above all infrastructures (railway networks, electricity grids), that spans several decades to a century, again slowing diffusion of new technologies.

Only in exceptional cases does the diffusion of new energy technologies proceed via the premature retiring of existing capital stock, as is the case in current cell phone markets or with information and communication technologies in general. In other words: if technological change in energy systems is on the policy agenda, one needs both a long breath and an early start.

2.1 Technology and Historical Energy Transitions

Two major transitions have shaped global energy systems since the onset of the Industrial Revolution (Nakicenovic et al., 1998). The first is characterised by the emergence of steam power relying on coal that helped to overcome the constraints of preindustrial energy systems (limited availability of mechanical power, low energy density, lack of ubiquitous and cheap transport systems, cf. Landes, 1969). This first energy transition took well over a century to fully unfold: from the late eighteenth century until the 1920s, when coal-based steam power constituted well over two-thirds of the global energy system. The second energy transition is characterised by the displacement of the previously dominating coal-based steam technology cluster by electricity (drives, light) and petroleum-based technologies (automobiles, aircraft, petrochemicals). This second energy transition is far from complete: some 2 billion people still lack access to modern energy services provided by electric appliances and end-use devices, as documented in the *Global Energy Assessment.*

These two historical energy transitions are both characterised by the four "grand" patterns of technological change in energy systems outlined earlier: clustering and spillovers, the dominance of performance over costs in the early phases of technology development, and of end-use applications over energy supply, as well as the long time constants of change.

Stationary steam engines were first introduced in the eighteenth century for dewatering coal mines. Stationary steam power subsequently spilled over to drive mechanisation in manufacturing (e.g., textiles) and agriculture (threshing) and also to mobile applications in the form of railways and steamships. Technology researchers have introduced the concept of "general purpose" technologies (e.g., Lipsey, Carlaw, and Bekar, 2005) to describe the cross-enhancing effects that arise if a technology is deployed in a variety of applications furthering knowledge spillovers and market growth (and corresponding economies of scale). Steam and electricity, hailed as the "greatest engineering achievement of the 20th century" (NAE, 2003), are prominent examples of general purpose technologies that assume their importance because of clustering and spillover effects. Information and communication technologies are a good current example.

Initial steam engines were by any standards inefficient and extremely expensive. The first atmospheric steam engines had thermal conversion efficiencies of one percent (consuming some forty-five pounds of coal per horsepower delivered, Ayres, 1989). It took a century to boost their thermal efficiency in a successive stream of innovations to some 20 percent (from Newcomen's atmospheric engine, to Watt's low pressure, to high-pressure engines that finally made railroads possible), and another century to reach the current efficiency of steam turbines of 40 percent. Crafts (2004) provides estimates of the costs of steam engines in the eighteenth and nineteenth centuries. Initial costs (by 1760) amounted to a phenomenal 12,000

US$(2003) per kW, akin to current fuel cells (often classified as prohibitively expensive). Yet modest performance and high costs provided sufficient incentive for their deployment, as alternatives were either not available where needed (e.g., water power) or did not provide the performance required in terms of power output and density (e.g., horsepower). After an extended period of experimentation, costs started to come down after the mid nineteenth century (i.e., 100 years after the introduction of steam engines), and by the beginning of the twentieth century costs had fallen to below $3,000 ($2003)/kW. A similar story of the adoption of new energy technologies despite initially extremely high costs can be told about the introduction of electricity and electric appliances for light and motive power (Devine, 1983; Smil, 2000).

Perhaps current concerns with the high costs of adopting climate-friendly technologies arise from the fact that, with a few exceptions (e.g., solar PV in remote off-grid applications), many current new energy technologies offer little comparative performance advantage in terms of services provided to consumers (apart from lower emissions whose benefits remain externalised). The conventional response to these concerns is subsidies, at least temporarily until costs can be "b(r)ought down." But these subsidies may have to be sustained over many decades rather than years (as evidenced by the innovation history of Brazilian ethanol). Conversely, the historical precedent – developing and introducing technologies whose decisive performance edge does not necessarily require subsidies for early niche market adopters – seems to receive little attention. Evidently such performance-driven niche markets can be supported and stimulated by public policy. Consider the case where critical public and technological infrastructures like communication networks would be required to have 100 percent back-up systems (as, for example, mandated in Switzerland) that could create a rather price-insensitive niche market for off-grid, decentralised energy system solutions.

Another robust finding from historical energy transitions is the importance, even dominance of end-use applications and markets for the introduction and development of energy technologies. As discussed in the steam engine example, stationary steam engines in industry and agriculture and mobile steam engines on ships and locomotives were by far the dominant markets of these new technologies compared to the energy supply sector (coal mines and coking and town gas plants that illustrate the emerging cluster of a complex coal-based chemistry and associated conversion technologies). In the case of electricity, it is no coincidence that the first innovation leaving Thomas Edison's R&D laboratory in Menlo Park was the incandescent light bulb. In the technology language of today: demand innovation (the electric light bulb) triggers supply-side innovations (electricity generation, transport, and distribution).

Given the multitude of energy technologies and applications, it is difficult to obtain a comprehensive picture of the entire energy technology landscape, so one needs to rely on a simple common metric (installed power) and the example of a country, the United States, where such data are available. Table 4.1 summarises the evolution of energy technologies for the United States since 1850 in terms of installed capacity (horsepower and Watts, converted to GW), differentiating between stationary and mobile end-use applications, as well as energy (supply) sector applications and three broad energy conversion categories: thermal (boilers, furnaces), mechanical (prime

Table 4.1. Energy technologies in the United States in the nineteenth and twentieth centuries by generic type and application (in GW, rounded numbers).

GW		1850	1900	1950	2000
stationary	thermal (furnaces/boilers)	300	900	1900	2700
end use	mechanical (prime movers)	1	10	70	300
	electrical (drives, appliances)	0	20	200	2200
mobile	animals/ships/trains/aircraft	5	30	120	260
end use	automobiles	0	0	3300	25000
stationary	boilers (power plants)	0	10	260	2600
supply	mechanical (prime movers)	<1	3	70	800
TOTAL	(numbers rounded)	300	1000	6000	34000

Notes: Italics indicate first-order estimates.
Sources: Data are derived from US DOC (1975, 2007); for other sources, see text.

movers like steam engines or electric turbine-generators), and electrical (appliances, lights, and other specific, non-substitutable electricity uses such as communication via radios, TVs, and computers).

Table 4.1 clearly shows the persistent dominance of end-use applications in the total installed energy technology base. Installed horsepower of prime movers in 2000 are based on 1992 values (US DOC, 2007), the most recent available. Thermal applications (boilers in end-use and power plants) have been inferred from final energy use statistics multiplied by representative load factors, or are derived from the rated thermal capacity of power plants. Electric applications are conservative estimates based on statistics of industrial electric drives of some 175 GW (US DOE, 1998) and residential electricity use statistics by application (EIA, 2001) multiplied by appliance specific load factors. The dominant estimated electric household technologies are small-scale motors in kitchen appliances and compressor pumps in residential air conditioning units.

By the 1850s, the beginnings of the U.S. steam age, the dominant energy technologies consisted of simple conversion devices to convert chemical energy (mostly fuel wood and some coal) into heat (ovens, boilers, furnaces). Horses constituted the dominant transport technology (conversion of chemical energy [feed] into mechanical energy) that surpassed the installed horsepower of the first stationary steam engines fivefold. By 1900, close to the peak of the steam-coal energy technology cluster, thermal conversion in boilers and furnaces accounted for 90 percent of the one TW installed energy conversion technologies of the United States. Nonthermal applications only constituted some sixty-three GW, with two-thirds represented by steam engines and the remainder accounted for by electricity applications (twenty GW lights and motors in end use and three GW generators in power plants). A century later, energy technologies in the United States add up to some thirty-four TW (120 kW per capita, ten times as large as in 1850), and the spectacular expansion of end-use technologies in electrified homes and industry as well as private transportation with a more than 1000-fold increase since 1900 is particularly striking (see Table 4.1). Automobile (car and truck) engines comprise nearly three-quarters of the total, roughly exceeding the thermal capacity of electric power plants by a factor of ten, illustrating the theoretical potential to entirely displace centralised generation (and utilities) by vehicle on-board electricity generation (e.g., via fuel cell vehicles, as first proposed by Lovins et al., 1996).

Table 4.1 provides a powerful summary illustration of the importance of energy end use for energy technologies in terms of market outlets and also explains why the largest efficiency improvement potentials are in energy end-use sectors and not in energy supply.

A historical perspective also allows an appreciation of the considerable time required for a turnover of the energy technology capital stock that ranges from many decades to well over a century (for a review, see, for example, Grubler, 1996; Grubler et al., 1999; IPCC, 2001). Crafts (2004) summarises the diffusion of steam power in the British economy: it took steam close to 100 years (until the 1860s) to gain a 50 percent market share in total installed horsepower, gradually displacing wind and waterpower. The electrification in U.S. industry provides another example: it took some forty years (to the 1920s) for electric drives to account for 50 percent of all prime movers in U.S. industry (Ausubel and Marchetti, 1997), with substantial (capital and labour) productivity effects arising only after that threshold was passed (Devine, 1983). The factors explaining this slow rate of diffusion of new technologies include their initial imperfections and high costs, the lack of corresponding organisational/institutional changes to realise their full productivity potential, and clustering and spillover effects that perpetuated the dominance of the old technology while taking considerable time to emerge with new technologies.

The dynamics of technological change of national and global energy systems have been increasingly well documented since Marchetti and Nakicenovic's early studies (1979), and confirm typical energy systems turnover times of 70–120 years at the global level. Figure 4.1 summarises the previously discussed two "grand" energy transitions (coal/steam replacing traditional renewables, and in turn modern energy technologies and carriers replacing steam/coal). Coal maintains its market share in primary energy of some 26 percent in 2008 primarily via coal-based electricity generation (based on BP, 2009). In terms of final energy use, that is, purchased energy forms that are compatible with the prevailing technologies in use by consumers, coal's share was less than 9 percent in 2007 (IEA, 2009), almost exclusively (some 80 percent) used by heavy, metallurgical industry that technologically is the last current testimony of the nineteenth-century steam/coal age.

The turnover times of these two grand transitions have historically been remarkably consistent. (Turnover time is measured as the number of years to grow from 1 percent to 50 percent, or from 10 percent to 90 percent market share). Historical market dynamics are approximated in Figure 4.2 by a set of coupled logistic equations over the 1850 to 1975 period (dashed lines) indicating turnover times of 120 years (phase out of traditional fuels and phase in of coal) and 80 years (growth of modern fuels/technologies, decline of coal-based steam applications) respectively. Note in particular the long period of initial slow market penetration of new technologies.

Another disconcerting trend since the mid 1970s is equally visible in Figure 4.1: a significant slowdown of historically observed (slow) energy market dynamics to near stagnation (model estimates shown as solid lines). In terms of turnover times, the post-1975 trends suggest that it might take almost 1,000 years to complete the historical energy "modernisation" transition – bad auspices indeed in view of the

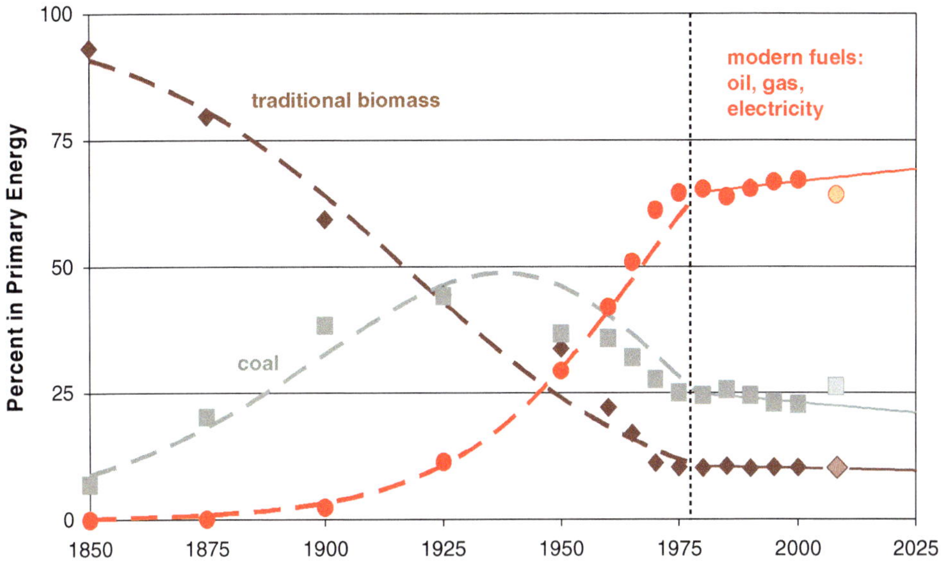

Figure 4.1. Two grand transitions in global energy systems measuring market shares in total primary energy use.
Notes: Traditional fuels (brown), coal (grey), and modern energy carriers (oil, gas, hydro, nuclear, and modern renewables, red). Historical market dynamics are approximated by a set of coupled logistic equations from 1850 to 1975 (dashed lines).
Sources: 1850 to 2000 data based on Grubler (2008); 2008 data (denoted by lighter shaded symbols) based on BP, 2009 and IEA, 2009.

increasingly recognised need for accelerated energy technology transitions in a climate-constrained world.

2.2 Technological Change in Future Scenarios

The role of technological change in future energy scenarios and in climate mitigation has been reviewed comprehensively in the IPCC's Fourth Assessment Report (Fisher et al., 2007; Halsnæs et al., 2007) and is also addressed in the *Global Energy Assessment*. Here we focus briefly on contrasting future scenarios with the historical perspective outlined previously, and also provide a simple illustration to show the levers of technological change in "bending the curve" in a transition to more sustainable energy systems. Such transitions often proceed along mutually exclusive, alternative technology pathways in the scenario literature and highlight the critical importance of near-term technology and policy decisions in structuring alternative long-term outcomes.

It is interesting to note that in contrast to the historical evidence, future scenarios portray almost exclusively technology transitions in energy supply systems, giving little detail on potentials and alternative pathways of technological change in energy end use. This conclusion does not suggest that the importance of end-use technologies goes unacknowledged by researchers and scenario modelling teams, but it rather reflects the current state of the art of modelling technological change in scenarios of energy transitions and climate stabilisation. Even the most detailed "bottom-up" models with detailed technological representations contain little detail at the level of energy end use. Rather technological change is commonly represented via aggregate indicators of energy use, that is, unspecified change is assumed to occur and is represented only in terms of energy demand outcomes in models. Reasons include both extreme paucity of technology-specific data as energy statistics are framed through the lens of economic activities and sectors rather than technologies, and the formidable challenge of deriving plausible and consistent scenario assumptions on the evolution of a potentially extremely large number of energy end-use applications (new transport and communication, manufacturing, and consumer applications) that would need to be modelled. To our knowledge, the scenario literature to date has included no example describing the potential evolution of alternative scenarios of new "general purpose" technologies like steam or electricity that have historically been the main technology drivers of energy transitions. Even scenarios of the possible emergence of a "hydrogen economy" (e.g., Barreto, Makihira, and Riahi, 2003) focus on technology substitutions of otherwise unchanged energy end-use service patterns. The best illustrations of the importance of technological change in future energy systems continue, therefore, to be those describing alternative transitions in energy supply systems.

Figure 4.2 gives an illustration from the IPCC Special Report on Emissions Scenarios (SRES, Nakicenovic et al., 2000) that became the ancestor of a wide range of derived scenarios following a similar scenario taxonomy. The SRES A1 "high growth" scenario family does not vary any salient scenario variables *except* technology. Yet the vast differences in terms of CO_2 emission outcomes are striking, and arise solely from differences in alternative technology development described in the A1FI and A1T scenario groups (and without any explicit climate policies).

This divergence in Figure 4.2 illustrates that technological change could cluster around alternative technological combinations that remain consistent and stable respectively through increasing returns to adoption that crowd out other alternatives. (It might be argued therefore that the intermediary, "balanced" A1B scenario is more a modelling/scenario construct than a development pathway consistent with historical experience; yet it could also be taken to represent the global aggregate of differentiated "locked-in" regional technology developments such as clean coal technologies in China versus renewable energy in Latin America.)

Another feature consistent with historical experience is that, despite the three scenario groups distinctively contrasting technology strategies and policies, emissions diverge only gradually (cf. the similar emission range across all scenarios by 2025). The seeds of long-term alternative outcomes have thus to be sown early on and translate into different environmental outcomes only slowly as new technologies replace gradually older technology vintages. It is this long-term leverage (change) effect of near-term technology policies that are of current interest, particularly in the climate change domain.

It is also possible to draw on the scenario literature to illustrate technological bifurcations with respect to individual technologies. Nakicenovic and Riahi (2002) have performed a comprehensive survey of the scenario literature with respect to energy technology projections for a wide range of energy supply technologies (for electricity generation as well as synfuels). As an example, Figure 4.3 shows scenario projections for solar PV that over the long term bifurcate into groups of scenarios of either comparatively low or alternatively high market deployment as a function of assumed technology characteristics, in particular investment costs, as

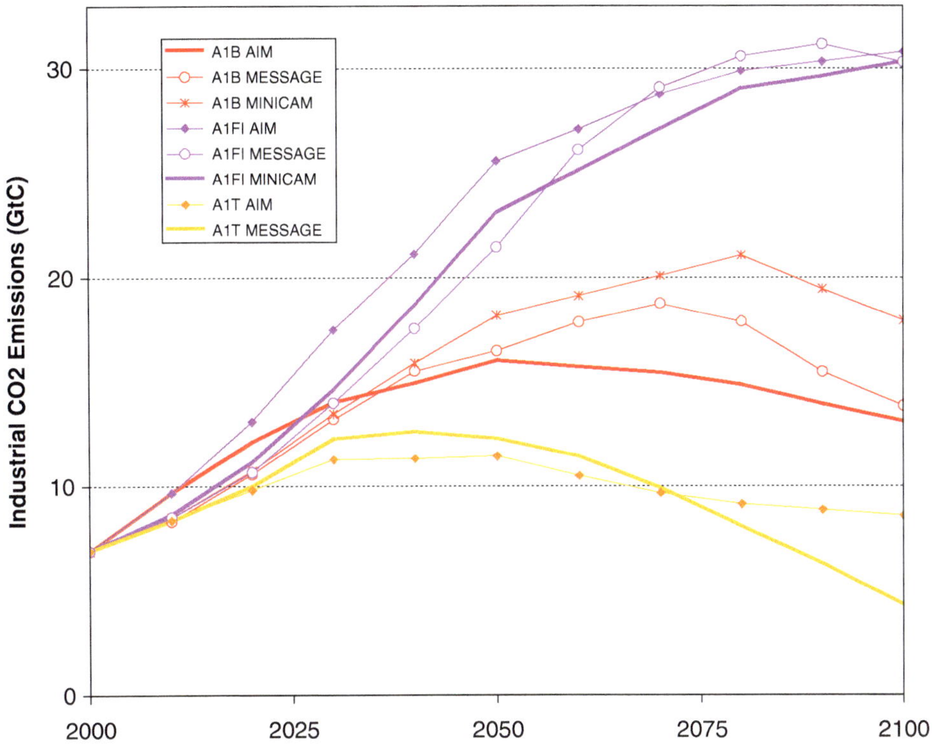

Figure 4.2. Industrial CO_2 emissions (GtC) for the IPCC SRES A1 "high growth" scenario family.
Notes: Scenarios share identical assumptions of the future, except technology development that lead alternatively to high emissions, fossil-fuel intensive (A1FI), or alternatively to low emissions (A1T) scenarios, with the "balanced" A1B scenarios taking an intermediary position. Bold lines denote so-called marker scenario; other scenarios illustrate modelling uncertainties in the representation of technological change even under shared overall assumptions of technology development trends across models.
Source: Nakicenovic and colleagues (2000).

well as future market deployment environments (absence versus existence of CO_2 emission constraints and their magnitude). Note in particular the bifurcation in PV deployment scenarios for 2100 as a function of assumed carbon constraints as well as the bimodal distribution in projected investment costs that differ by a factor of ten.

The scenario literature provides two important policy lessons. First, the temporal dimensions of technological bifurcations in energy systems are extremely long. Market deployment scenarios for 2020 (not shown) suggest only modest, niche market inroads into the global energy system, and even the most ambitious scenarios do not project more than sixty EJ PV electricity equivalent by 2050, with a technological bifurcation only expected well after that date. By 2100, scenarios cluster either around small (0–60 EJ) or large (100–180 EJ) solar PV markets, and their respective scenarios can be clearly identified as either assuming no CO_2 constraints and high investment costs (low PV scenario cluster) or alternatively stringent CO_2 constraints combined with low investment cost projections (high PV scenario cluster).

Second, the scenarios illustrate well the needed leveraging of both "supply push" and "demand pull" policies in triggering such a policy-induced long-term technological bifurcation. R&D efforts and improved designs and "debugging"

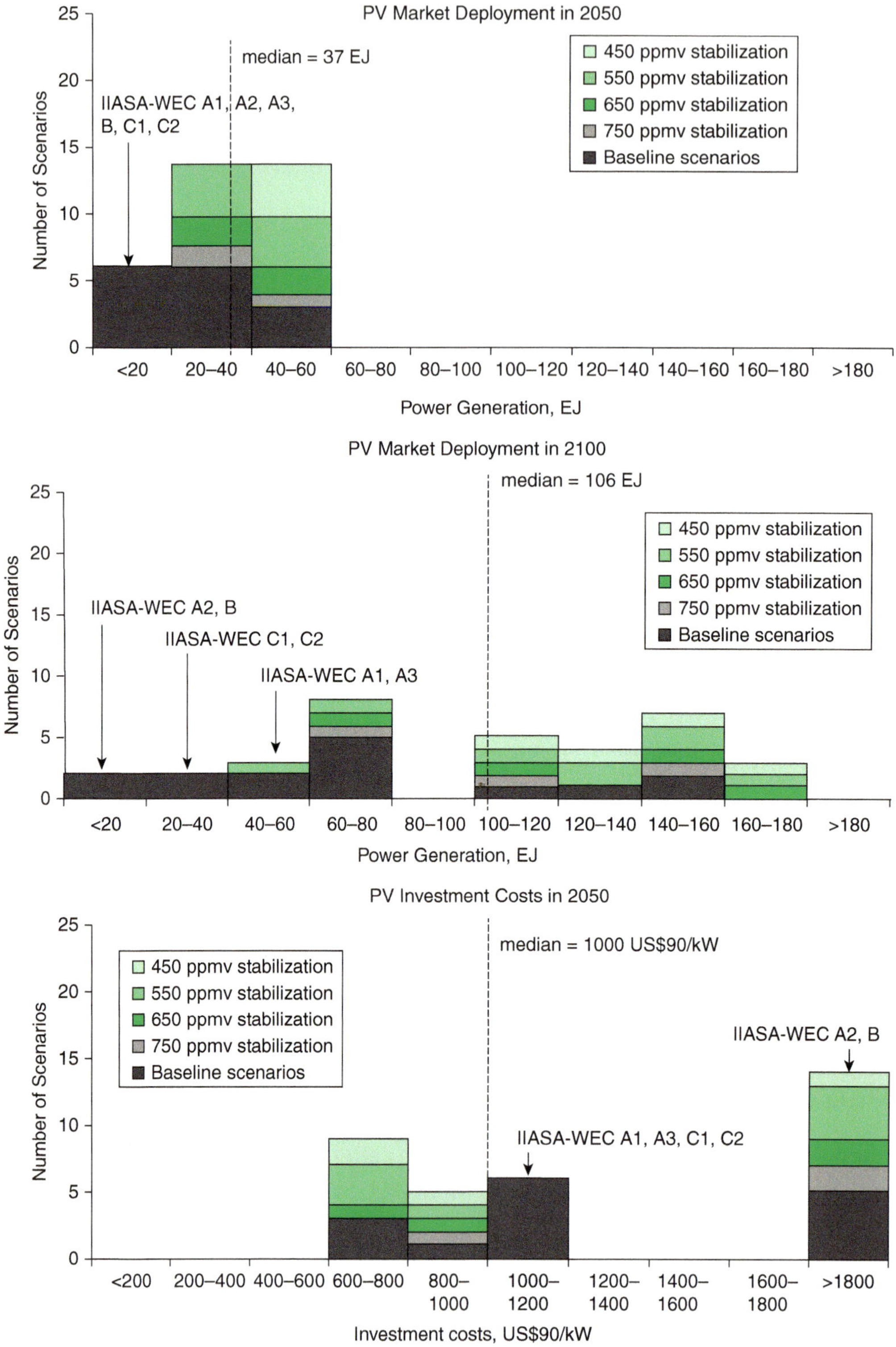

Figure 4.3. Synthesis of PV market deployment scenarios (top panels) in the scenario literature as a function of investment costs (bottom panels) and CO_2 emission constraints.
Notes: The market environment is represented through the absence (baseline scenarios, black) or existence of various CO_2 emissions constraints (ppm stabilisation targets, shades of green). Panels summarise the scenario literature for 2050 and 2100.
Source: Nakicenovic and Riahi (2002).

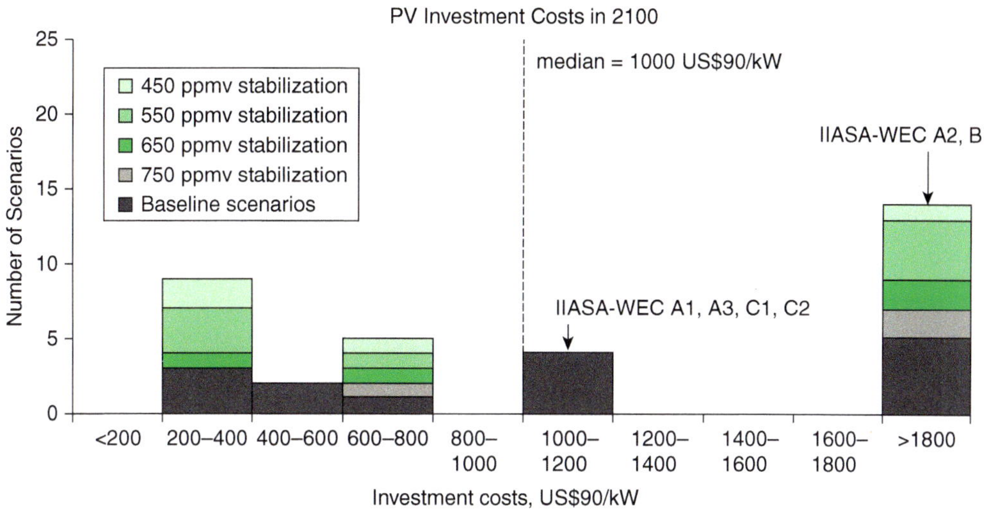

Figure 4.3. *(continued)*

through niche market application feedbacks are underlying the alternative trends in projected PV technology characteristics (most notably costs that reflect other technology characteristics such as conversion efficiencies). Conversely, changes in relative prices in scenarios of CO_2 constraints constitute the necessary complementary "demand pull" policies that are needed to yield marked differences in long-term technology outcomes.

3 Conclusions

In conclusion, certain generalisable patterns (with related policy implications) can be observed in the dynamics of energy transitions historically and in future scenarios.

First, the demand for better, different, and cheaper energy services and their associated end-use technologies have driven supply-side transformations. Falls in the effective costs of energy service provision lead to dramatic rises in the level of energy service demand. However, energy services and end-use technologies are relatively poorly represented in scenario studies of future technological change.

Second, innovations attract end users initially through their performance advantages, not lower costs. These end users constitute market niches that are protected from full cost competition and allow experimentation, learning, and other processes to improve, adapt, and reduce the costs of technologies as a basis for widespread diffusion. Improved environmental performance is alone insufficient to support technologies through this process unless the pricing of environmental externalities affects the cost competitiveness of the energy services provided.

Third, spillovers, clustering, interdependencies, and infrastructures result in strongly path-dependent technological change. Exacerbated by the longevity of much energy capital stock, this means that the time constants of change in the energy system are long, measured in decades, not years. As a consequence of this path dependency, near-term choices define long-term outcomes though divergence emerges only gradually over the medium term as existing capital stock is retired. Both "technology push" drivers such as falling costs as a result of sustained R&D

investments and "market pull" drivers such as externality pricing are necessary and complementary for supporting low carbon technological change.

FURTHER READING

For a more detailed version of this case study with additional empirical material, see Wilson and Grubler (2011).

REFERENCES

Arthur, B. W., 1989. Competing Technologies, Increasing Returns, and Lock-In by Historical Events. *The Economic Journal*, 99(394): 116–31.

Ausubel, J. H. & Marchetti, C., 1997. Elektron: Electrical Systems in Retrospect and Prospect. In Ausubel, J. H. & Langford, H. D. (eds.) *Technological Trajectories and the Human Environment*. Washington, DC: National Academy Press, 115–40.

Ayres, R. U. 1989. *Technological Transformations and Long Waves*. Laxenburg, Austria: International Institute for Applied Systems Analysis (IIASA).

Barreto, L., Makihira, A., & Riahi, K., 2003. The Hydrogen Economy in the 21st Century: A Sustainable Development Scenario. *International Journal of Hydrogen Energy*, 28(3): 267–84.

BP, 2009. *BP Statistical Review of World Energy June 2009*, London, UK: BP.

Crafts, N., 2004. Steam as a General Purpose Technology: A Growth Accounting Perspective. *Economic Journal*, 114(495): 338–51.

Devine, W. D., 1983. From Shafts to Wires: Historical Perspective on Electrification. *Journal of Economic History*, 43(2): 347–72.

EIA, 2001. *End-Use Consumption of Electricity 2001*, Washington, DC: Energy Information Administration (EIA).

Fisher, B. S., Nakicenovic, N., Alfsen, K., Corfee Morlot, J., de la Chesnaye, F., Hourcade, J.-C., Jiang, K., Kainuma, M., La Rovere, E., Matysek, A., Rana, A., Riahi, K., Richels, R., Rose, S., Van Vuuren, D., & Warren, R., 2007. Issues Related to Mitigation in the Long Term Context. In Metz, B., Davidson, O. R., Bosch, P. R., Dave, R., & Meyer, L. A. (eds.) *Climate Change 2007: Mitigation. Contribution of Working Group III to the Fourth Assessment Report of the Inter-governmental Panel on Climate Change*. Cambridge, UK: Cambridge University Press, 169–250.

Freeman, C. & Perez, C., 1988. Structural Crises of Adjustment, Business Cycles and Investment Behaviour. In Dosi, G., Freeman, C., Nelson, R., Silverberg, G., & Soete, L. (eds.) *Technical Change and Economic Theory*. London: Pinter Publishers, 38–66.

Grubler, A., 1996. Time for a Change: On the Patterns of Diffusion of Innovation. *Daedalus*, 125(3): 19–42.

1998. *Technology and Global Change*, Cambridge, UK: Cambridge University Press.

Grubler, A., 2008. Energy Transitions. In Cleveland, C. J. (ed.) *Encyclopedia of Earth*. Washington, DC: Environmental Information Coalition, National Council for Science and the Environment.

Grubler, A., Nakicenovic, N., & Victor, D. G., 1999. Modeling Technological Change: Implications for the Global Environment. *Annual Review of Energy and the Environment*, 24: 545–69.

Halsnæs, K., Shukla, P., Ahuja, D., Akumu, G., Beale, R., Edmonds, J., Gollier, C., Grubler, A., Ha Duong, M., Markandya, A., McFarland, M., Nikitina, E., Sugiyama, T., Villavicencio, A., & Zou, J., 2007. Framing issues. In Metz, B., Davidson, O., Bosch, P., Dave, R., & Meyer, L. (eds.) *Climate Change 2007: Mitigation*. Cambridge, UK: Contribution of Working Group III to the Fourth Assessment Report of the Intergovernmental Panel on Climate Change, Cambridge University Press, 117–67.

IEA, 2009. *Energy Balances of Non-OECD Countries*, Paris: International Energy Agency (IEA).

IPCC, 2001. *Climate Change 2001: Mitigation. Contribution of Working Group III to the Third Assessment Report of the Intergovernmental Panel on Climate Change.*, Cambridge, UK: Cambridge University Press.

Landes, D. S. 1969. *The Unbound Prometheus: Technological Change and Industrial Development in Western Europe From 1750 to the Present*. Cambridge, UK: Cambridge University Press.

Lipsey, R. G., Carlaw, K. I., & Bekar, C. T., 2005. *Economic Transformations: General Purpose Technologies and Long-term Economic Growth*, Oxford, UK: Oxford University Press.

Lovins, A., Brylawski, M., Cramer, D., & Moore, T., 1996. *Hypercars: Materials, Manufacturing, and Policy Implications*. Snowmass, CO: Rocky Mountain Institute.

Marchetti, C. & Nakicenovic, N., 1979. *The Dynamics of Energy Systems and the Logistic Substitution Model*. Laxenburg, Austria: International Institute for Applied System Analysis.

NAE, 2003. *A Century of Innovation: 20 Engineering Achievements that Transformed our Lives*. Washington, DC: National Academy of Engineering (NAE).

Nakicenovic, N., Alcamo, J., Davis, G., de Vries, B., Fenhann, J., Gaffin, S., Gregory, K., Grubler, A., Tae, Y. J., Kram, T., La Rovere, E. L., Michaelis, L., Mori, S., Morita, T., Pepper, W., Pitcher, H., Price, L., Riahi, K., Roehrl, A., Rogner, H.-H., Sankovski, A., Schlesinger, M., Shukla, P., Smith, S., Swart, R., Van Rooijen, S., Victor, N., & Zhou, D., 2000. *Special Report on Emissions Scenarios*, Cambridge, UK: IPCC and Cambridge University Press.

Nakicenovic, N. & Riahi, K., 2002. *An Analysis of Technological Change across Selected Energy Scenarios*. Laxenburg, Austria: International Institute for Applied Systems Analysis.

Rosenberg, N., 1994. *Exploring the Black Box: Technology, Economics, and History*. Cambridge, UK: Cambridge University Press.

Smil, V., 1994. *Energy in World History*, Boulder, CO: Westview Press.

2000. Energy in the Twentieth Century: Resources, Conversions, Costs, Uses, and Consequences. *Annual Review of Energy & Environment*, 25: 21–51.

Solow, R. M., 1957. Technical Change and the Aggregate Production Function. *The Review of Economics and Statistics*, 39(3): 312–20.

US DOC, 1975. *Historical Statistics of the United States: Colonial Times to 1970*. Washington, DC: U.S. Department of Commerce (US DOC).

2007. *Statistical Abstract of the United States*. Washington DC: U.S. Department of Commerce (US DOC).

US DOE, 1998. *United States Industrial Electric Motor Systems Market Opportunities Assessment December 2002*. Washington, DC: U.S. Department of Energy (US DOE).

Wilson, C. & Grubler, A., 2011. Lessons from the History of Technological Change for Clean Energy Scenarios and Policies. *Natural Resources Forum*, 35: 165–84.

5 Historical Diffusion and Growth of Energy Technologies

Charlie Wilson

Author's Summary

Over the course of the twentieth century, a range of energy supply and energy end-use technologies has diffused widely throughout the energy system. Unit capacities of many energy technologies have also increased. The industry-level growth (diffusion) and the unit-level growth (up-scaling) varies spatially as technologies move out of the initial markets in which they are first commercialised to spread through later-adopting regions. Analysing these historical growth dynamics at industry and unit levels reveals general patterns that appear robust across very different energy technologies.

At the industry level, the extent to which a technology's installed capacity grows is consistently related to the time duration of that growth. The consistency of this extent-duration relationship holds as technologies diffuse spatially, although the relationship steepens in later-adopting regions as diffusion accelerates. At the unit level, increases in the unit size of a technology follow a lengthy formative phase of experimentation with many smaller-scale units. Subsequent up-scaling is most characteristic of technologies with clear economies of scale servicing homogeneous market segments. There is no evidence, however, to support an acceleration in this up-scaling phase as technologies diffuse spatially. These observed patterns have important implications, not least in striking a cautionary note on pushing for significant jumps in unit size before the successful completion of a "formative phase" of experimentation with smaller-scale units.

Editors' Guide

Relationship of Case Study to Energy Technology Innovation System Framework

The dynamics of technological diffusion and up-scaling explored in this case study provide a broad context for understanding the more detailed country-specific and technology-specific case studies that follow. The case study emphasises two important innovation system outcomes: increases in the total installed capacity base of a technology, and changes in the size of the units that make up that capacity. The patterns of change observed are consistent with many of the elements and processes of the energy technology innovation system framework.

First, the *knowledge* generated by experimenting with many small-scale units during a technology's formative phase is the necessary precursor to widespread growth. This broadly corresponds to the *market formation* stage of the innovation life cycle during which *feedbacks* from end users and other market actors help shape the refinement, improvement, and adaptation of the technology to market demands.

Second, the duration of this formative phase is often extended over a period of twenty or more years, a period that precedes up-scaling and initiates widespread diffusion that itself lasts many decades. Time – or more aptly, patience – is therefore a *resource* needed for the successful functioning of innovation systems, but one that is not substitutable by financial or human capital. Generating the necessary *knowledge* and developing the necessary *institutions* to support rapid up-scaling takes time.

Third, up-scaling is characteristic of many energy technologies and consistent with the capturing of *scale economies* associated with larger unit sizes. The extent to which up-scaling occurs depends on the interactions between scale economies as a *technology-push* driver of up-scaling (to reduce costs), and end-user demands in relevant market segments as a *market-pull* driver of up-scaling (to underwrite demand and reduce risk). Both technology-push and market-pull drivers interact to determine the respective rates of up-scaling at the unit scale frontier and in terms of average capacities.

Fourth, *knowledge spillovers* from early-adopting to later-adopting regions are evidenced in the accelerating rates at which technologies diffuse in later markets, although this is traded off against relatively lower extents of diffusion. Analogous effects of knowledge spillovers in the accelerating rate of up-scaling in later markets is not, however, evidenced, pointing again to the non-substitutability of time in the development of the necessary institutions to support technology up-scaling.

Insights for Energy Technology Innovation Policy

- The formative phase plays a central role in developing the knowledge and institutions necessary for technologies to upscale successfully (increases in unit size). The formative phase involves experimentation with many (smaller-scale) units. Policy efforts to "force" early and rapid up-scaling should be considered highly risky.
- Successful experimentation, improvements, and learning characteristic of early diffusion are associated with granular or small-scale, dispersed technological units. Investments in testing these small-scale units bear lower financial and innovation failure risks compared to the uncertain deployment of large-scale units. Experimentation can be supported by technology portfolios encouraging diversity and granularity.

- Patience is a virtue, particularly in innovation systems. Pervasive change takes time; experimentation and formation take time. Time is an important resource, but a non-substitutable one. The potential for large investments in financial resources to short-circuit or accelerate the innovation processes behind the observed diffusion patterns appears limited.
- Technology-push and market-pull drivers interact, here most clearly in the case of up-scaling. This is consistent with a "necessary but not sufficient" characterisation of any singular technology policy emphasis.

1 Analysing Cross-Technology Diffusion Dynamics

1.1 Introduction

Over the past 100 years, global primary energy consumption has increased almost tenfold (see also Smil, 2000: 229). In the 1960s, roughly one coal-fired steam turbine unit averaging 125 MW in capacity was installed every other day, and around three in four of these were in OECD countries alone. In the 1990s, Boeing and Airbus' combined production was about three commercial jet aircraft every other day carrying the equivalent of around 150 MW of power plant.

The ever-expanding capacity of the energy system to convert primary energy into energy carriers into useful services (and thence into human welfare) is thus driven by increasing numbers and increasing sizes of energy technologies: more coal power plants and jet aircraft; larger-capacity coal power plants and jet aircraft.

These wholesale transformations of the energy system over the twentieth century, on both supply and demand sides, are typically characterised at the industry level. As a current example, frequent reference is made to the double digit growth rates of the wind or solar photovoltaic industries (IEA, 2008). The headline growth in these technology industries in turn comprises, or is supported by, growth in related industries of materials, components, control systems, installers, business services, and so on – all part of interdependent technology clusters (Grubler, 1995).

For many energy technologies, growth at the industry level has been complemented by increases in the size or capacity of the technological unit itself. These unit-level changes have been referred to as "up-scaling" (Luiten and Blok, 2003). Figure 5.1 uses coal power and passenger jet aircraft as examples (see also Gardiner, 1983: 229). Steam turbine units in coal-fired power plants in OECD countries rarely exceeded 200MW for the first fifty years of their commercial deployment, before up-scaling rapidly in the 1950s and 1960s to the 1,000 MW range. In the case of jet aircraft, Boeing's defining 707–100 model, certified for commercial flight in September 1958, carried 110 to 140 passengers a range of around 6,700 km. The introduction of the Airbus A380 in 2007 was the most recent up-scaling of aircraft capacity, carrying 555 to 822 passengers a range of more than 15,000 km. The corresponding increase in engine capacity, expressed in MW terms, has been from 70MW to around 300MW. (For details of the kN thrust to MW conversion, and full details of all data sources, see Wilson, 2009.)

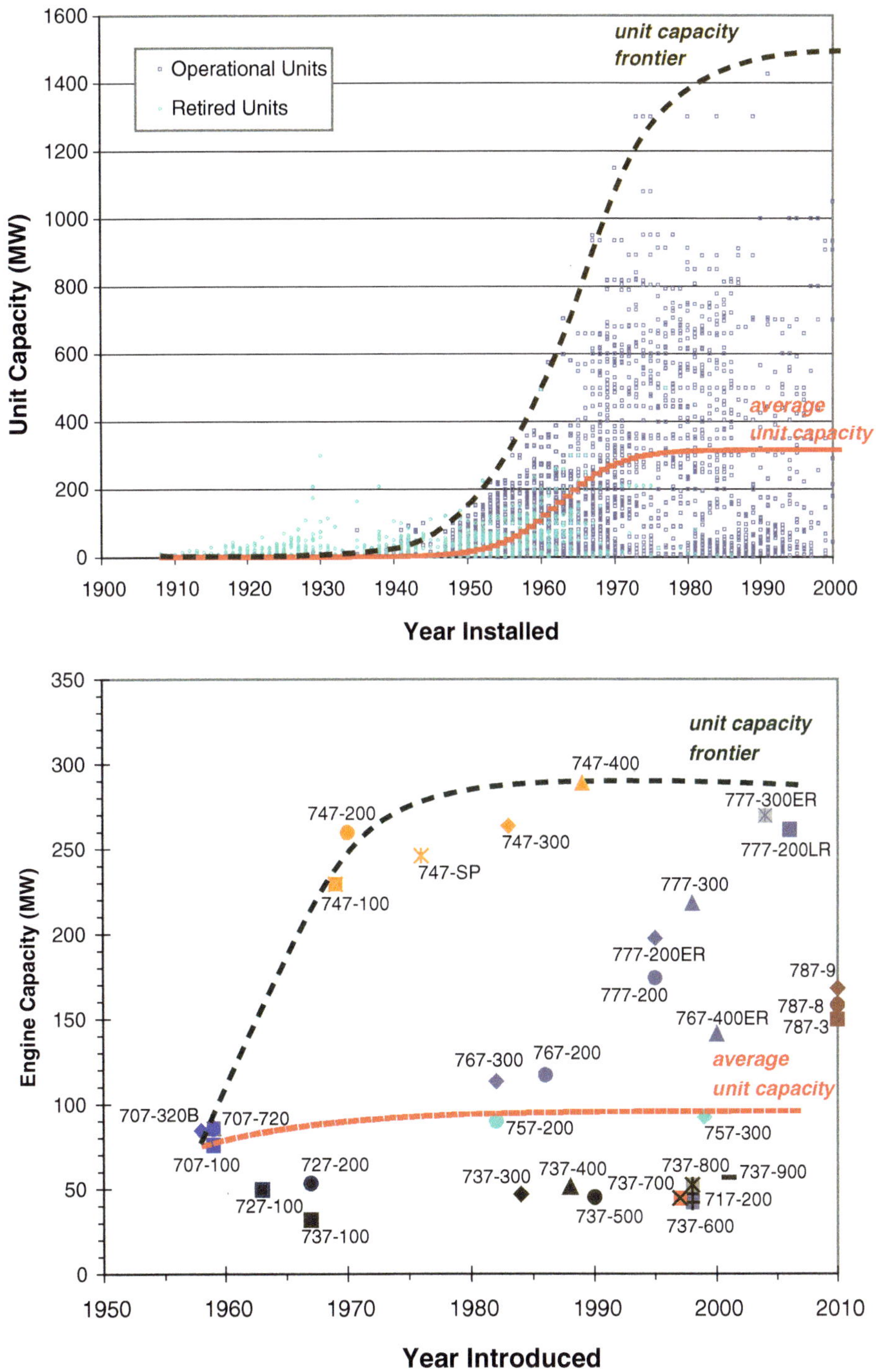

Figure 5.1. Unit-level growth or "up-scaling" of energy technologies: Steam turbine units in coal-fired power plants, 1908–2000, OECD (top panel); jet aircraft, 1958–2010, Global (bottom panel).
Notes: Main data sources include Jackson (1998); Platts (2005); see also Wilson (2009) for details.

Table 5.1. Wind power growth distinguishing unit, plant, industry, and system levels.

Level	Wind power example	Observed growth dynamics over the past thirty years
Unit	wind turbine	Maximum turbine capacities installed each year have increased from the kW range to 3–5 MW, with further increases anticipated as installations move into the emerging offshore market niche.
Plant	wind farm	Wind farms combining turbines with balance of system components (e.g., transformers, grid interconnects) have increased in maximum capacity from the low MW range to many hundred MW arrays.
Industry	wind industry	Wind industry growth, measured as the annual % change in total installed capacity, is consistently double digit. Behind this capacity growth is an increasingly consolidated and globalised sector of turbine manufacturers, component suppliers (e.g., generators, gearboxes), and service industries (e.g., insurers, financiers) with localised assembly operations.
System	electricity system	Despite rapid growth, the wind industry is a small niche within continually expanding electricity systems comprising centralised and decentralised generation, transmission and distribution infrastructures, and proliferating numbers of electrical end-use technologies.

1.2 Aim and Method

This case study examines the historical record for regularities in the industry-level growth (diffusion) and unit-level growth (up-scaling) of a representative sample of energy technologies. Of particular interest is the relationship between industry-level and unit-level growth and how this may change as technologies diffuse out of their initial markets. Common patterns observed should allow forward-looking inferences to be drawn about energy technologies and related policies.

Table 5.1 provides a simple illustration of the industry and unit levels using wind power as an example. The plant and system levels are included for the complete picture.

1.3 Data

Historical time series data on industry-level diffusion and unit-level up-scaling were collected for a sample of energy technologies, ranging from centralised, capital-intensive energy supply technologies to distributed, low-cost technologies directly providing useful services to end users (see Table 5.2). Energy supply technologies included refineries, large-scale power plants (coal, nuclear, hydro, natural gas), and small- to medium-scale power plants (wind, hydro). End-use technologies included jet aircraft, cars, and compact fluorescent light bulbs. Additional technologies were sampled (solar PV, mobile phones) but excluded from the analysis as their relatively early stage of diffusion meant overall dynamics were not yet observable.

Energy conversion capacity expressed in MW was used as a common metric of both industry-level and unit-level growth to allow cross-technology comparisons. In all cases, capacity data meaningfully related to the service provided by the technologies (e.g., the horsepower of cars and the thrust of jet engines relate to

Table 5.2. Historical energy technologies analysed.

Technology	Time series	Regions with logistic growth *(included in analysis)*				Regions with non–logistic growth[a] *(excluded from analysis)*
		Global	Core	Rim	Periphery	
Oil Refineries	1940–2000	Global	OECD +FSU	Middle East, Asia (ex. China), L.America	China, Africa	–
Coal Power	1908–2000	Global	OECD	(1) FSU	Middle East+ L.America+Africa	Rim
Nuclear Power	1956–2000	Global	OECD	(1) FSU	–	Rim, Periphery
Hydro Power	1900–2000	Global	OECD	(1) FSU	–	Rim, Periphery
Natural Gas Power[b]	1903–2000	Global	OECD	(1) FSU(2) Asia	Midde East+ L.America +Africa	
Wind Power	1977–2008	–	Denmark	–	–	Global, Rim, Periphery
Passenger Jet Aircraft[c]	1958–2007	Global	Boeing	Airbus	–	Periphery
Passenger Cars	1900–2005	Global	United States	(1) FSU (2) OECD ex. US	–	Periphery
Compact Fluorescent Light Bulbs	1990–2003	–	North America +Western Europe	–	–	Global, Rim, Periphery

[a] Either growth still in exponential phase reducing reliability of logistic fit, or insufficient data.
[b] For technologies with distinct, sequential phases of growth, logistic functions were fitted to the "first phase" of growth if it evidenced a clear plateau.
[c] Boeing sales used as proxy for core region; Airbus sales for rim region; Boeing and Airbus combined for global data.
Acronyms: OECD = Organization of Economic Cooperation and Development (corresponding to developed countries); FSU = former Soviet Union (corresponding to economies in transition); L.America = Latin America.
Notes: See Wilson (2009, 2012) for details and data sources.

the mobility service provided). Industry-level growth was expressed in terms of cumulative total capacity; unit-level growth was expressed in terms of both average and maximum unit capacities.

Table 5.2 shows the time series data used in the cross-technology analysis, ordered from high to low unit capacity. In all cases, the time series data begins with the first commercial application of the technology through to the present. Detailed explanations and plots of all the data can be found in Wilson (2009). (All data are also available online at http://www.iiasa.ac.at/web/home/research/researchPrograms/TransitionstoNewTechnologies/Scaling-Dynamics-of-Energy-Technologies1.en.html.)

1.4 Logistic Growth Functions

Growth function parameters were used to compare the extents and durations of industry-level growth between different technologies. Logistic or S-shaped growth functions were fitted to the data subject to criteria of accuracy (minimum $R^2 = 95\%$) and reliability (minimum % of estimated asymptote reached by data = 60%). Details of the logistic function are shown in the box below.

Time series data that were not logistic in form and so were excluded from the analysis are shown in Table 5.2. For full details of the logistic function fitting, and its rationale, see Wilson (2009, 2012).

Three-parameter logistic function describing S-shaped growth.

$$y = K / (1+e^{-b(t-t0)}) \text{ and } \Delta t = \log 81.b^{-1}$$

with:
K = asymptote (saturation level)
b = diffusion rate (steepness)
Δt (delta t) = time period over which y grows from 10% to 90% of K
$t0$ = inflection point at $K/2$ (maximal growth)

We used two logistic function parameters in our analysis (see box):

- K, the asymptote parameter, as a measure of the *extent* of growth;
- Δt, the time between 10 percent and 90 percent of the asymptote, as a measure of the *duration* of growth.

1.5 Spatial Disaggregation

Global data were disaggregated into four regions to account for the diffusion of technologies from their initial market of first commercial application through to subsequent and then final markets. These regions were defined for each technology by the sequence of spatial diffusion in the data. These regions are labelled "core," "rim," and "periphery" in that sequence (following Grubler, 1998). Thus, for example, Denmark is the core region for wind power; the United States is the core region for cars; the OECD is the core region for natural gas power. As the former Soviet Union presented a particular case historically with the sequence of diffusion relative to Western markets being either broadly concurrent (e.g., nuclear power) or markedly subsequent (e.g., solar PV), the "rim" region was split into "rim1," corresponding to the former Soviet Union and Eastern European countries, and "rim2," corresponding to the remainder (see Table 5.2).

1.6 Overview of Analysis

The analysis is divided into three sections. The first section observes how rapidly and how pervasively energy industries have grown historically using energy conversion capacities as a metric. (For further details on both method and findings, see Wilson et al., 2012.) The second section considers the relationship between unit-level growth and this industry-level growth, focussing on the characteristic "formative phase" of experimentation and learning that precedes the main industry growth phase. The third section characterises the unit-level growth or up-scaling in more detail, looking at changes in both average and maximum unit capacities. (For further details on both method and findings, see Wilson, 2009, 2012.) The first and third sections also explicitly consider whether patterns observed are robust as technologies diffuse spatially out of their markets of first commercialisation.

2 Findings: Industry-Level Growth

2.1 Consistent Cross-Technology Relationship between Extent and Duration of Diffusion

Figure 5.2 (left panel) shows the historical data series of cumulative total capacity for the nine energy technologies analysed (note log-scale y-axis). These data describe the capacity expansion of each technology in its core region or initial market of commercial deployment (see Table 5.2 for details). Fitted logistic functions to these cumulative total capacity data then allow the extent and duration of industry-level growth to be compared for different technologies. Figure 5.2 (right panel) shows the normalised K as a measure of *extent* of diffusion plotted against Δt as a measure of *duration* of diffusion (note log scale y-axis). K is normalised to account for changes in the size of the energy system into which each technology diffused. (For details of the method, see Wilson et al., 2012.)

In general, a technology should take longer to diffuse to a greater extent notwithstanding the many factors that affect growth (Grubler, Nakicenovic, and Victor, 1999). Figure 5.2 supports this finding, demonstrating a consistent extent-duration relationship between historical energy technologies of markedly different characteristics and vintage.

The extent-duration relationship of growth in cumulative total capacity describes the energy system's ability to accumulate energy conversion potential over time through the manufacture and installation of capital stock. Refineries, power plants, jet aircraft, cars, and light bulbs are characterised by different energy service demands and have distinctive costs and efficiencies, capital intensiveness, turnover rates, regulatory contexts, implications for infrastructure, and so on. Yet observed consistencies between extents and durations of capacity growth historically suggest

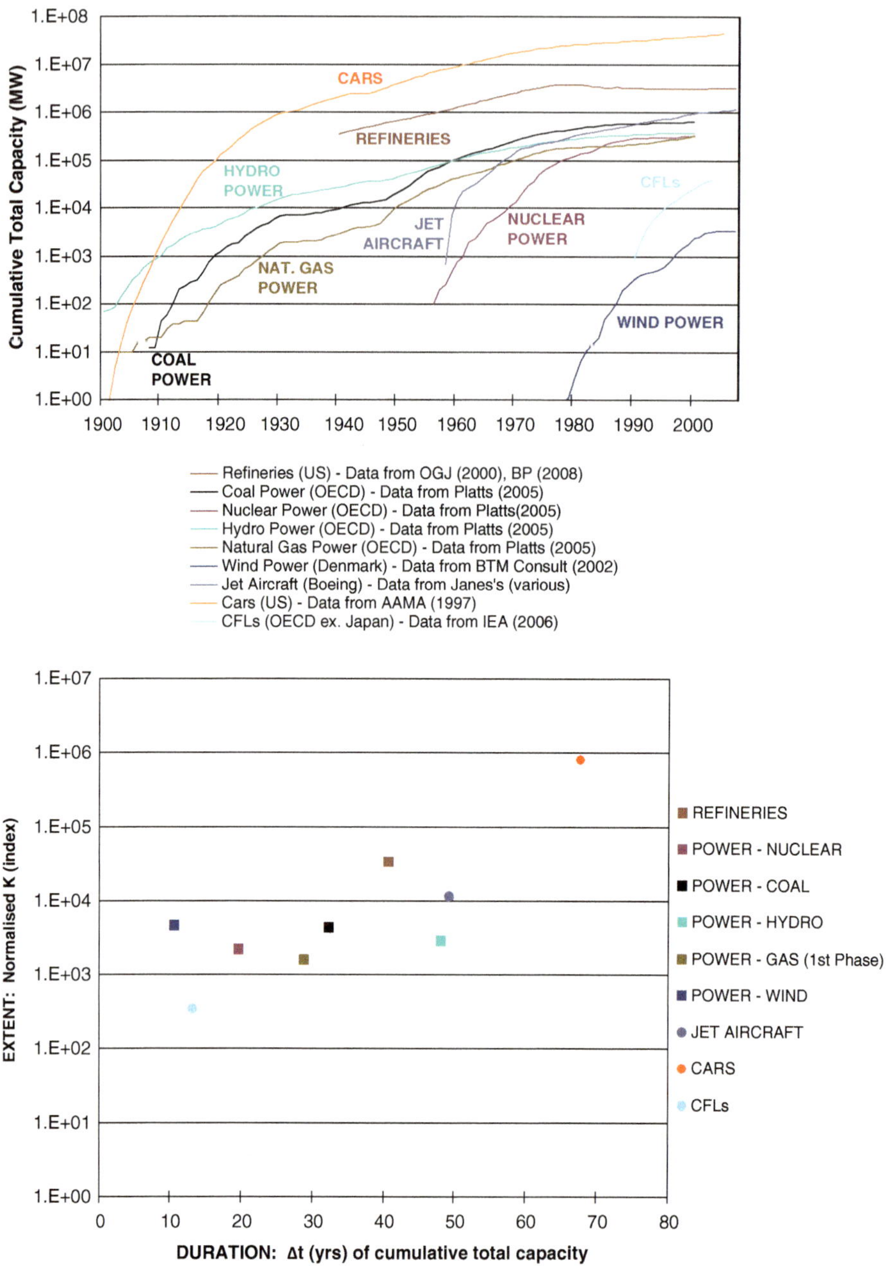

Figure 5.2. Capacity growth of nine energy technologies in their respective core regions. *Notes*: Time series data (top panel); Logistic function parameters K versus Δt (bottom panel). See figure legend and Wilson and colleagues (2012) for data sources.

influences on diffusion act *proportionately* on both extent and duration, so preserving the relationship seen in the extent-duration relationship plots (bottom panel, Figure 5.2). This cross-technology consistency in the historical relationship between extent and duration of industry-level growth describes the inherent inertia of a large, complex, interrelated system of technologies, infrastructures, and institutions (Unruh, 2000; Wilson and Grubler, 2011).

2.2 Shorter Duration but Less Extensive Growth as Technologies Diffuse Spatially

As technologies diffuse spatially from initial (core) into subsequent (rim, periphery) markets, knowledge spillovers can lead to an increase in the speed of diffusion. However, this tends to be associated with a decrease in the pervasiveness of diffusion, measured, for example, as lower saturation densities. This general pattern of spatial diffusion being associated with more rapid but less pervasive growth is known as Schmidt's Law (Grubler, 1990).

Figure 5.3 shows Δts (as a measure of the duration of diffusion) for the nine energy technologies analysed as they diffuse spatially through the sequence from core, rim1, and rim2 to periphery regions. Omitted bars mean data were either unavailable or data described exponential growth that was not yet clearly identifiable as logistic (see Table 5.2). As logistic growth is symmetrical about the inflection point at which growth rates are maximal, the duration of diffusion is inversely related to the rate of diffusion. Thus longer durations (high Δts) indicate slower rates.

The basic pattern observed in Figure 5.3 confirms Schmidt's Law, and by implication, the effect of knowledge spillovers as technologies diffuse spatially. Durations of industry-level growth tend to decrease (or rates of growth increase) from core to rim to periphery regions.

The accelerating rates of growth are shown more clearly in Figure 5.4, which plots logistic growth functions fitted to the data for each region on the same y-axis scale, by indexing each function's asymptote parameter (K) to equal 1.00. The rate (steepness) and timing (position of curve on the x-axis) can be readily compared for core, rim1, rim2, and periphery regions. Three technologies are illustrated: refineries, natural gas, and passenger cars. In the cases of refineries and natural gas, the logistic functions describe a first phase of growth that culminated in a period of saturation (slow / no growth) following the oil shocks in the 1970s. For each of the three technologies, diffusion in the core region (blue line) is less rapid (flatter curve) than diffusion in the later regions (red, green, purple lines).

With the cumulative total capacity data, it was not possible to test whether these increases in rates (decreases in durations) as technologies diffused spatially were also associated with a decrease in the pervasiveness of diffusion as the core, rim, and periphery regions are of different sizes. However, Figure 5.5 does show a steepening or "acceleration" of the extent-duration relationship (shown previously in Figure 5.2) from core to rim to periphery regions. This is consistent with the general finding of more rapid but less pervasive growth as technologies diffuse spatially.

Figure 5.5 plots the extent of diffusion (normalised K) against the duration of diffusion (Δt) for all the data points available. The number of data points decreases from core to rim to periphery as capacity growth is more likely to be still exponential and so not describable by logistic growth functions. However, the steepening of the relationship between K and Δt is clearly evidenced, with increasing exponents in the best fit lines for the data points in the core, rim1, and rim2 regions. (No best fit line is shown for the periphery region, which has only three data points.) As technologies diffuse spatially, a given extent of growth requires a shorter duration of growth.

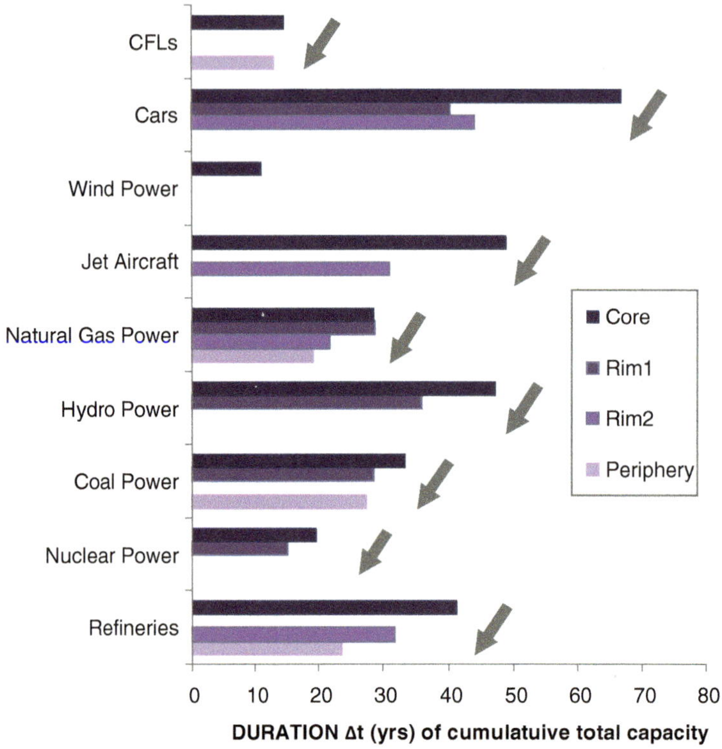

Figure 5.3. Duration of capacity growth as technologies diffuse spatially.
Notes: Bars show Δts of industry-level growth in each technology's core, rim1, rim2, and periphery regions. Omitted bars mean data were unavailable or did not describe logistic growth. See Wilson (2009) for details.

3 Findings: Industry-Level and Unit-Level Growth

3.1 Experimentation in the Formative Phase of a Technology's Life Cycle

The industry-level growth dynamics shown in the previous section conceal a series of phases in a technology's overall life cycle from origination and development through to deployment, widespread diffusion, and ultimately, senescence. Here, the emphasis is on the commercial deployment of technologies that begins with a "formative phase" during which a technology is iteratively tested, refined, and adapted to market conditions (Jacobsson and Lauber, 2006).

The formative phase of a technology's commercial deployment describes the critical period between the development of an innovation and the emergence of positive feedbacks or "cumulative causation" that sustain its diffusion (Jacobsson and Bergek, 2004). The formative phase is characterised by experimentation, "an iterative process of understanding what doesn't work and what does," encompassing both success and failure (Thomke, 2003: 2). Ruttan calls this "debugging" through a process of "designing-by-experience" (2001).

Experimentation thus involves repeated and iterative testing and modification of technologies to improve their performance, reduce their cost, and adapt them to specific market demands based on feedback from end users. This typically takes place in market niches that offer some protection from competitive pressures

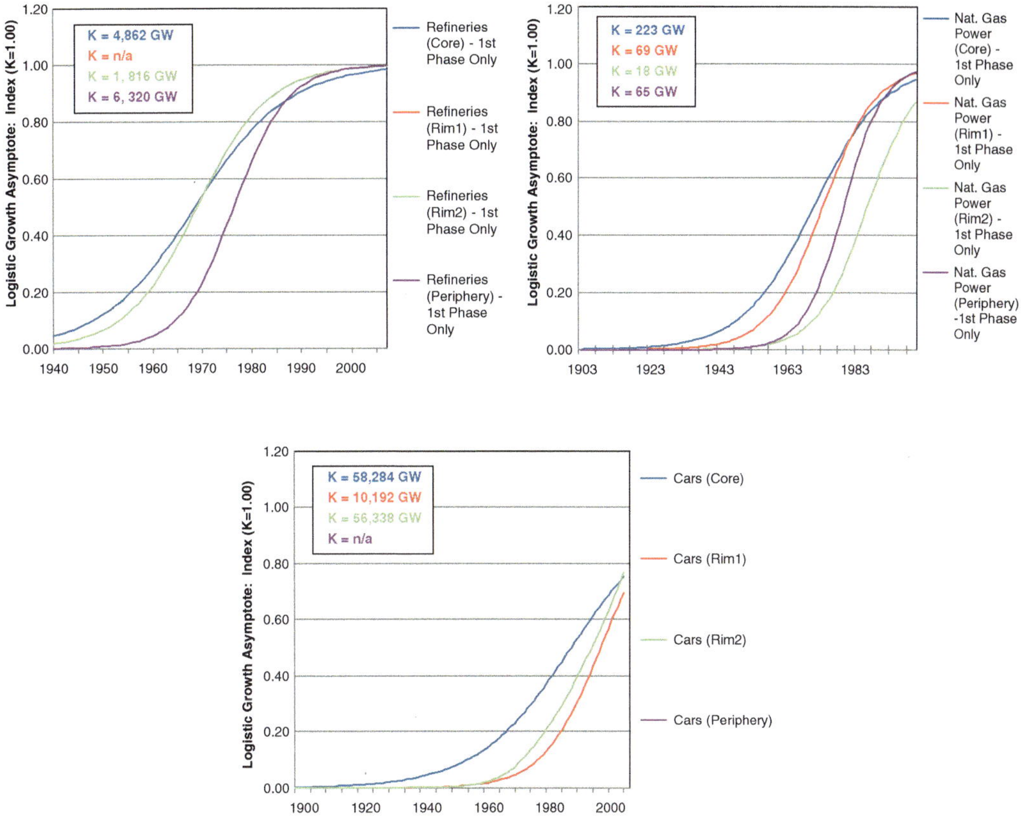

Figure 5.4. Rate and timing of capacity growth as technologies diffuse spatially.
Notes: Logistic growth functions fitted to cumulative total capacity data and indexed so each asymptote (K) = 1.00 (absolute values of K shown in boxes). Data where available for all regions (core, rim1, rim2, periphery) for refineries (top left panel), natural gas power (top right panel), passenger cars (bottom panel). See Wilson (2009) for details.

(Kemp, Schot, and Hoogma, 1998). Market niches for low-carbon or efficient end-use technologies are often created or sustained by public policy (Schot and Geels, 2008). The formative phase is also important for building expectations for a technology's future diffusion trajectory, providing a means of "articulating" the designs, markets, policies, and end-user demands of a technology (Kemp et al., 1998).

3.2 Formative Phase Growth in Unit Numbers but Not Unit Sizes

Figure 5.6 explores changes at the unit level during the formative phase of energy technologies' diffusion. Specifically, changes in the *numbers* of units over time are contrasted with changes in the *sizes* of units over time. Figure 5.6 shows data for both steam turbine units in coal-fired power plants (left panel) and wind turbines (right panel) as examples of the general pattern observed. The blue lines show changes in the numbers of additional units installed each year; the red and green lines show changes in the average and maximum capacities of those units respectively. In both cases, the data describe diffusion in each technology's core region: the OECD in the case of coal power; Denmark in the case of wind power. For wind power, the

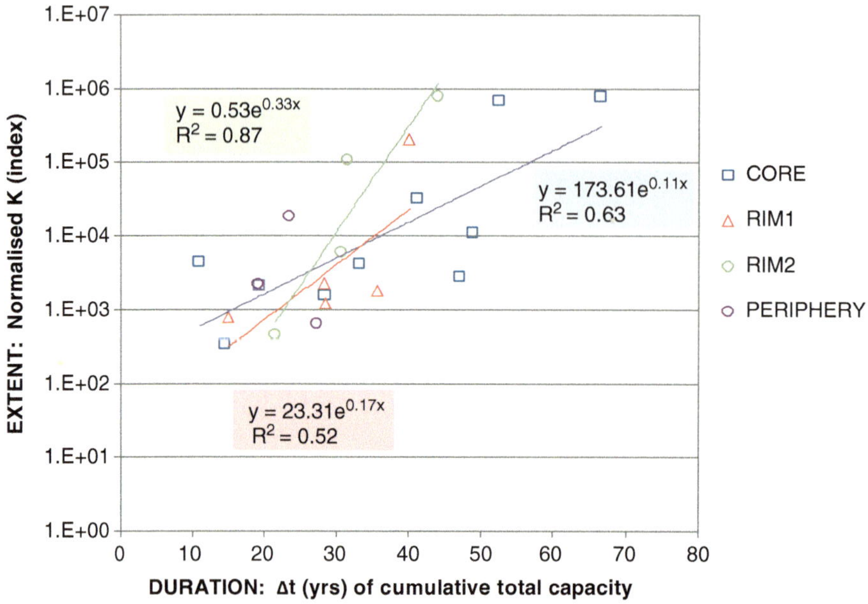

Figure 5.5. Extent-duration relationship of capacity growth as technologies diffuse spatially. *Notes*: Data points show logistic function parameters K versus Δt for all available data in each technology's core, rim1, rim2, and periphery regions. Omitted points mean data were unavailable or did not describe logistic growth.

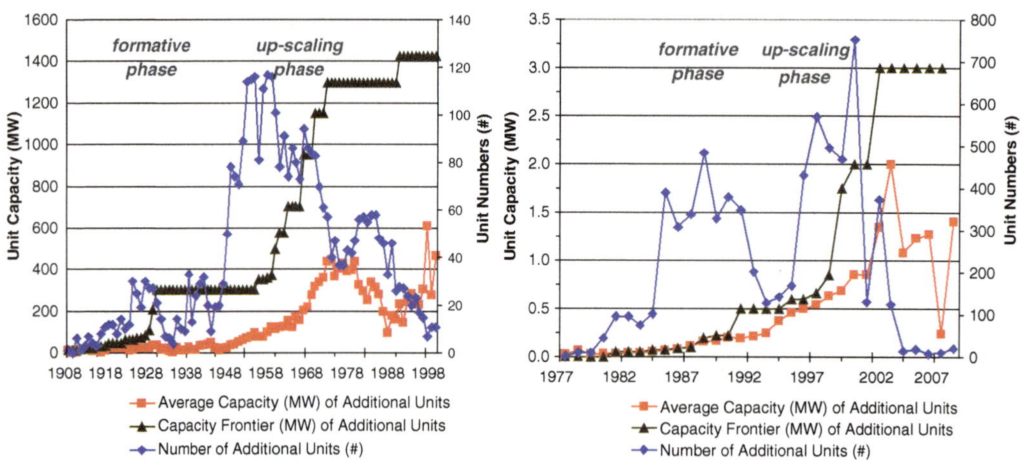

Figure 5.6. Changes in unit numbers and unit capacities for coal power (left panel) and wind power (right panel).
Notes: Data for coal power from Platts (2005); Data for wind power from Danish Energy Agency (2008).

commercial history of new wind turbine models developed by Vestas, the leading Danish (and global) manufacturer, is used as an approximation of maximum unit capacities. The observable up-scaling of turbine capacity is still far from saturation, particularly in the offshore market segment for which 5MW and larger unit capacities are envisaged (GWEC, 2008).

Figure 5.6 shows how the initial period of diffusion at the industry level is characterised by only small increases in unit capacities (red lines, right-hand y-axis) but steady, albeit fluctuating growth in the numbers of units installed (blue lines, left-hand y-axis). The processes of experimentation described earlier occur, therefore, using increasing numbers of small-scale units. This formative phase *precedes* the up-scaling or increase in unit capacities of technologies in commercial settings: building many before building big. This sequence from the formative phase to the up-scaling phase is indicated on the graphs. The up-scaling phase is characterised by a rapid and sustained increase in unit size, particularly when considered at the scale frontier (green lines, left-hand y-axis).

The process of experimentation with many small-scale units is further illustrated in Table 5.3, which compiles available data from the energy technologies analysed in their core regions (which vary geographically and in size). Technologies are ordered from top to bottom according to how long ago they began diffusion. The right-hand column shows the number of years and number of units built during a formative phase, which is measured from first commercial deployment to the point at which new units reach 10 percent of the eventual unit scale frontier.

An extended period of experimentation with many small-scale units should mean these formative phases last several decades and involve the deployment of many units. This is indeed the case. Nuclear power is the outlier with a relatively short formative phase and relatively few numbers of units built prior to up-scaling. But in fact, this exception supports the generalisable rule. The unit-scale frontier of nuclear power increased fivefold in the decade that followed commissioning of the first 50MW commercial reactor in 1956. Ultimately, these rapid increases in unit size were a contributing factor to the rising complexity that created diseconomies of scale and constrained further growth of the industry in the late 1970s (Grubler, 2010; Lovins et al., 2003).

4 Findings: Unit-Level Growth

4.1 Up-Scaling Driven by Economies of Scale at the Unit Level

"Up-scaling" is an increase in the capacity of an individual technological unit to convert energy into a useful service. Historical examples of up-scaling were given in Figure 5.1. Up-scaling is typically associated with economies of scale that describe reductions in average unit costs as the size of individual units ("unit"-scale economies) or the volume of total production ("manufacturing"-scale economies) increase over the long run, that is, assuming all production inputs are variable.

Figure 5.6 illustrates the up-scaling phases for coal power and wind power in their respective core regions (see also Figure 5.1). After a fifty-year formative phase, the maximum size of steam turbine units in coal-fired power plants in the OECD increased by an order of magnitude from 300MW to 1,300MW over an eighteen-year period from 1955 to 1973. (The average size of new units increased from 100MW to 400MW over the same period.) This followed a formative phase of around five decades during which maximum unit capacities did not exceed the 300MW unit installed in 1930.

Wind power makes an interesting comparative case as, like coal power, economies of scale at the unit level might be expected to drive significant and rapid increases in unit capacities following the maturation of the formative phase. Larger turbines

Table 5.3. Formative phases of energy technologies: durations & numbers of units.

Technology	Initial market	First commercial capacity installed	10 percent of maximum unit scale reached	Formative phase: no. of years | No. of units
Hydro Power	OECD	<1900s	1934	40 years | 3,500 units
Natural Gas Power	OECD	1900s	1948	50 years | >400 units
Coal Power	OECD	1900s	1950	50 years | >775 units
Nuclear Power	OECD	1950s (1940s*)	1963	10 years | 25 units
Wind Power	Denmark	1970s (1880s*)	1987	15–100 years | >1,400 units
Refineries**	US	1860s–70s	(1948 – average capacity only)	(80–90 years | >500 units?**)

* First nuclear installations on submarines date to 1940s; first wind power generators date to 1880s, but from 1970s in their modern form.

** Refineries data are indicative only. Saturation capacity measured in terms of average rather than maximum capacities; number of units are a rough estimate.

allow longer blades with more than proportional increases in power output, and the further benefit of stronger, more laminar winds at higher hub heights. Despite this economic incentive for up-scaling, the right panel of Figure 5.6 shows how the unit-scale frontier of wind turbines in Denmark increased very little during a formative phase lasting two decades from the late 1970s. The period of rapid up-scaling was then concentrated in a six-year period from 1997 to 2002 during which the largest Vestas turbines increased in capacity from 0.7MW to 3MW, and the average capacities of all turbines installed in the Danish electricity system increased from 0.5MW to around 1MW.

As documented in the case study on wind power, in contrast to this Danish experience, countries like Germany, Sweden, and Netherlands placed early emphasis on rapidly increasing turbine capacities to capture unit-scale economies (Åstrand and Neij, 2006). Relative to the Danish case, this premature move to the up-scaling phase failed to build an enduring industry (Meyer, 2007). As Heymann puts it: "The problems in wind technology development [in Germany and the United States] demonstrate that the testing, design improvement, and maturation of complex technologies require much practical experimentation and at least as much time, money, and effort as do the initial design and construction" (1998: 667). In the successful Danish case, up-scaling occurred only once the fundamental design issues with the technology, which were addressed during the formative phase, were settled.

4.2 Up-Scaling Influenced by Market Demand for Different Size Units

Figure 5.7 shows the up-scaling dynamics for the full sample of energy technologies analysed distinguishing growth in average unit capacities (top panel) from growth in the unit capacity frontier (bottom panel). Each line describes the changes over

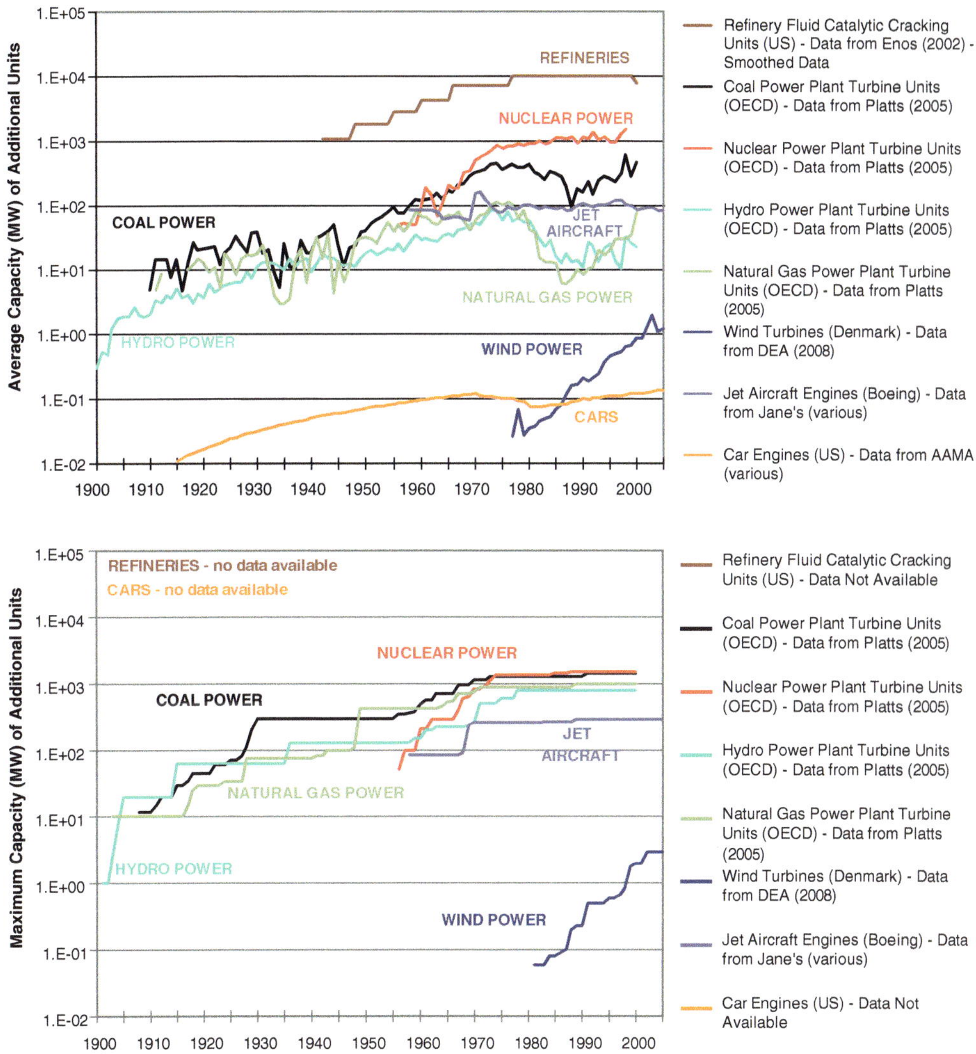

Figure 5.7. Up-scaling of energy technologies since 1900: Average unit capacities (top panel); maximum unit capacities (bottom panel).
Notes: See figure legend and Wilson (2012) for data sources.

time of the capacity in MW of newly installed "units": steam turbine units in coal, gas, and nuclear power plants; turbine units in hydro power plants; wind turbines in wind farms; jet engines in passenger aircraft; internal combustion engines in cars; and compact fluorescent light bulbs in lighting systems.

Increases in the maximum size of additional units (Figure 5.7, bottom panel) show the relatively short duration of the up-scaling phases of many of the technologies, but in particular nuclear power. The unique issues associated with managing nuclear fuel cycles coupled with the need to reduce capital costs drove rapid up-scaling from the mid 1960s to the mid 1970s following a relative short formative phase (see Table 5.3). For refineries, only average capacity data were available: up-scaling was concentrated during the several decades following World War II. Increases in unit capacities largely saturated by the 1970s. As the largest capacity end-use technology, jet aircraft also exhibited rapid and early up-scaling between the period of first commercial deployment (the Boeing 707–100 in 1958) to the introduction of the Boeing 747–100 in 1969. The recent introduction of the larger Airbus A380 has only slightly extended the unit scale frontier.

Increases in the average size of additional units (Figure 5.7, top panel) show the market demand for units of different sizes. For technologies with homogeneous market demand – that is, providing a particular function in a single market niche or segment – increases in the average and maximum sizes of additional units should be similar. In other words, up-scaling characterises all or most new units being installed. This is clearest for nuclear power, wind power, and to a lesser extent, coal power.

For technologies with heterogeneous market demand – that is, providing different functions in many market niches or segments – increases in the average size of additional units should be slower than increases in maximum size. In other words, up-scaling characterises only some of the new units installed in those market niches that demand larger units (e.g., because of the availability of economies of scale). This is clearest for natural gas power, hydro power, and to a lesser extent, jet aircraft.

4.3 No Evidence for "Leapfrogging" of Up-Scaling as Technologies Diffuse Spatially

"Leapfrogging" describes the adoption by a developing country of a less-polluting or more efficient technology from a developed country, thus avoiding a polluting or inefficient part of the developed country's development pathway (Goldemberg, 1998). The evidence for leapfrogging is mixed, and as a result the concept is contested (Gallagher, 2006; Perkins, 2003). Figure 5.8 explores the evidence for leapfrogging relating to up-scaling in two different ways using a set of three power generation technologies (coal, nuclear, natural gas).

First, leapfrogging may imply an acceleration in the rates of up-scaling (steepness of the curves in Figure 5.8) as technologies diffuse spatially. This would be consistent with the spilling over of knowledge related to up-scaling generated in initial markets to later markets in which available scale economies could be more readily and more rapidly captured. Second, leapfrogging may imply an increase in the extents of up-scaling (maximum unit capacities shown as K values in the boxes in Figure 5.8) as technologies diffuse spatially. This would similarly be consistent with knowledge spillovers from initial to later markets, but in this case enabling larger-capacity units to be developed and deployed (to the extent these are demanded in particular market segments – see earlier in this chapter for discussion).

Neither form of leapfrogging relating to up-scaling was evidenced in the data. Up-scaling rates do not accelerate from core to rim to periphery regions (blue to red / green to purple lines in Figure 5.8). Nor do up-scaling extents increase from core to rim to periphery regions (K values in boxes in Figure 5.8). Maximum unit sizes in core regions are consistently higher in core regions than in rim2 and periphery regions (excluding the rim1 data for the former Soviet Union). For example, the typical maximum unit capacity of coal power in the OECD (core region) lies somewhere around 1,000 MW, in Asia (rim region) around 800 MW, and in the rest of the world (periphery region) around 700 MW.

This absence of leapfrogging relating to up-scaling reinforces the earlier finding that an initial and often extended formative phase is important for generating the

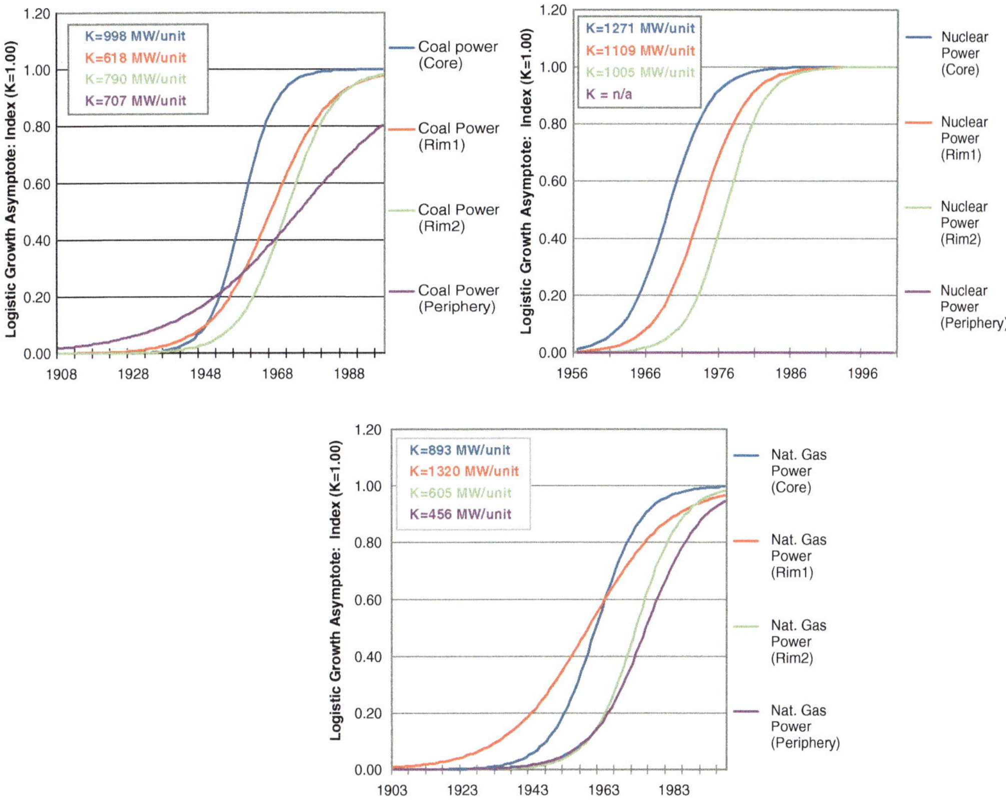

Figure 5.8. Rate and timing of up-scaling as technologies diffuse spatially.
Notes: Logistic growth functions fitted to data on maximum capacity of additional units and indexed so each asymptote (K) = 1.00 (absolute values of K shown in boxes). Data where available for all regions (core, rim1, rim2, periphery) for coal power (top left panel), nuclear power (top right panel), and natural gas power (bottom panel). See Wilson (2009) for details.

knowledge and institutions needed to design, develop, and increase the unit size of energy technologies. Such knowledge is not fully transferrable from initial market to later markets through spillovers or other forms of knowledge exchange. This implies that too rapid or too early up-scaling in later markets may be infeasible or inappropriate given the absence of necessary supporting conditions developed locally through experience, experimentation, and learning (Dahlman et al., 1987; Gallagher, 2006).

5 Conclusions

Table 5.4 summarises the common patterns or growth dynamics observed at both industry and unit levels for the sample of energy technologies analysed.

The consistent relationship between the extent and the duration of industry-level growth implies a trade-off faced by accelerated diffusion that is likely to result in less extensive or less pervasive diffusion (i.e., lower market penetration). The extended length of the formative phase as a precursor to subsequent unit-level growth (up-scaling) as well as industry-level growth (diffusion) similarly cautions against over-exuberant efforts to compress the timescales needed to ready large-scale energy

Table 5.4. Common growth dynamics observed historically in energy technologies.

Category	Empirical finding
Industry-Level Growth (Diffusion)	There is a consistent relationship between the extent and duration of diffusion measured in terms of cumulative total capacity. The relationship between the extent and duration of diffusion accelerates as technologies diffuse spatially out of their initial markets: the time taken to reach a given level of cumulative total capacity decreases.
Industry-Level and Unit-Level Growth	Diffusion begins with an often extended formative phase, involving experimentation with many small-scale units. Knowledge generated during the formative phase underwrites subsequent increases in unit size (up-scaling).
Unit-Level Growth (Up-Scaling)	Up-scaling is more rapid for technologies with clear economies of scale. Up-scaling in terms of average unit capacities is less rapid for technologies that service different market demands. There is no evidence for "leapfrogging" relating to up-scaling as technologies diffuse spatially. Knowledge spillovers do not enable later markets to up-scale energy technologies more rapidly.

technologies for widespread deployment. Extensive diffusion takes time; so do the knowledge generation processes that underwrite the capturing of economies of scale at the unit level.

These observations from core markets broadly hold as technologies diffuse spatially into later-adopting markets. Over the sequence from core to rim to periphery regions, the relationship between the extent and duration of diffusion remains consistent while steepening: knowledge spillovers enable faster diffusion in later markets. Up-scaling, however, does not similarly accelerate. Later-adopting regions need time to develop the knowledge, capabilities, and institutions to support increases in the unit sizes of energy technologies.

The findings of this analysis of the historical diffusion and growth of a sample of nine energy technologies are inherently general and are predicated on technologies that have successfully diffused and that are mature enough to have exhibited signs of saturation (and so logistic growth). Given also the limited scope of the historical data set, generalising findings beyond successful, mature technologies in core markets should be cautious.

FURTHER READING

This case study extends a body of research begun in Wilson (2009) and more recently applied to learning and formative phases historically (Wilson, 2012), and to the evaluation of future scenarios of technological diffusion (Wilson et al., 2012). For a more general overview of the dynamics of technological change and diffusion in the energy system, see Grubler and colleagues (1999).

REFERENCES

Åstrand, K. & Neij, L., 2006. An Assessment of Governmental Wind Power Programmes in Sweden – Using a System Approach. *Energy Policy*, 34: 277–96.

Dahlman, C. J., Ross-Larson, B., & Westphal, L. E., 1987. Managing Technological Development: Lessons from the Newly Industrializing Countries. *World Development*, 15(6): 759–75.
Danish Energy Agency, 2008. *Wind Turbine Master Data Register*. Copenhagen, DK: Danish Energy Agency.
Gallagher, K. S., 2006. Limits to Leapfrogging in Energy Technologies? Evidence from the Chinese Automobile Industry. *Energy Policy*, 34(4): 383–94.
Gardiner, J. P., 1983. Robust and Lean Designs with State-of-the-Art Automotive and Aircraft Examples. In Freeman, C. (ed.) *Design, Innovation & Long Cycles in Economic Development*. London, UK: Design Research Publications, 215–48.
Goldemberg, J., 1998. Leapfrog Energy Technologies. *Energy Policy*, 2(10): 729–41.
Grubler, A., 1990. *The Rise and Fall of Infrastructures, Dynamics of Evolution and Technological Change in Transport*. Heidelberg, Germany: Physica-Verlag.
1995. *Industrialisation as an Historical Phenomenon*. Laxenburg, Austria: IIASA.
1998. *Technology and Global Change*. Cambridge, UK: Cambridge University Press.
2010. The Costs of the French Nuclear Scale-Up: A Case of Negative Learning by Doing. *Energy Policy*, 38(9): 5174–88.
Grubler, A., Nakicenovic, N., & Victor, D. G., 1999. Dynamics of Energy Technologies and Global Change. *Energy Policy*, 27(5): 247–80.
GWEC, 2008. *Global Wind Energy Outlook*. Brussels, Belgium: Global Wind Energy Council.
Heymann, M., 1998. Signs of Hubris – The Shaping of Wind Technology Styles in Germany, Denmark and the United States 1940–1990. *Technology and Culture*, 39(4): 641–70.
IEA, 2008. *Energy Technology Perspectives 2008 – Scenarios and Strategies to 2050*. Paris: Organization for Economic Cooperation & Development.
Jackson, P. (ed.) 1998. *Jane's All the World's Aircraft:1998–99*. Surrey: Jane's Information Group.
Jacobsson, S. & Bergek, A., 2004. Transforming the Energy Sector: The Evolution of Technological Systems in Renewable Energy Technology. *Industrial and Corporate Change*, 13(5): 815–49.
Jacobsson, S. & Lauber, V., 2006. The Politics and Policy of Energy System Transformation – Explaining the German Diffusion of Renewable Energy Technology. *Energy Policy*, 34(3): 256–76.
Kemp, R., Schot, J., & Hoogma, R., 1998. Regime Shifts to Sustainability through Processes of Niche Formation: The Approach of Strategic Niche Management. *Technology Analysis & Strategic Management*, 10(2): 175–98.
Lovins, A., Kyle Datta, E., Feiler, T., Rabago, K. R., Swisher, J., Lehmann, A., & Wicker, K., 2003. *Small is Profitable: The Hidden Economic Benefits of Making Electrical Resources the Right Size*. Snowmass, CO: Rocky Mountain Institute.
Luiten, E. E. M. & Blok, K., 2003. Stimulating R&D of Industrial Energy-Efficient Technology; the Effect of Government Intervention on the Development of Strip Casting Technology. *Energy Policy*, 31(13): 1339–56.
Meyer, N. I., 2007. Learning from Wind Energy Policy in the EU: Lessons from Denmark, Sweden and Spain. *European Environment*, 17(5): 347–62.
Perkins, R., 2003. Environmental Leapfrogging in Developing Countries: A Critical Assessment and Reconstruction. *Natural Resources Forum*, 27(3): 177–88.
Platts, 2005. *World Power Plant Database*. London, UK: Platts.
Ruttan, V., 2001. *Technology, Growth, and Development: An Induced Innovation Perspective*. Oxford, UK: Oxford University Press.
Schot, J. & Geels, F. W., 2008. Strategic Niche Management and Sustainable Innovation Journeys: Theory, Findings, Research Agenda, and Policy. *Technology Analysis & Strategic Management*, 20(5): 537–54.
Smil, V., 2000. Energy in the Twentieth Century: Resources, Conversions, Costs, Uses, and Consequences. *Annual Review of Energy & Environment*, 25: 21–51.
Thomke, S. H., 2003. *Experimentation Matters: Unlocking the Potential of New Technologies for Innovation*. Cambridge, MA: Harvard Business School Press.

Unruh, G. C., 2000. Understanding Carbon Lock-In. *Energy Policy*, 28(12): 817–30.

Wilson, C., 2009. *Meta-Analysis of Unit and Industry Level Scaling Dynamics in Energy Technologies and Climate Change Mitigation Scenarios*. Laxenburg, Austria: International Institute for Applied Systems Analysis (IIASA).

2012. Up-Scaling, Formative Phases, and Learning in the Historical Diffusion of Energy Technologies. *Energy Policy*, 50: 81–94.

Wilson, C. & Grubler, A., 2011. Lessons from the History of Technological Change for Clean Energy Scenarios and Policies. *Natural Resources Forum*, 35: 165–84.

Wilson, C., Grubler, A., Bauer, N., Krey, V., & Riahi, K., 2012. Future Capacity Growth of Energy Technologies: Are Scenarios Consistent with Historical Evidence? *Climatic Change, 118(2)*: 381–95.

6 Input, Output, and Outcome Metrics for Assessing Energy Technology Innovation

Charlie Wilson

Author's Summary

A variety of quantitative methods have been used to analyse energy technology innovation. These range from efforts to identify and isolate the effects of particular policies or innovation processes such as research and development, to studies of the broader social benefits of energy technology innovation. Innovation metrics can thus describe innovation inputs, outputs, or outcomes. This case study presents the commonly used metrics in each of these categories, together with the main issues associated with their application. The discussion then identifies broader issues with the use of quantitative metrics to analyse energy technology innovation, including the lack of integrated metrics, problems with quality control, and difficulties with cross-country and cross-technology comparisons.

Editors' Guide

Relationship of Case Study to Energy Technology Innovation System Framework

Many of the elements of the innovation system are observable by proxy through particular quantitative metrics.

Metrics of innovation inputs predominantly describe the financial *resource mobilisation* that underpins the accumulation of knowledge, the strengthening of actors and institutions, and the improving performance of technologies (i.e., the other three dimensions of the innovation system). Financial resource input metrics – whether investments or expenditures – can be biased towards early *innovation stages* (particularly R&D) as data are more readily available and the processes and consequences of *knowledge generation* become clearer. Labour resources are another important input, and offer a proxy for tacit knowledge that is embodied in skilled workers.

Metrics of innovation outputs and outcomes predominantly describe the *adoption and use* of technologies and their consequences. Examples include the charting of improvements in the *relative advantage* of new technologies (e.g., falling costs) or of the up-scaling of technologies (or whole industries) to capture available *economies of scale*. Certain output metrics such as publications and patents can be used as proxies for *knowledge codification*. Such metrics can also formally involve *exchange and interaction* between innovation system actors, as in the case of scientific conferences or symposia.

Several metrics are designed explicitly to assess the overall performance of functioning of the innovation system by linking different elements. Weighting patents by citations links innovation inputs (*knowledge generated* by other actors as measured by citation counts) to innovation outputs (*knowledge codified* in patents). Learning curves that express cost reductions as a function of cumulative experience with a technology also link innovation inputs (*learning*) with innovation outputs (improving *relative advantage*). Cost-benefit analyses of the returns on innovation investments link innovation inputs (financial *resources mobilised*) with innovation outcomes (the consequences of the *adoption and use* of technologies in terms of broader economic, environmental, or social benefits). Modelling studies seek to quantify the relative influence of *technology-push* and *demand-pull* drivers of innovation outputs such as technology cost (*relative advantage*) or market penetration (*adoption and use*).

However, none of the available metrics cover all elements of the innovation system, and so all offer a partial view. In particular, elements in the *actors and institutions* dimension tend to be poorly observable by proxy through the use of metrics. *Knowledge depreciation* and *knowledge spillovers* are also difficult to measure robustly. The prevalence of *knowledge spillovers* from energy innovation to and from other, non-energy domains, and the issues that confound cross-country comparisons (such as labour cost differentials, cultural norms towards peer-reviewed publication, or the institutional nature of intellectual property rights), raise questions about how to draw analytical boundaries around the energy technology innovation system.

Insights for Energy Technology Innovation Policy

- Metrics offer a useful window into the functioning of the energy technology innovation system. Each metric is particularly associated with specific elements of the innovation system and/or specific innovation stages. Multiple metrics are needed for a compound view of the innovation system.
- The use of single, simple metrics to capture innovation processes is therefore problematic. Learning curve models have become widely used in public policy both to assess and to justify the magnitude of expenditure and policy initiatives to subsidise demand for new energy technologies. Yet such models omit key variables, have limited predictive power, and capture a limited set of the causal mechanisms linking dependent and independent variables within the innovation system.
- The ultimate outcomes of energy innovation are many and varied, ranging from emission reduction and energy security to cost reduction and welfare provision. Policy objectives should be clearly stated so appropriate metrics can be used to evaluate progress (and so contribute to policy learning). Multiple

objectives may require multiple metrics that allow synergies and trade-offs to be transparently identified.

- Public policy plays an important role in shepherding technologies through the "valley of death" between innovation and successful diffusion (focussing particularly on the demonstration and market formation stages), yet this is poorly characterised by available metrics.

1 Quantitative Analysis of Energy Technology Innovation

A variety of quantitative methods have been used to analyse energy technology innovation. These range from efforts to identify and isolate the effects of particular policies or innovation inputs such as R&D investments, to studies of the broader social benefits of energy technology innovation.

One area of interest has been the relative influence of technology-push and demand-pull factors on innovation outcomes including (as examples) technology costs or market penetration. Some researchers have focussed on particular energy technologies including wind power (e.g., Klaassen et al., 2005) and solar PV (Watanabe, Wakabayashi, and Miyazawa, 2000). Findings have proven sensitive to assumptions about knowledge depreciation and about the time lags between policy actions and the responses of innovation system actors. Both these factors are inherently difficult to estimate empirically.

Other researchers have sought to generalise underlying innovation mechanisms in highly stylised reduced form models (e.g., Kouvaritakis, Soria, and Isoard, 2000; Miketa and Schrattenholzer, 2004). These include learning or experience curves that describe cost reductions as experience with a technology accumulates (Junginger, Van Sark, and Faaij, 2010; Wene, 2000). Resulting models of learning are of direct interest to policy makers concerned with the magnitude and outcomes of public investments or policy initiatives to subsidise the demand for energy technology innovations (Duke and Kammen, 1999). But again, this approach is not without its problems. Learning curve models have been critiqued on the basis of their atheoretical specification (e.g., assumptions of perfect substitutability of R&D and learning by doing as sources of technological knowledge) as well as on statistical grounds (e.g., problems of collinearity in model parameter estimation). Moreover, observed discontinuities in learning rates, perhaps resulting from omitted variables, as well as observed high variability of learning rates across technologies and market environments (Nemet, 2009b), limit the reliability and predictive power of models for specific policy interventions.

Shifting from the measurement and modelling of innovation inputs (e.g., R&D investments) and outputs (e.g., cost reductions), cost-benefit analyses have also been used to quantify the broader impacts and outcomes of energy technology innovation. The National Research Council in the United States conducted one such study to measure the overall value of federal energy R&D programmes (NRC, 2001) and a follow-up study that applied the same methodology prospectively to analyse expected social returns on investment under a range of assumptions including future carbon prices (NRC, 2007). The scope of these cost-benefit analyses is inherently broader than that of a modelling-based analysis of particular innovation processes or specific technology case studies. Yet insights remain contingent on various assumptions, including those related to counterfactual innovation outcomes in the event that federal R&D investments were not made.

Quantitative analysis is not only retrospective. Decision analytic techniques, commonly used by R&D managers, have been used to elicit the subjective judgement of experts familiar with particular technologies (Clemen and Kwit, 2001; Peerenboom, Buehring, and Joseph, 1989). In the United States, this is now recommended practice for helping guide and evaluate public energy technology innovation investments (NRC, 2007). Findings can be used to parameterise models or to quantify roadmaps of future technology trajectories. Recent examples include studies of solar PV and carbon capture technologies (Baker, Chon, and Keisler, 2009).

2 Innovation Metrics

The most commonly used quantitative metrics of innovation relating to energy technology can be divided into three types. *Input* metrics describe financial, labour, and other inputs to the innovation system and innovation processes (like R&D). *Output* metrics describe defined products of the innovation system and innovation processes. These extend from the demonstrated technical feasibility of prototype innovations, the successful scaling up of technological options either at the level of individual technologies (e.g., larger wind turbines) or of entire industries (e.g., solar PV market deployment), or reaching desired prespecified technology performance targets (in terms of efficiency or costs). *Outcome* metrics describe broader energy sector or economy-wide impacts of the successful diffusion of innovations into the marketplace (e.g., carbon emission reductions, jobs created, social returns on innovation investments). These metric proxies of innovation inputs, outputs, and outcomes can be either intangible (e.g., knowledge stock, practical problems and ideas) or tangible and human (e.g., investments needed, scientists, laboratories) (Freeman and Soete, 2000). The following three sections provide an overview of commonly used innovation metrics and their attendant issues, drawing on Gallagher and colleagues (2006) and Sagar and Holdren (2002) unless otherwise noted.

2.1 Input Metrics

Table 6.1 summarises commonly used input metrics applied to energy technology innovation. Common issues associated with the practical application of each metric are also noted.

2.2 Output Metrics

Table 6.2 summarises commonly used output metrics applied to energy technology innovation. Again, common issues associated with the practical application of each metric are also noted.

2.3 Outcome Metrics

Table 6.3 summarises commonly used outcome metrics applied to energy technology innovation with frequent issues associated with the practical application of each metric also noted.

Table 6.1. Commonly used metrics of innovation inputs.

Metric	Description	Issues
R&D Expenditure	• Public, private, or total expenditure on R&D (e.g., real $). • Can also include expenditure on demonstration projects (i.e., R,D&D) • Expressed either in absolute terms or as "R&D intensities" normalised for total output (GDP), total investment, etc.	• Data on public R&D expenditure typically available; time series data allows trends to be analysed.[1] • Private R&D data, particularly in nonlisted companies, can be difficult to obtain; if available, data are often highly aggregated so difficult to isolate expenditures specific to energy R&D.
Investment	• Public, private, or total investment in innovation (e.g., real $). • Includes R&D expenditure (see above) but also investments in demonstration, early deployment, and diffusion. • Narrow investment metrics can also be normalised (e.g., early stage venture capital as % of total venture capital).[2]	• Similar issues to R&D expenditures (see above) but broader categorisation of investment can avoid disaggregation issues (e.g., of aggregated investment figures in corporate accounts). • Investment data tend to underrepresent later stage innovation activities (see text below) and conflate R&D and demonstration stages. • Some databases compile private investments into specific technology sectors by investor type (e.g., venture capital, private equity); but investment targets are usually start-up companies rather than innovation per se, and databases mainly cover industrialised markets.[3]
Human Resources	• Number of scientists and engineers engaged in R&D. • Can be weighted by education (e.g., highest degree attained) or type of training. • Expressed in absolute terms, by sector, or per capita.	• Use as a proxy for "tacit" knowledge embodied in labour input to innovation process. • Simplified metric does not account for quality or efficiency, nor differences in research infrastructure and capital equipment (so difficult to assess R&D labour productivity). • As with investment metrics, difficult to isolate labour input specific to energy innovation, particularly in diversified companies.

[1]. For U.S. R&D trends and analysis, see Nemet and Kammen (2007); Margolis and Kammen (1999) and for detailed data, see Gallagher and Anadón (2009); for selected global data, see Sagar and Van der Zwaan (2006).

[2]. Innovation ratios measure the proportion of high-risk private capital invested in early stage (or high tech) technologies as a proxy measure of the extent to which innovation systems support capital constrained entrepreneurial firms through the "valley of death" between innovation and commercial diffusion. See Rin, Nicodano, and Sembenelli (2006) for an example. In principle, innovation ratios could be used to measure different success factors of an innovation system: for example, total venture capital investments as a proportion of total investments would capture the successful leveraging of risk-taking private capital premised on broad market acceptance and diffusion.

[3]. Examples include the European Venture Capital Association yearbooks (for the EU) and New Energy Finance (in the United States).

Table 6.2. Commonly used metrics of innovation outputs.

Metric	Description	Issues
Publications	• Number of peer-reviewed articles. • Can be weighted by citations or impact factors. • Can also include other research publications (reports, books, evaluations, etc.) • Workshops and conferences	• Readily available information, but English-language bias. • Difficult to define clear system boundaries for energy technology innovations: for example, should articles on complementary technologies such as catalysts, materials, and control systems be included? • Useful metric for programme evaluation if quality or impact weighted.
Patents	• Number of patents filed or granted. • Can be weighted by citations.	• Similar issues to publications: readily available, but difficult to define system boundaries. Greatest validity if used at low level of aggregation.[1] • Biased towards industrialised countries and towards industrial sectors with higher propensity to patent. • Patents generally relate to R&D rather than later stage innovation activities, and are not necessarily good predictors of successful commercialisation.
Technologies	• Number of technologies commercialised. • Can be in terms of plants, production lines, product variants, process improvements, companies, turnover, etc.	• Most visible measure of ultimate success of innovation process. • Difficult to define clear system boundaries for what constitutes a technology, particularly for complex multicomponent systems (e.g., aircraft). • Fails to capture increases in learning and know-how for technologies based on tacit or noncodified knowledge (e.g., energy-efficient building design).
Technology Characteristics	• Ratios of technical to service characteristics.	• Change in ratios of technical characteristics to performance or service characteristics indicate directionality and variety of innovations, as well as their proximity to the technological frontier.[2] • Technology specific: not possible to use in meta-analyses.

[1]. For an overview of issues relating to patents as an output metric, see Basberg (1987); OECD (2009). For studies using patent citations to assess environmental and energy technology innovation, see Nemet (2009a); Popp (2005). For a recent application of patent count data to assess the effectiveness of different types of environmental policy on renewable energy innovation, see Johnstone and colleagues (2010), and for a similar study but on energy-efficient building technology, see Noailly (2012).

[2]. For applications of this approach to helicopters, see Saviotti and Trickett (1992); to aircraft, see Frenken and Leydesdorff (2000); to refineries, see Nguyen and colleagues (2005).

Table 6.3. Commonly used metrics of innovation outcomes.

Metric	• Description	• Issues
Market Penetration	• Number or capacity of technologies sold or used. • Market share (or penetration rate) is alternative measure normalising for size of market or economy, hence describing the extent of substitution into – or capture of – a specific market.	• Generally available data. • Extensive empirical literature on diffusion dynamics, market penetration, and substitution effects.[1] • Bias towards "successful" innovations that have diffused widely. Best suited to retrospective historical analysis.
Learning Rates	• Rate of cost reduction of a technology. • Conventionally measured as the % reduction in unit cost per doubling of cumulative production. • Can be measured for production plants, organisations or technologies.	• Learning rates emphasise the commercialisation phase of a technology but substantive cost reductions may also occur in the earlier innovation stages. • Production and cost (or price) data generally available, and mechanisms for learning effects have been extensively researched.[2] • Learning rates can vary widely between variants of the same technology and between plants producing the same technology; learning rates are also sensitive to timing and data-fitting issues.[3]
Economic Benefits	• Cost-benefit analysis. • Can be aggregated as net social benefit, or left disaggregated in multiple "accounts" (e.g., labour, environment, knowledge, etc.).	• (Investment) costs generally easier to quantify than benefits, which can include environmental and energy security externalities, knowledge stocks and spillovers, option values of technology portfolios, as well as more conventional net employment, tax, and consumer surplus benefits. • Cost-benefit analyses widely used as core component of programme evaluations.[4]
Energy / Emissions Intensity	• Primary energy (GJ), electricity (GWh), or emissions (e.g., tCO_2 tSO_x) per unit of GDP. • Normalisation can also be more tightly defined, for example, per sector, or per power plant.	• Readily available data; meaningful as part of time series trend. • Aggregate impact of innovation only, and potentially confounded by structural changes to economic activity, nonprice-induced changes, and inter-fuel substitution.
Project / Programme Evaluation	• Size and number of programmes in terms of employees, turnover, investment, outputs, etc.	• Difficult to assess quality, so need to complement with case study or survey research (see text below). • Similar issues with tacit knowledge as for technologies (see above under outputs).

[1]. For a theoretical overview and empirical treatment of energy and environmental technologies globally, see Grubler (1998).

[2]. In general terms, the experience of "doing" or producing a good gives rise to incremental improvements in design, process techniques, product efficiency, labour productivity, and so on as a basis for reductions in unit cost. For reviews related to manufacturing, see Argote and Epple (1990) and to energy innovation, see Grubler and colleagues (1999). For empirical analyses of energy technology learning rates, see IEA (2000); Junginger and colleagues (2010); McDonald and Schrattenholzer (2001).

[3]. For an analysis of learning rate variability between different specification coal power plants, see Joskow and Rose (1985). For issues with data fitting with examples for solar PV and wind power, see Nemet (2009b).

[4]. For a comprehensive evaluation of two major energy technology innovation programmes, see NRC (2001), and for the same social return on investment framework applied prospectively, see NRC (2007). For an example of an econometric evaluation of programme costs relating to demand-side management with a subsequent critique, see Auffhammer, Blumstein, and Fowlie (2008); Loughran and Kulick (2004).

3 Discussion

3.1 Lack of Integrated Metrics for Energy Technology Innovation

With respect to high-tech companies, Hagedoorn and Cloodt (2003) discuss the absence of a generally accepted metric of innovation despite the long tradition of using single metrics to study innovation, including R&D, patents, and new product announcements. In response they propose a composite innovation metric to describe companies' overall "innovative performance."

It is similarly salient that no single metric dominates assessments of energy technology innovation. As shown in Tables 6.1, 6.2, and 6.3, this can be explained partly by issues of data availability, reliability, and definition that mean each metric has its shortcomings. Defining what constitutes an energy technology is also far from straightforward. Energy assessments typically focus on the supply side only, but many activities in the industrial, agricultural, and residential and commercial buildings sectors are also fundamentally connected to energy technology innovation.

A more general observation, however, is that no single metric captures the *overall* success of the energy technology innovation endeavour, linking inputs to outputs and broader outcomes. Learning rates are one exception as they are a hybrid metric that links inputs (investments in production) to outputs (cost per unit production), but they are typically confined to the diffusion phase of technologies that have succeeded in the market. As such, they omit the uncertain and complex interrelationships among innovation processes during the early innovation stages that generate technological novelty.

Other fields of innovation – for example, agriculture, pharmaceuticals, aerospace – are similarly characterised by a mixed, staged involvement of both public and private resources, and a complex set of time-lagged processes linking innovation inputs to outputs. In these fields, innovation assessments commonly use econometric or regression techniques to evaluate either net social returns on R&D or productivity gains. In agriculture, for example, researchers have long used such techniques to examine the effectiveness of particular structures and roles in the innovation process (Evenson, Waggoner, and Ruttan, 1979), and to account for the often long time lags between innovation inputs and outputs (Alston et al., 2000).

Certainly, the use of econometric methods to evaluate innovation is not without its issues (Griliches, 1987). However, innovation assessments in the energy field have not similarly converged on a broadly accepted set of analytical approaches. Many, if not all, of the data and methodological limitations raised in Table 6.1 equally hamper the assessment of innovation in other fields. Neither is energy unique in the capital intensiveness of innovation outcomes (cf. pharmaceuticals) and the shape of returns in high-volume, low-margin regulated markets (cf. agriculture).

With respect to the evaluation of research, "problems arise because not everyone perceives the same purpose" (Alston et al., 2000: 14). Is this a particular problem with assessments of energy technology innovation? Agricultural innovation may emphasise productivity outcomes; pharmaceutical innovation may emphasise health outcomes. In contrast, there are many, varied, and oft-competing objectives for energy technology innovation. Does successful innovation reduce emissions in absolute terms? reduce energy or carbon intensity? reduce unit costs? reduce the cost of final service provision? increase option values in diverse portfolios? improve security, reliability, flexibility? ensure backstops are available? increase accessibility? And so on. The most appropriate dependent variable of an energy technology innovation assessment may also be technology specific. This creates problems for meta-analyses and dampens convergence on a set of common assessment techniques and indicators.

3.2 Underrepresentation of Demonstration and Deployment Activities

Although the overall innovation process is characterised by multiple feedbacks between stages, a general distinction can be made between research, development, demonstration, market formation, and diffusion activities. Early deployment during the market formation stage is a precursor (but overlapping) stage to diffusion in which barriers to widespread commercialisation are targeted (e.g., through "demand-pull" policies) (Sagar and Van der Zwaan, 2006). R&D expenditures and investment data, as well as patent and publication data, tend to relate to the earlier research and development phases. For the financial metrics, this is driven by data availability associated with government's more central role in these higher-risk activities with public good characteristics (e.g., nonappropriable knowledge generation). Outcome metrics inherently relate to the diffusion phase. This leaves the demonstration and early deployment phases underrepresented by innovation metrics (Sagar and Gallagher, 2004). Both phases are critical to avoiding the so-called valley of death between innovation and successful diffusion if technologies still carry preventively high risks for private sector capital. Metrics of demonstration activities can be extracted from demonstration project databases for specific technologies (Harborne and Hendry, 2009). However, available statistics (e.g., IEA (2011) data on public R,D&D expenditures) are extremely limited with respect to technology demonstration activities. Assessing deployment activities is also difficult to do outside of specific programme evaluations (see Table 6.1). For examples on clean coal and buildings, see Sagar and Gallagher (2004).

3.3 Cross-Country Comparison of Innovation Metrics

Cross-country comparisons of technological capabilities and innovation processes are complicated by the absence of standardised data. For a detailed discussion of issues, see Archibugi and Coco (2005). As noted in Table 6.1, both R&D expenditure and investment data vary as to the system boundaries or definition of what is included (e.g., sources and targets) and to the coverage of the innovation life cycle (Sagar and Holdren, 2002). Comparisons of innovation metrics across countries also need to control for confounding factors. For example, comparisons of human resource metrics need to account for labour cost differentials, and comparisons of peer-reviewed papers may be biased by different cultural attitudes towards the importance of publications, and by access to English-language journals (Archibugi and Coco, 2005). The same is broadly true of patents as the institutional nature of intellectual property rights varies widely between countries both in terms of propensity to patent, as well as the quality and accessibility of the patent system (Basberg, 1987). Cross-country comparisons of energy and emissions intensity are confounded by numerous factors including the structural weighting of economic activity towards certain types of production (e.g.,

Alcantara and Duro, 2004), relative energy prices (Miketa, 2001), climatic factors, urban form, resource endowments, and so on. Comparison of within-country time trends are informative through decomposition analyses to isolate the influence of technological change (IEA, 2004).

3.4 Quality Control of Quantitative Metrics

As with any quantitative metric, quality control is important to ensure results are meaningful. Tables 6.1, 6.2, and 6.3 include some examples. Numbers of publications or patents can be weighted by subsequent citations or some other measure of impact. Changes in the ratios between certain technical and performance characteristics offer another indication of qualitative changes in a technology's design (Saviotti and Metcalfe, 1984).

In general, however, complementary qualitative assessments of innovation processes are needed. Methods include interviews, process tracing, surveys of innovation process participants, and detailed case studies of innovation programmes (as in this book). Important qualities of energy technology innovations to assess include contribution to portfolio diversity and innovation system management. Creating and maintaining diverse knowledge stocks can enhance system resilience and preserve option value with respect to long-term climate change mitigation goals (Sanden and Azar, 2005). Management capabilities, as well as sufficient infrastructure and collaborative relationships to support knowledge generation, are important structural features of innovation systems and influence their successful functioning. (For an overview of innovation systems at a sectoral level, see Malerba (2002); for applications to renewable energy, see Hekkert and Negro (2009), and to biofuels, see Jacobsson (2008).)

4 Conclusion

The growing body of quantitative analysis and modelling using input, output, and outcome metrics of energy technology innovation has greatly improved our understanding of innovation processes. Systematic renderings of the whole energy technology innovation system are, however, rare. Wilson and colleagues (2012) compare a range of different metrics of inputs, outputs, and outcomes into global energy technology innovation (see also Figure 6.1). They find directed innovation efforts to clearly privilege energy supply technologies even though end-use technologies dominate innovation system outputs and desired innovation outcomes in terms of climate change mitigation or social rates of return. This is but one potential application of a compound set of metrics to assess the overall functioning of the energy technology innovation system.

However, as a final cautionary note, it is important to recognise the inherent limitations of quantitative metrics that emphasise particular elements of the energy technology innovation system and risk offering overly mechanistic explanations of causality. Quantitative assessments are also limited by data availability, as is evidenced by the relative dearth of empirical work on energy end-use technologies compared to energy supply technologies, and on developing country transitions compared to developed country innovation systems. Descriptively as well as analytically rich

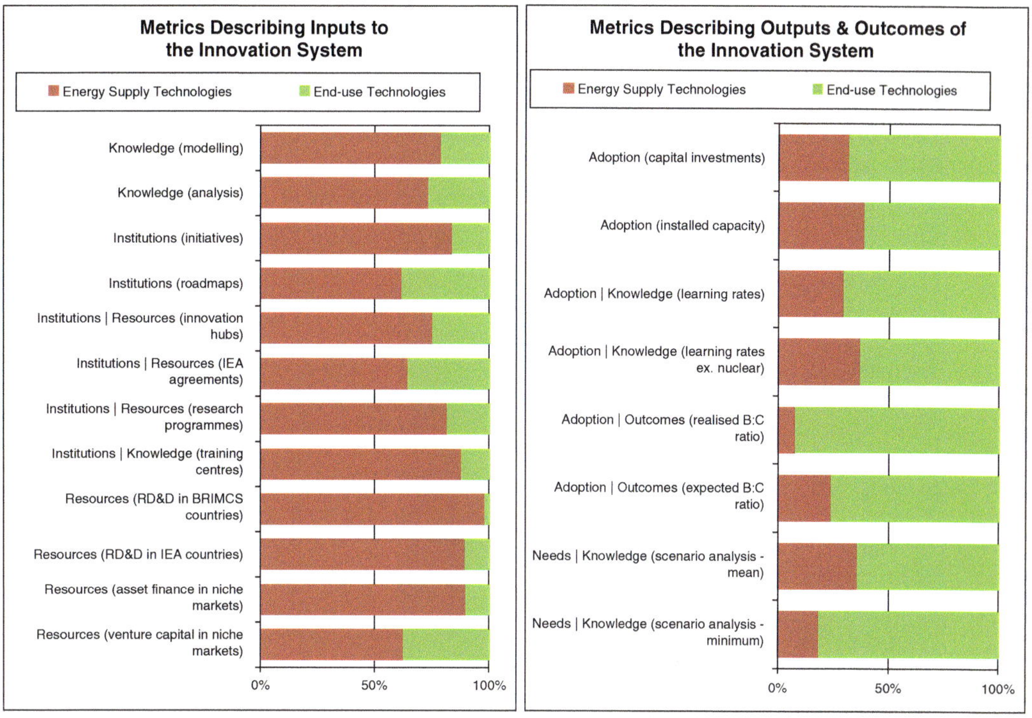

Figure 6.1. Using input, output, and outcome metrics to assess the technological emphasis of directed innovation efforts.
Notes: Bars show relative emphasis on energy supply (red) and energy end-use technologies (green) using representative indicators for different innovation inputs (left panel) and outputs and outcomes (right panel).
Source: Wilson and colleagues (2012).

case studies of energy technology innovation can help to counter some of these limitations.

5 Further Reading

For a good overview of issues associated with assessing innovation in energy technologies, see Gallagher, Holdren, and Sagar (2006). For an interesting recent application of one widely used metric, patent counts, see Johnstone, Haščič, and Popp (2010). For a comparison of innovation input and output metrics in the context of the energy system, see: Wilson and colleagues (2012).

REFERENCES

Alcantara, V. & Duro, J. A., 2004. Inequality of Energy Intensities across OECD Countries: A Note. *Energy Policy*, 32(11): 1257–60.

Alston, J. M., Chan-Kang, C., Marra, M. C., Pardey, P. G., & Wyatt, T., 2000. *A Meta-Analysis of Rates of Return to Agricultural R&D. Washington, DC:* International Food Policy Research Institute (IFPRI).

Archibugi, D. & Coco, A., 2005. Measuring Technological Capabilities at the Country Level: A Survey and a Menu for Choice. *Research Policy*, 34(2): 175–94.

Argote, L. & Epple, D., 1990. Learning Curves in Manufacturing. *Science*, 247(4945): 920–24.

Auffhammer, M., Blumstein, C., & Fowlie, M., 2008. Demand-Side Management and Energy Efficiency Revisited. *The Energy Journal*, 29(3): 91–104.

Baker, E., Chon, H., & Keisler, J., 2009. Advanced Solar R&D: Combining Economic Analysis with Expert Elicitations to Inform Climate Policy. *Energy Economics*, 31: S37–S49.

Basberg, B. L., 1987. Patents and the Measurement of Technological Change: A Survey of the Literature. *Research Policy*, 16(2–4): 131–41.

Clemen, R. T. & Kwit, R. C., 2001. The Value of Decision Analysis at Eastman Kodak Company, 1990–1999. *Interfaces*, 31(5): 74–92.

Duke, R. & Kammen, D. M., 1999. The Economics of Energy Market Transformation Programs. *The Energy Journal*, 20(4): 15–64.

Evenson, R. E., Waggoner, P. E., & Ruttan, V. W., 1979. Economic Benefits from Research: An Example from Agriculture. *Science*, 205(4411): 1101–1107.

Freeman, C. & Soete, L., 2000. *The Economics of Industrial Innovation*. Cambridge, MA: MIT Press.

Frenken, K. & Leydesdorff, L., 2000. Scaling Trajectories in Civil Aircraft (1913–1997). *Research Policy*, 29(3): 331–48.

Gallagher, K. S. & Anadón, L. D., 2009. *DOE Budget Authority for Energy Research, Development, & Demonstration Database*. Cambridge, MA: Belfer Center for Science and International Affairs, Harvard University.

Gallagher, K. S., Holdren, J. P., & Sagar, A. D., 2006. Energy-Technology Innovation. *Annual Review of Environment and Resources*, 31(1): 193–237.

Griliches, Z., 1987. R&D and Productivity: Measurement Issues and Econometric Results. *Science*, 237(4810): 31–35.

Grubler, A., 1998. *Technology and Global Change*. Cambridge, UK: Cambridge University Press.

Grubler, A., Nakicenovic, N., & Victor, D. G., 1999. Dynamics of Energy Technologies and Global Change. *Energy Policy*, 27(5): 247–80.

Hagedoorn, J. & Cloodt, M., 2003. Measuring Innovative Performance: Is There an Advantage in Using Multiple Indicators? *Research Policy*, 32: 1365–79.

Harborne, P. & Hendry, C., 2009. Pathways to Commercial Wind Power in the US, Europe and Japan: The Role of Demonstration Projects and Field Trials in the Innovation Process. *Energy Policy*, 37(9): 3580–95.

Hekkert, M. P. & Negro, S. O., 2009. Functions of Innovation Systems as a Framework to Understand Sustainable Technological Change: Empirical Evidence for Earlier Claims. *Technological Forecasting and Social Change*, 76(4): 584–94.

IEA, 2000. *Experience Curves for Energy Technology Policy*. Paris: OECD/IEA.

2004. *Oil Crises & Climate Challenges: 30 Years of Energy Use in IEA Countries*. Paris: OECD/IEA.

2011. *Energy Technology RD&D, Online Data Base*, 2011 edition. Paris: International Energy Agency (IEA).

Jacobsson, S., 2008. The Emergence and Troubled Growth of a "Biopower" Innovation System in Sweden. *Energy Policy*, 36(4): 1491–1508.

Johnstone, N., Haščič, I., & Popp, D., 2010. Renewable Energy Policies and Technological Innovation: Evidence Based on Patent Counts. *Environmental and Resource Economics*, 45(1): 133–55.

Joskow, P. L. & Rose, N. L., 1985. The Effects of Technological Change, Experience, and Environmental Regulation on the Construction Cost of Coal-Burning Generating Units. *The RAND Journal of Economics*, 16(1): 1–27.

Junginger, M., Van Sark, W., & Faaij, A. (eds.) 2010. *Technological Learning in the Energy Sector: Lessons for Policy, Industry and Science*. Cheltenham, UK: Edward Elgar.

Klaassen, G., Miketa, A., Larsen, K., & Sundqvist, T., 2005. The Impact of R&D on Innovation for Wind Energy in Denmark, Germany and the United Kingdom. *Ecological Economics*, 54(2–3): 227–40.

Kouvaritakis, N., Soria, A., & Isoard, S., 2000. Modelling Energy Technology Dynamics: Methodology for Adaptive Expectations Models with Learning by Doing and Learning by Searching. *International Journal of Global Energy Issues*, 14(1): 104–15.

Loughran, D. S. & Kulick, J., 2004. Demand-Side Management and Energy Efficiency in the United States. *The Energy Journal*, 25(1): 19–43.
Malerba, F., 2002. Sectoral Systems of Innovation and Production. *Research Policy*, 31(2): 247–64.
Margolis, R. M. & Kammen, D. M., 1999. Evidence of Under-Investment in Energy R&D in the United States and the Impact of Federal Policy. *Energy Policy*, 27(10): 575–84.
McDonald, A. & Schrattenholzer, L., 2001. Learning Rates for Energy Technologies. *Energy Policy*, 29(4): 255–61.
Miketa, A., 2001. Analysis of Energy Intensity Developments in Manufacturing Sectors in Industrialized and Developing Countries. *Energy Policy*, 29(10): 769–75.
Miketa, A. & Schrattenholzer, L., 2004. Experiments with a Methodology to Model the Role of R&D Expenditures in Energy Technology Learning Processes; First Results. *Energy Policy*, 32(15): 1679–92.
Nemet, G. F., 2009a. Demand-Pull, Technology-Push, and Government-Led Incentives for Non-Incremental Technical Change. *Research Policy*, 38(5): 700–709.
2009b. Interim Monitoring of Cost Dynamics for Publicly Supported Energy Technologies. *Energy Policy*, 37(3): 825–835.
Nemet, G. F. & Kammen, D. M., 2007. U.S. Energy Research and Development: Declining Investment, Increasing Need, and the Feasibility of Expansion. *Energy Policy*, 35(1): 746–55.
Nguyen, P., Saviotti, P. P., Trommetter, M., & Bourgeois, B., 2005. Variety and the Evolution of Refinery Processing. *Industrial and Corporate Change*, 14(3): 469–500.
Noailly, J., 2012. Improving the Energy Efficiency of Buildings: The Impact of Environmental Policy on Technological Innovation. *Energy Economics*, 34(3): 795–806.
NRC, 2001. *Energy Research at DOE: Was it Worth it? Energy Efficiency and Fossil Energy Research 1978 to 2000*. Washingon, DC: Board on Energy and Environmental Systems, National Research Council (NRC).
2007. *Prospective Evaluation of Applied Energy Research and Development at DOE (Phase Two)*. Washington, DC: National Research Council (NRC).
OECD, 2009. *Patent Statistics Manual*. Paris: Organisation for Economic Cooperation and Development (OECD).
Peerenboom, J. P., Buehring, W. A., & Joseph, T. W., 1989. Selecting a Portfolio of Environmental Programs for a Synthetic Fuels Facility. *Operations Research*, 37(5): 689–99.
Popp, D., 2005. Lessons from Patents: Using Patents to Measure Technological Change in Environmental Models. *Ecological Economics*, 54(2–3): 209–26.
Rin, D. M., Nicodano, G., & Sembenelli, A., 2006. Public Policy and the Creation of Active Venture Capital Markets. *Journal of Public Economics*, 90(8–9): 1699–1723.
Sagar, A. D. & Gallagher, K. S., 2004. Energy Technology Demonstration and Deployment. In National Committee on Energy Policy (ed.) *Ending the Energy Stalemate: A Bipartisan Strategy to Meet America's Energy Challenges*. Washington, DC: National Committee on Energy Policy.
Sagar, A. D. & Holdren, J. P., 2002. Assessing the Global Energy Innovation System: Some Key Issues. *Energy Policy*, 30(6): 465–69.
Sagar, A. D. & Van der Zwaan, B., 2006. Technological Innovation in the Energy Sector: R&D, Deployment, and Learning-By-Doing. *Energy Policy*, 34(17): 2601–2608.
Sanden, B. A. & Azar, C., 2005. Near-Term Technology Policies for Long-Term Climate Targets – Economy Wide Versus Technology Specific Approaches. *Energy Policy*, 33(12): 1557–76.
Saviotti, P. P. & Metcalfe, J. S., 1984. A Theoretical Approach to the Construction of Technological Output Indicators. *Research Policy*, 13(3): 141–51.
Saviotti, P. P. & Trickett, A., 1992. The Evolution of Helicopter Technology, 1940–1986. *Economics of Innovation and New Technology*, 2: 111–30.
Watanabe, C., Wakabayashi, K., & Miyazawa, T., 2000. Industrial Dynamism and the Creation of a "Virtuous Cycle" between R&D, Market Growth and Price Reduction – The Case of Photovoltaic Power Generation (PV) Development in Japan. *Technovation*, 20(6): 299–312.

Wene, C. O., 2000. *Experience Curves for Energy Technology Policy*. Paris: International Energy Agency.

Wilson, C., Grubler, A., Sims Gallagher, K., & Nemet, G. F., 2012. Marginalization of End-Use Technologies in Energy Innovation for Climate Protection. *Nature Climate Change*, 2(11): 780–88.

7 Technology Portfolios: Modelling Technological Uncertainty and Innovation Risks

Arnulf Grubler, Sabine Fuss, with contributions from David McCollum, Volker Krey, and Keywan Riahi

Authors' Summary

The case study reviews a range of methodologies and model-based applications that can assist policy decisions under irreducible technology innovation uncertainties. One approach uses scenario analysis. A scenario study that explored a range of salient future uncertainties identified energy efficiency and conservation as the single most important and also most robust technology option for climate mitigation. Another approach uses portfolio theory to quantify the benefits from diversified technology portfolios in the framework of risk-averse decision making. Such studies suggest risk aversion leads to higher adoption rates of currently higher-cost energy technology options such as modern biomass, renewables, and carbon capture and storage. Diversification not only reduces the mean of risk exposure but also drastically lowers the tails of extreme undesirable outcomes. Portfolio theory and scenario analysis can also be combined, as illustrated by a study of three different energy system objectives (energy security, air pollution, climate change) within a multi-criteria optimisation framework. Scenarios in which objectives were met individually, and then all together, under a range of uncertainties again demonstrated the benefits of portfolio diversification with greater emphasis on energy efficiency and "general purpose" energy conversion technologies. The examples discussed in this case study show that formal tools and approaches, including scenario analysis, portfolio theory, and multi-criteria optimisation, are increasingly available to move technology portfolio decisions onto a more rational ground.

Editors' Guide

Relationship of Case Study to Energy Technology Innovation System Framework

Innovation outcomes are irreducibly uncertain. Innovation investments and expectations should be robust to the possibility of failure. Analysing innovation uncertainties and anticipating the consequences of innovation failures, as well as successes, is an integral part of the *knowledge generation* dimension of the innovation system. Salient uncertainties, ranging from future costs and other performance attributes (as examples of *adopter preferences*) to energy service demands or resource availabilities, can be identified, quantified, and systematically assessed.

Such analysis invariably concludes that diverse innovation portfolios outperform hopes pinned on silver bullet technologies or innovation breakthroughs. Portfolio diversification links to the importance of *entrepreneurial risk taking*, enhanced experimentation to support *learning*, and earlier *niche market* investments supported or protected from commercial competition by *public policy*. Portfolio diversification can thus be regarded as an institution that brings to life all four dimensions of the innovation system.

Analysis of innovation outcomes under conditions of uncertainty also invariably concludes that efficiency improvements and the subsequent *adoption and use* of energy-using technologies should be an essential, even dominant part of innovation portfolios. End-use efficiency outperforms energy supply technologies across a wide range of criteria, is more robust to uncertainty, provides a more diverse set of benefits, and relaxes the constraints on allowable energy supply mixes, which in turn makes the energy system more resilient to technology-specific innovation failures. This also illustrates well why a *systemic perspective on innovation* is so important. Energy technology innovation systems have been, and continue to be, heavily weighted towards energy supply technologies. A systemic analysis against explicit criteria (such as contribution to climate change mitigation) reveals this stark bias. The *knowledge generated* can thus guide innovation decisions particularly in terms of *public policy*.

A range of formal tools are available to inform the rational design of energy innovation portfolios and *public policy* decisions to *mobilise resources* in support of particular technologies. These tools, which include scenario analysis, portfolio theory, and multi-criteria optimisation, have particular strengths so their application should be tailored to particular decision needs. These tools do not remove the need for judgement; rather, they make the evidence base on which such judgements are made more transparent, and the criteria against which portfolios are designed more clear. Portfolio design criteria can apply to energy technologies (e.g., lowest cost), to decision maker characteristics (e.g., aversion to mean or extreme risks), to energy system outcomes (e.g., climate change mitigation, energy security), or to portfolio characteristics (e.g., robustness to salient uncertainties), as well as to the relative weighting of these various aspects.

Insights for Energy Technology Innovation Policy

- Diverse innovation portfolios with a strong emphasis on efficient end-use technologies can help make future energy systems more robust to uncertainties, including the possibility of technology-specific innovation failures.
- A diverse range of energy technologies supported by innovation policy is associated with experimentation and learning in diverse market niches. Given absolute resource constraints, an innovation strategy centred on diversity implies granularity in technology choices. Single, "lumpy," large-scale

investments in technologies at scale (invariably energy supply technologies) exhaust innovation resources and expose resulting portfolios to the risk of failure.

- To date, innovation portfolios (public R&D investments), as well as the many institutions comprising the innovation system, have been strongly biased towards energy supply technologies (particularly nuclear fission) at the expense of end-use efficiency. A systemic perspective on energy innovation urges the rebalancing of innovation portfolios towards efficient end-use technologies based on analysis using transparent and clear portfolio design criteria (e.g., climate change mitigation potentials).
- General purpose technologies with inherent flexibility, versatility, or substitutability in both resource inputs and energy service "outputs" are a central feature of optimal innovation portfolios. Electricity generation has been a historic exemplar, and is expected to play an ever more important future role. Innovation policy should prioritise other potential general purpose technologies.

1 Introduction

Most large-scale energy assessments rely on deterministic energy models. Technological uncertainty is either not represented at all (e.g., IEA, 2008) or is described through a set of scenarios that vary the most important salient technological uncertainties such as availability, costs, and resource constraints, as well as changing market environments like relative prices (e.g., Nakicenovic et al., 1998, 2000).

Such approaches can offer only limited guidance for investments under technological uncertainty. Deterministic models tend to promote technologies that are least costly under the assumed parameterisation and under a perfect-foresight decision-making paradigm. Under slightly different circumstances, however, such choices can turn out to be considerably more expensive (e.g., if the technologies considered "optimal" do not develop at the projected rates) or even dangerous (e.g., if safety risks are ignored). It is, therefore, inherently important to include uncertainty into any energy system modelling and specifically in applications designed to guide policy making. Fortunately, modelling techniques have greatly improved over the last years. This is in part due to increased computing power that allows for more complex models and analysis, but also due to advances on the theoretical side as well as in modelling tools from other fields including finance.

Energy systems modelling can incorporate innovation uncertainty in two main ways: scenario analysis and portfolio theory. Both approaches are reviewed and illustrated in this case study. A combination of both approaches using a multi-criterion optimisation framework is also discussed. The case study concludes by drawing some generalisable insights for crafting technology portfolios.

2 Scenario Analysis

Scenario analysis is a technique widely used to explore salient uncertainties of the future. As opposed to a (singular) deterministic forecast, a scenario is a set of conditional "what if ... then" assumptions integrated into formal energy or climate policy models to explore the outcomes of exogenously assumed contingencies. Examples of such contingencies include high or low energy or carbon prices, or a moratorium on nuclear power or its expansion. In order to investigate how the optimal energy mix changes under different assumptions and alternative developments, scenario analysis and model sensitivity analyses have proven useful and insightful tools. Relaxing certain assumptions (e.g., that prices remain stable in the long run) or imposing additional restrictions (e.g., because of the policy goal of achieving a less emission-intensive fuel mix) can lead to quite different strategies for energy technology investments over time.

Kann and Weyant (2000) present scenario analysis (along with sensitivity analysis) as a way to identify key uncertain parameters, which can then be modelled stochastically (see, for example, Zapert, Gaertner, and Filar, 1998) using simplified versions of integrated assessment models. The simplest procedure for propagating uncertainty through a model is by sampling from a joint distribution across input values (such as future technology costs derived from scenario studies), even though in practice such joint distributions are difficult to specify and also involve a large degree of subjectivity in deriving probabilistic uncertainty distributions from a wide range of scenarios whose probability of occurrence remains unknowable.

A large-scale scenario study (Riahi, Grubler, and Nakicenovic, 2007) addressed explicitly the question of how the portfolio of GHG mitigation technologies changes as a function of the representation of different salient uncertainties including energy demand, resource constraints, availability and costs of technologies, and finally, the magnitude of GHG emissions constraints. All these uncertainties were varied together in twenty-two scenarios. The scenarios were quantified using the bottom-up, linear programming model MESSAGE that lends itself particularly to this type of analysis because of its detailed representation of energy technologies. For the quantification of the impact of scenario uncertainty, the metric of cumulative future GHG emission reductions under alternative scenario specifications was adopted, deploying the concept of mitigation "wedges" (Pacala and Socolow, 2004). A mitigation wedge is simply the contribution to the cumulative emission reduction over the period 2000–2100 that a particular technological option provides compared to a baseline scenario.

First, three baseline scenarios without GHG emissions constraints were compared to corresponding hypothetical baselines that assume a "frozen" state of technology of the year 2000 (i.e., no technological change/improvements at all). Then for each baseline scenario a range of increasingly stringent GHG emissions constraint scenarios was calculated (constraints varied from as low as 480 ppm CO_2-equivalent GHG concentration by 2100 up to 670 ppmv-equivalent). Additional scenarios were performed to quantify the effects of innovation failures, that is, assuming a particular technology (e.g., nuclear, or carbon capture and sequestration, CCS) would not be available. The concept of mitigation wedges is illustrated in Figure 7.1 for an illustrative, high-emissions scenario (A2r), its corresponding "frozen technology" baseline, and a GHG constraint resulting in 670 ppmv-equivalent concentration by 2100.

The calculated technology-specific GHG mitigation wedges are summarised in Table 7.1 showing mean as well as minima/maxima across all scenarios analysed. The

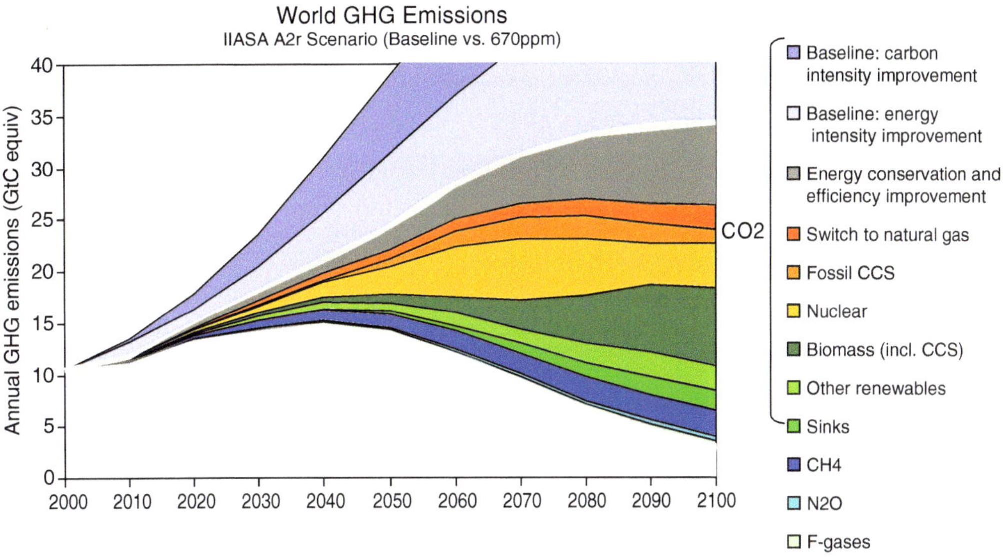

Figure 7.1. Greenhouse gas mitigation wedges for climate stabilisation.
Notes: A wedge is the integral in terms of cumulative emissions reductions over the time period 2000–2100 compared to respective baseline scenarios. Wedges are calculated as the difference in emissions between a hypothetical baseline with frozen 2000 technology (top two "baseline" wedges) or a scenario achieving atmospheric stabilisation at 670 ppmv-equivalent by 2100 (all other wedges) compared to baseline scenario without GHG constraints.

ranking of different mitigation options is quite robust across the scenarios explored (even including larger scenario samples than reported here) with energy efficiency and conservation the single most important option with typically >50 percent contribution to cumulative 2000–2100 emission reduction.

Riahi and colleagues (2007) propose an additional "robustness" measure to describe different mitigation options, expressed as the dispersion between minima/maxima across all scenarios calculated. For instance, energy efficiency turns out to be very robust as all scenarios rely to a large degree on that option whereas nuclear exhibits low robustness as its deployment varies widely across the scenarios (*high* in scenarios of high demand, lack of technology progress in renewables, and stringent climate constraints; *low* in low demand scenarios with high shares of renewables and medium climate constraints).

Table 7.1 also compares the calculated future mitigation potentials of technological options to cumulative past and current R&D expenditures, summarised for the total of all IEA countries (that account for the majority of global public energy R&D). This reveals a highly significant bias in past and current R&D portfolios in favour of nuclear and at the detriment of energy efficiency and conservation. The scenarios suggest that more than half of future GHG emission reductions rely on improvements in energy efficiency and conservation, compared to some 10 percent for nuclear. This compares to cumulative nuclear R&D expenditures well above 50 percent of the entire public energy R&D budget of all IEA countries over the period 1974–2008 and a share of less than 10 percent for energy efficiency and conservation. To put the latter numbers into an absolute perspective: IEA countries' cumulative public R&D into energy efficiency totalled some 38 billion US$2008 in PPP terms, which is *lower*

Table 7.1. Technologies needed for future greenhouse gas mitigation versus cumulative public energy R&D in IEA countries.

	Cumulative emission reduction (2000–2100)				Cumulative R&D (1974–2008)	
	(mean all scenarios)		Min	Max	10e9	
	GtC	%			$2008	%
Energy efficiency	1695	59.2	666	3008	38	9.1
Fossil Fuels	177	6.2	19	415	54	12.8
Renewables	520	18.2	64	917	36	8.7
Nuclear	243	8.5	64	425	225	53.8
Others	229	8.0	72	361	65	15.5
Total	2864	100.0	885	5126	417	100.0

Notes: Mitigation expressed as cumulative 2000–2100 emission reduction (GtC-equivalent) with mean and min/max across scenarios describing future uncertainties (Riahi et al., 2007); R&D in billion US$2008 in PPP-terms (IEA, 2009).
Source: Grubler and Riahi (2010).

than total cumulative expenditure into fusion energy (41 billion), a highly uncertain option having demonstrated neither technical feasibility nor economic viability and social and environmental acceptability.

Scenario analysis can thus help to reveal such apparent technology innovation biases and assist decision makers in crafting technology portfolios. It helps systematically understand the magnitudes of salient uncertainties and the corresponding risks involved across the range of possible outcomes. However, the choice of scenarios can never be fully comprehensive of all future uncertainties and contingencies. In addition, the scenario approach does not capture the full impact that uncertainty has on decisions of technology adoption. In a deterministic perfect foresight model formulation, scenarios describing technology investment trajectories represent cost minima in terms of discounted systems costs. Uncertainty and associated (e.g., financial investment) risks can have a significant influence on the behavioural patterns governing the composition of the energy mix (e.g., in delaying needed upfront investments).

In effect, decisions depend critically on the risk preferences of a decision maker whether or not he or she will trade more risk for a higher expected return on his or her investment and vice versa. For instance, a decision maker might decide to bet the entire energy strategy on the breakthrough of a *potentially* low-cost, clean-energy technology (e.g., fusion), which constitutes a considerable risk in cases when the anticipated technology promise does not materialise (e.g. fusion turns out to be either infeasible or far too expensive). Diversifying the technology innovation and investment portfolio is therefore a useful strategy to hedge against risks, but the extent of diversification and thus the composition of the resulting portfolio depends on both risk attitudes (premiums) of the decision maker as well as on a correct representation of the risks associated with technological (innovation) uncertainty (e.g., failure or cost overruns of new technologies).

3 Portfolio Theory

Addressing these issues associated with decision making is at the heart of portfolio theory (Markowitz, 1952, 1959). Originally designed to find the optimal mix of financial assets to keep the risk at a prespecified level, or to minimise risk for a fixed rate of return, portfolio theory has recently also been applied to energy sector investment in a number of applications (see the review in Bazilian and Roques, 2008). Earlier work includes Bar-Lev and Katz (1976); Awerbuch and Berger (2003); Awerbuch (2006); Van Zon and Fuss (2006); Krey and Zweifel (2006); and Roques, Newberry, and Nuttall (2008), among others. (Real options theory is another tool adapted from finance – see box.)

Box. Real Options Theory

Another decision-making tool adapted from finance is real options theory (Dixit and Pindyck, 1994). Real options analysis values an investment opportunity as an option, which will only be exercised when the immediate benefits of doing so exceed the value of waiting and making a decision later. With sufficiently large uncertainty, this continuation value can be so large that the adoption of a new technology might be postponed until information becomes available that enables the investor to make a better decision. Note that, while portfolio theory allows for a top-down view of the energy mix, the real options approach is more suited to represent decision making at the plant and investment project level. Recent applications to energy investment include Madlener, Kumbaroglu, and Ediger (2005); Fleten, Maribu, and Wangensteen (2007); Laurikka and Koljonen (2006); Szolgayová, Fuss, and Obersteiner (2008); Yang and colleagues (2008); and Fuss and Szolgayová (2009). The focus of the review here is on large-scale energy systems modelling, however, for which the use of real options approaches is practically limited by issues of computational practicality. See also Kann and Weyant (2000) on this topic.

Core to both scenario analysis and portfolio theory is the required representation of salient uncertainties. Relevant references here are Messner, Golodnikov, and Gritsevskyi (1996); Krey, Martinsen, and Wagner (2007); Grubler and Gritsevskyi (2002); and Gritsevskyi and Nakicenovic (2000). These studies not only focus on stochastic energy prices, but also on technological uncertainties, in particular uncertain future technology costs and how these respond to alternative innovation and investment strategies (i.e., uncertain increasing returns to technology adoption).

Portfolio theory has the advantage of explicitly capturing the benefits from diversification in the framework of risk-averse decision making. Krey and Riahi (2009), for example, address a very important problem in current large-scale energy systems modelling: conventional analysis suggests that optimal investments should focus on technologies projected to have the least cost in the future. However, this ignores the potential of currently high-cost alternatives providing a more acceptable risk profile. Krey and Riahi (2009) implement an endogenous risk (premium) formulation into a stochastic variant of the technology-explicit MESSAGE model of the global energy system in which all salient model variables are treated as uncertain. Their model runs show that optimal investment strategies based on conventional

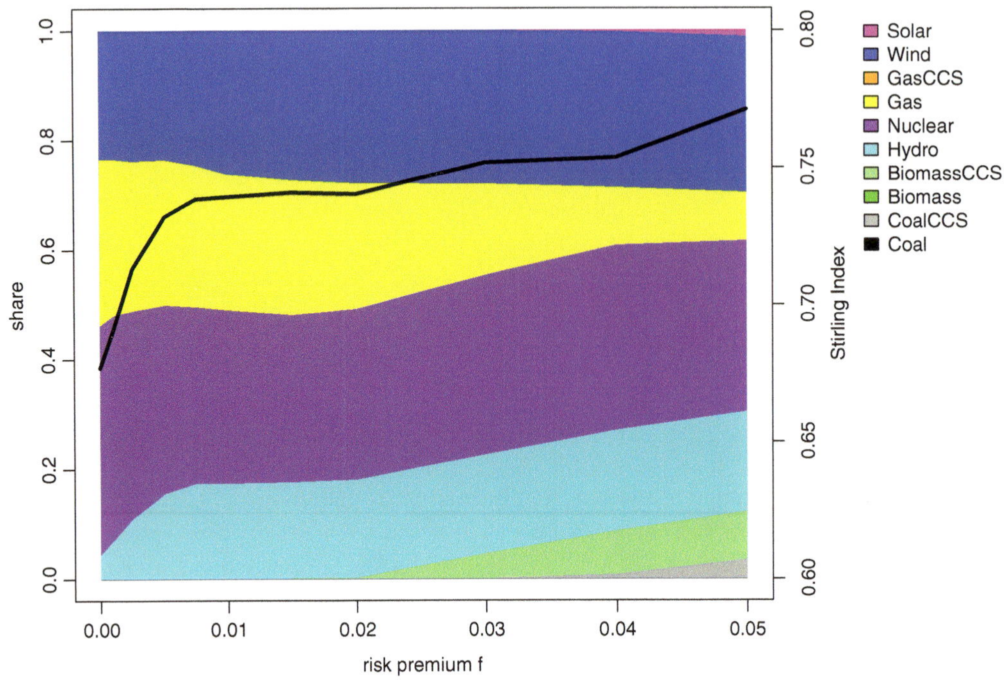

Figure 7.2. Technology shares (coloured blocks, left axis) in 2050 for a global energy system and corresponding Stirling diversification index (line, right axis) versus increasing risk premium (fraction of total discounted systems costs).
Notes: A zero risk premium corresponds to a deterministic model solution with all uncertain variables set at mean values. Positive risk premiums (maximum of 5 percent in above figure) reflect the decision makers' willingness to pay for portfolio diversification to hedge against risks arising from uncertainty represented as lognormal distributions around mean values. Modelled uncertainties include future technology availability and costs as well as uncertain carbon prices. Higher risk aversion leads to a more diversified technology portfolio (as represented by the higher Stirling diversification index and changing shares in technologies deployed).
Source: Krey and Riahi (2009).

modelling (ignoring uncertainty) can be very costly. Risk aversion leads to higher adoption rates of currently higher-cost technology options such as modern biomass and renewables and carbon capture and storage (CCS) in order to hedge against inevitable innovation uncertainty (Figure 7.2). Thus, the model suggests higher short-to-medium term investments into advanced (but currently costlier) technologies under risk aversion.

Krey and Riahi (2009) also conduct sensitivity analysis to identify risk-hedging strategies, which lead to improved robustness of energy investment decisions across a wide range of future uncertainties, including future carbon prices. Energy end-use and efficiency improvements were not modelled in the study so the results describe optimal technology portfolio diversification of energy supply options as a function of the willingness to pay to hedge against technological uncertainty (represented by the "risk premium" expressed as a percentage of total discounted energy systems costs). CO_2 emission reductions prove larger under optimal risk management of technological uncertainty.

The authors do not only consider variance as a risk measure, but also take into account the risk of high-impact tail events. For example, a risk premium of only about

1 percent of total energy expenditures is found to decrease the value of the ninety-ninth percentile of economic risk by more than a factor of two. A "risk ignorant" investment strategy thus imposes extremely high energy system costs if anticipated future cost reductions do not materialise. Diversification thus not only reduces the mean of risk exposure but especially drastically lowers the tails (and economic consequences) of extreme events (undesirable innovation outcomes).

Portfolio theory and scenario analysis are evidently not two mutually exclusive ways to incorporate uncertainty into energy and climate policy modelling and the ensuing crafting of technology strategies. In particular, portfolios can be designed to be robust across scenarios, so that scenario analysis can provide the basis for better-informed portfolio optimisation. In this context, it is also important to consider different measures of risk. The choice of the risk metric depends on whether the underlying distributions have fat tails or not and whether the decision maker is averse to the amount that can potentially be lost in the tails rather than against the variance of his or her costs (e.g., variance versus "conditional value-at-risk" – see Fortin and colleagues (2008) for an application).

Another important question relates to the cost and risk associated with strategies that are robust across a wider set of scenarios, so that the optimal portfolio would perform best even if the worst scenario materialised. This has, for instance, been implemented by Fuss and colleagues (2009) using mini-max criteria in the objective function. This method also benefits from being essentially distribution free insofar as all scenarios are treated as equally likely (avoiding the problem of assigning inherently subjective scenario probabilities).

4 Multi-Criteria Optimisation

These illustrative modelling applications provide important insights and policy guidance under conditions of uncertainty, but nonetheless are restricted by the use of a single policy (and hence mathematical optimisation) criterion, be it minimising future GHG emissions or minimising financial risks (costs) associated with insufficient technology portfolio diversification. In reality, however, public policy as well as energy sector investments should consider a range of different policy objectives that need to be factored in when designing technology portfolios. As well as climate change mitigation, energy security, affordability, and the health impacts of energy systems are among the examples of such a range of sometimes conflicting policy objectives and decision criteria. In this context, multi-criteria optimisation approaches can shed additional insights into technology portfolio strategies (for reviews see, for example, Figueira, Salvatore, and Ehrgott, 2005; Stewart, 1992; Wierzbicki, Makaovski, and Wessels, 2000).

A recent modelling study by McCollum, Krey, and Riahi (2011) explored the implications of pursuing three alternative policy objectives (improving energy security, lessening traditional air pollution and the ensuing health impacts of energy systems, as well as meeting climate change targets) first individually and then in an integrated fashion. The study combines the scenario technique for describing the future uncertainties that lead to a wide range of technology deployment scenarios within a multi-criteria optimisation framework. In this case, future uncertainties are represented by the varying degrees of stringency of adhering to the three policy objectives modelled, from weak to intermediate and stringent.

A significant finding from the modelling study is that optimising energy systems for the policy objectives of energy security and traditional air pollutants yields limited co-benefits ("policy spillovers") on the climate change policy objective. Conversely, using climate change mitigation as a policy entry point yields significant co-benefits on the other two policy objectives of energy security and air pollution/health.

Another important finding was that the additional costs of meeting all three policy objectives simultaneously are comparatively modest compared to the costs of meeting any policy objective alone, and that the aggregate costs of an integrated policy approach (optimising all three criteria simultaneously) are significantly below the sum of the costs incurred when pursuing each policy objective individually (McCollum et al., 2011).

The results from the study can also be examined with respect to its implications for technology portfolios using cumulative (2010–50) energy sector investments as a metric (which links to the potential for new business models, jobs, and overall macroeconomic multiplier effects). Figure 7.3 shows illustrative results for the policy scenarios modelled that meet each criteria individually at the "stringent" policy fulfilment levels, as well as their combined multi-objective optimisation (with all three policy objectives at their "stringent" policy fulfilment level).

The dominance of energy end use in total energy system investments and the only modestly higher investments needed to achieve all three policy objectives simultaneously are both striking. In the original study, only price/policy-induced investments into energy efficiency were modelled (i.e., excluding current and future baseline investments without additional policies). To account for energy investments comprehensively, a range of current energy end-use investments from 300 to 3,500 billion US$ per year was added based on empirical estimates developed for the *Global Energy Assessment* (Grubler et al., 2012). This number is (conservatively) considered to remain constant over the modelling time horizon 2010–50. The range in energy end-use investments reflects the different system boundaries that can be drawn to define the end-use technology (broadly) or just its energy-using component (narrowly). See Wilson and Grubler (2011) and the case study on end-use investments for more details.

To interpret Figure 7.3, it is important to realise the magnitude of the "baseline" energy sector investments (i.e., excluding policy-induced investments) estimated to be about 200 trillion US$2005 over the period 2010 to 2050, or some 5 trillion US$2005 per year on average. Meeting various policy objectives requires additional investments that increase across the three policy objectives modelled, from energy security and air pollution, to climate change mitigation. More important is to realise the very small extra investments required to meet all three policy objectives simultaneously, compared to a scenario in which only one objective is met. For instance, the integrated policy scenario entails cumulative investments of some 260 trillion US$2005, compared to the "climate protection only" policy scenario with some 253 trillion US$2005, a difference of only some 3 percent.

It is also instructive to examine the different implications for technology portfolios among the alternative policy formulations. Singular policy objectives also tend to increase investments into single-purpose technologies (e.g., high investments into end-of-pipe pollution control technologies in the "air pollution only" policy scenario). Conversely, under an integrated policy paradigm, investments into

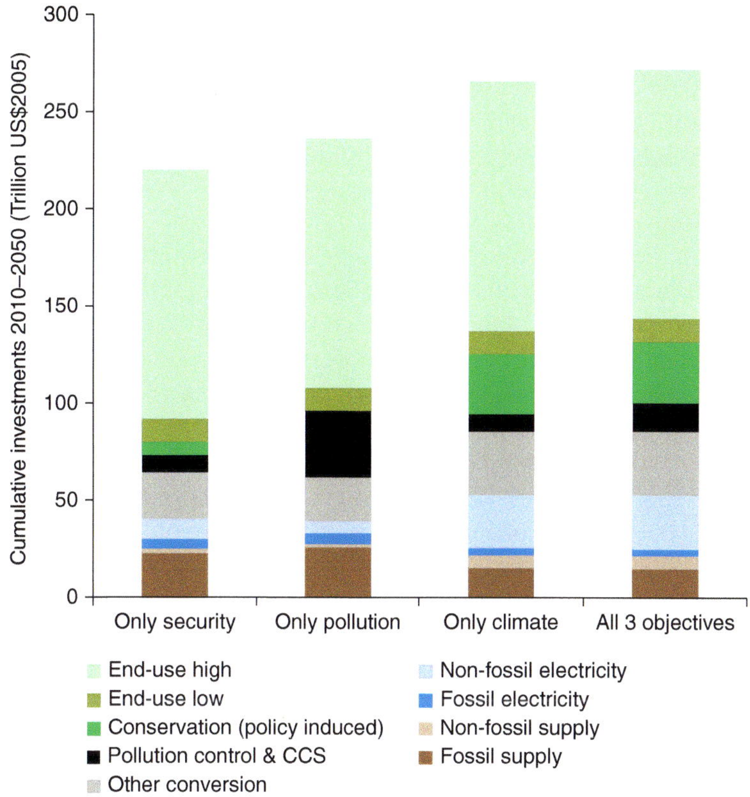

Figure 7.3. Cumulative global energy technology investments in three alternative policy objectives, and a fourth policy integration scenario using multi-criteria optimisation.
Notes: Investments shown in trillion US$2005 during 2010–50.
Source: McCollum and colleagues (2011).

"general purpose" technologies, most notably increased energy efficiency, increase significantly. End-use investments account for up to 61 percent of all energy sector investments in such a policy integration scenario.

Equally noteworthy is the increasing diversification of energy supply and infrastructure investments with more nonfossil fuel supply and electricity generation investments, in addition to stepped up investments into "general purpose" energy conversion technologies and infrastructures such as syngases and hydrogen that can be supplied from a variety of fossil and nonfossil sources. These changes in long-term technology portfolios under a changed paradigm of an integrated policy framework provides useful guidance for innovation policies that have to date insufficiently supported energy end use and efficiency as well as general purpose energy conversion technologies.

5 Conclusions

This case study provided a brief summary of methods and examples of modelling applications available to assist policy makers in crafting better technology innovation portfolios. The main conclusion is that formal tools in the form of scenario analysis, portfolio theory, and multi-criteria optimisation are increasingly available to move technology policy decisions onto a more rational footing. The choice of a particular tool will evidently depend on the policy question at hand and is of secondary importance compared to the need to arrive at technology portfolio decisions in a reproducible and transparent fashion. Formal tools and public disclosure are good instruments to assure transparency and reproducibility, and to date technology policy decisions have been falling short on these criteria.

Modelling studies in the field of energy technologies reveal a general pattern of portfolio diversification and of enhanced experimentation and earlier niche market investments under technological and innovation uncertainty, the extent of which depends on the (user specified) degree of risk aversion in addition to the underlying innovation uncertainty distributions. A second robust finding from the modelling studies is a much greater emphasis on energy end-use efficiency in technology portfolios responding to a range of risks and policy objectives. Greater energy end-use efficiency provides multiple benefits and also reduces risks and increases flexibility in supply-side energy technology portfolios.

These qualitative findings are in stark contrast to what appears to be a systemic energy technology innovation bias (Wilson et al., 2012) extending from public and private sector R&D, technology programmes and roadmaps, analysis and modelling, through to private venture capital and niche market deployment incentives. Invariably, few selected energy supply technologies (nuclear in public R&D, solar and wind electricity in niche market incentives) obtain the lion's share of funding and policy attention, with energy end-use and efficiency falling significantly short. And yet improved energy efficiency has been shown to dominate the desired outcomes of innovation efforts including mitigation portfolios, social rates of return, or learning and cost reduction potentials. Technology portfolio analyses can reveal such innovation biases and misalignments in policies and incentives and thus help in starting to address them through changed innovation policies.

FURTHER READING

Fuss and colleagues (2009) provide a useful application of portfolio theory in relation to climate change mitigation. Awerbuch (2006) is another interesting and detailed demonstration of portfolio design in an energy context.

REFERENCES

Awerbuch, S., 2006. Portfolio-Based Electricity Generation Planning: Policy Implications for Renewables and Energy Security. *Mitigation and Adaptation Strategies for Global Change*, 11: 693–710.

Awerbuch, S. & Berger, M., 2003. *Energy Security and Diversity in the EU: A Mean-Variance Portfolio Approach*. Paris: International Energy Agency.

Bar-Lev, D. & Katz, S., 1976. A Portfolio Approach to Fossil Fuel Procurement in the Electric Utility Industry. *Journal of Finance*, 31: 933–47.

Bazilian, M. & Roques, F., 2008. *Analytic Methods for Energy Diversity and Security. Applications of Mean Variance Portfolio Theory. A Tribute to Shimon Awerbuch*. London: Elsevier.

Dixit, A. & Pindyck, R., 1994. *Investment under Uncertainty*. Princeton, NJ: Princeton University Press.

Figueira, J., Salvatore, G., & Ehrgott, M., 2005. *Multiple Criteria Decision Analysis: State of the Art Surveys*. New York: Springer.
Fleten, S. E., Maribu, K., & Wangensteen, I., 2007. Optimal Investment Strategies in Decentralized Renewable Power Generation under Uncertainty. *Energy Economics*, 32: 803–15.
Fortin, I., Fuss, S., Hlouskova, J., Khabarov, N., Obersteiner, M., & Szolgayova, J., 2008. Integrated CVaR and Real Options Approach to Investments in the Energy Sector. *Journal of Energy Markets*, 1: 61–85.
Fuss, S., Johansson, D., Szolgayová, J., & Obersteiner, M., 2009. Impact of Climate Policy Uncertainty on the Adoption of Electricity Generating Technologies. *Energy Policy*, 37(2): 733–43.
Fuss, S. & Szolgayová, J., 2009. Fuel Price & Technological Uncertainty in a Real Options Model for Electricity Planning. *Applied Energy*, 87(9): 2938–44.
Gritsevskyi, A. & Nakicenovic, N., 2000. Modeling Uncertainty of Induced Technological Change. *Energy Policy*, 28: 907–21.
Grubler, A., Aguayo, F., Gallagher, K., Hekkert, M., Jiang, K., Mytelka, L., Neij, L., Nemet, G., & Wilson, C., 2012. Chapter 24 – Policies for the Energy Technology Innovation System (ETIS). *Global Energy Assessment – Toward a Sustainable Future*. Cambridge, UK and New York, NY: Cambridge University Press and Laxenburg, Austria: The International Institute for Applied Systems Analysis.
Grubler, A. & Gritsevskyi, A., 2002. A Model of Endogenous Technological Change through Uncertain Returns on Innovation. In Grubler, A., Nakicenovic, N., & Nordhaus, W. (eds.) *Technological Change and the Environment*. Washington, DC: Resources for the Future Press, 280–319.
Grubler, A. & Riahi, K. 2010. "Do Governments Have the Right Mix in Their Energy R&D Portfolios?" *Carbon Management*, 1(1): 79–87.
IEA, 2008. *Energy Technology Perspectives 2008 – Scenarios and Strategies to 2050*. Paris: Organization for Economic Cooperation & Development.
Kann, A. & Weyant, J., 2000. Approaches for Performing Uncertainty Analysis in Large-Scale Energy/Economic Policy Models. *Environmental Modeling and Assessment*, 5: 29–46.
Krey, B. & Zweifel, P., 2006. *Efficient Electricity Portfolios for Switzerland and the United States*. Zurich, Switzerland: University of Zurich.
Krey, V., Martinsen, D., & Wagner, H.-J., 2007. Effects of Stochastic Energy Prices on Long-Term Energy-Economic Scenarios. *Energy*, 32: 2340–49.
Krey, V. & Riahi, K., 2009. *Risk Hedging Strategies under Energy System and Climate Policy Uncertainties – A Stochastic Scenario Analysis using the SMS Modeling Framework*. Laxenburg, Austria: International Institute for Applied Systems Analysis.
Laurikka, H. & Koljonen, T., 2006. Emissions Trading and Investment Decisions in the Power Sector – A Case Study of Finland. *Energy Policy*, 34: 1063–74.
Madlener, R., Kumbaroglu, G., & Ediger, V., 2005. Modelling Technology Adoption as an Irreversible Investment under Uncertainty: The Case of the Turkish Electricity Supply Industry. *Energy Economics*, 27: 39–163.
Markowitz, H. M., 1952. Portfolio Selection. *Journal of Finance*, 7: 77–91.
1959. *Portfolio Selection: Efficient Diversification of Investments*. New York: Wiley.
McCollum, D., Krey, V., & Riahi, K., 2011. An Integrated Approach to Energy Sustainability. *Nature Climate Change*, 1: 428–29.
Messner, S., Golodnikov, A., & Gritsevskyi, A., 1996. A Stochastic Version of the Dynamic Linear Programming Model Message III. *Energy*, 21(9): 775–84.
Nakicenovic, N., Alcamo, J., Davis, G., de Vries, B., Fenhann, J., Gaffin, S., Gregory, K., Grubler, A., Tae, Y. J., Kram, T., La Rovere, E. L., Michaelis, L., Mori, S., Morita, T., Pepper, W., Pitcher, H., Price, L., Riahi, K., Roehrl, A., Rogner, H.-H., Sankovski, A., Schlesinger, M., Shukla, P., Smith, S., Swart, R., Van Rooijen, S., Victor, N., & Zhou, D., 2000. *Special Report on Emissions Scenarios*. Cambridge, UK: IPCC and Cambridge University Press.
Nakicenovic, N., Grubler, A., & McDonald, A. (eds.) 1998. *Global Energy: Perspectives*. Cambridge: Cambridge University Press.

Pacala, S. & Socolow, R., 2004. Stabilization Wedges: Solving the Climate Problem for the Next 50 Years with Current Technologies. *Science*, 305(5686): 968–72.
Riahi, K., Grubler, A., & Nakicenovic, N., 2007. Scenarios of Long-Term Socio-Economic and Environmental Development under Climate Stabilization. *Technological Forecasting and Social Change*, 74(7): 887–935.
Roques, F. A., Newberry, D. M., & Nuttall, W. J., 2008. Fuel Mix Diversification Incentives in Liberalized Electricity Markets: A Mean-Variance Portfolio Theory Approach. *Energy Economics*, 30: 1831–49.
Stewart, T., 1992. A Critical Survey on the Status of Multiple Criteria Decision Making Theory and Practice. *OMEGA International Journal of Management Science*, 205(5/6): 569–86.
Szolgayová, J., Fuss, S., & Obersteiner, M., 2008. Assessing the Effects of CO2 Price Caps on Electricity Investments – A Real Options Analysis. *Energy Policy*, 36: 3974–81.
Van Zon, A. & Fuss, S., 2006. *Irreversible Investment under Uncertainty in Electricity Generation: A Clay-Clay-Vintage Portfolio Approach with an Application to Climate Change Policy in the UK.*
Wierzbicki, A., Makaovski, M., & Wessels, L., 2000. *Model-based Decision Support with Environmental Applications*. Dordrecht, the Netherlands: Kluwer Academic.
Wilson, C. & Grubler, A., 2011. *A Comparative Analysis of Annual Market Investments in Energy Supply and End-Use Technologies*. Laxenburg, Austria: International institute for Applied Systems Analysis.
Wilson, C., Grubler, A., Sims Gallagher, K., & Nemet, G. F., 2012. Marginalization of End-Use Technologies in Energy Innovation for Climate Protection. *Nature Climate Change*, 2(11): 780–88.
Yang, M., Blyth, W., Bradley, R., Bunn, R., Clarke, C., & Wilson, T., 2008. Evaluating the Power Investment Options with Uncertainty in Climate Policy. *Energy Economics*, 30: 1933–50.
Zapert, R., Gaertner, P. S., & Filar, J. A., 1998. Uncertainty Propagation within an Integrated Model of Climate Change. *Energy Economics*, 205: 5–6.

PART III

KNOWLEDGE IN THE ENERGY TECHNOLOGY INNOVATION SYSTEM

8 Solar Water Heater Innovation in the United States, China, and Europe

Gregory F. Nemet

Author's Summary

Solar water heaters provide an example of three decades of innovation in an end-use technology. In this case, the diffusion of the technology underwent an extremely rapid boom and bust that was centered in California in the early 1980s. The technology was largely abandoned in the United States for two decades, although diffusion has been much more rapid and consistent in China and to a lesser extent in Europe.

Solar water heaters use a working fluid to absorb sunlight and provide heating and hot water in residential and commercial settings. The biggest technical improvement was the advent in the 1970s of selective coatings, which could absorb more sunlight. Since then, the technology has been rather stable, with some improvements in lifetime and reliability. The technology is currently cost-effective, especially in large installations with high demand for hot water. China is currently by far the world's largest market for solar water heaters.

The key period of R&D investment was in the 1970s at national laboratories and universities in the United States. Market subsidies in the 1980s were not monitored and were rampantly abused, with many installations that leaked and caused extensive damage to structures far in excess of the cost of the solar water heater itself. In response, the technology was largely abandoned in the United States for two decades; the industry went from sales of US$1billion/year in 1982 to US$30million/year in the late 2000s, and hundreds of firms went out of business. As a result, much of the learning gained in the period of rapid deployment was lost. And perceptions of poor reliability persist and have proven difficult to overcome – bad news lasts. An exception to this general policy failure has been a program in Hawaii that makes consumer rebates contingent on an inspection that occurs one year after installation.

In the United States, therefore, the history of solar water heaters is one of policy inconsistency, stagnant costs, and the implications of bad outcomes in early stages that are not easily forgotten and can have substantial spillover effects on other technologies. This presents a challenge to the need to support experimentation and intelligent failures.

Editors' Guide

Relationship of Case Study to Energy Technology Innovation System Framework

The early history of solar water heaters clearly illustrates the interdependencies between the various innovation system processes relating to *knowledge*, and between the different dimensions of the innovation system as a whole. In the United States, the rapid growth in installations driven by market incentives in the early 1980s led to *learning* improvements in the performance of the technology, particularly in terms of its reliability. These improvements and associated R&D activities that led to the *codification* of knowledge in patents and research papers helped some of the accumulated knowledge stock endure the market collapse during which the number of firms and personnel shrank drastically. But this boom and bust in market activity did mean a *depreciation* of the tacit, practical knowledge of the installers. This exacerbated problems with poor-quality installations that left a lasting *legacy* of failure surrounding the industry in the United States. The apparent absence of robust linkages between different groups of *actors* – particularly technicians and researchers on one hand and installers operating in the market on the other – implied the necessary *exchange and interaction* between the developers and users of the technology was insufficient to maintain a strong base of support as public resources dried up.

Nor had the technology become sufficiently cost competitive. Cost reductions as manufacturing experience and scale increased were offset by, among other things, the rising cost of material inputs illustrating the contingency of learning effects. The rapid shift in emphasis from R&D efforts (*technology-push*) to market deployment (*demand-pull*) at a still relatively early stage of the technology's deployment points to the absence of a systemic approach. This contributed to subsequent relative failure.

Some of these lessons have been learned in the evident success in various European countries at steadily growing the installed capacity base of solar water heaters using a suite of policy measures that target and develop different innovation system processes together. China also demonstrates the potential for regulatory formation of *niche markets* (e.g., hotels in urban areas) rather than financial incentives to stimulate market deployment, while flagging up the risks inherent to early and rapid *scaling up* of an industry prior to product standardization and policy harmonization.

Insights for Energy Technology Innovation Policy

- A large, high-profile failure in the early stage of the U.S. solar water heater industry has proven extremely difficult to overcome. Trust in the technology's performance and potential was very slow to rebuild, and also undermined policy support.
- Boom-and-bust cycles in policy support and market activity lead to knowledge depreciation through the loss of experience and tacit knowledge. Codifying

knowledge to the extent possible can mitigate this. So can more diverse networks and relationships between actors. In this case, the lack of involvement of utilities and builders appears to have hurt the industry.

- Verification of deployed technologies is inexpensive and essential to sustain legitimacy and market development. Inspection and verification have proven successful in Hawaii.
- Both R&D and market incentives placed excessive emphasis on solar collector units rather than on system integration. The overambitious demand-pull policy framework that failed to also address continuous improvements and cost reductions left the industry vulnerable to policy discontinuity.

1 Technology and Market Overview

Solar water heaters (SWH) expose a working fluid to solar radiation to raise the fluid's temperature. The fluid circulates and can be used indirectly for space heating or directly as hot water. In direct use, heated potable water is typically stored in a well-insulated tank. In most cases in developed countries, SWH systems work as hybrid systems, in which heat from the combustion of natural gas or electric resistance supplements the solar heated water. The configuration of the SWH and the integration with conventional forms of water heating is one of several design elements that distinguish SWH systems. For example, passive systems use temperature gradients to circulate the working fluid while active systems use electrically powered pumps. In direct systems, the working fluid is potable water and leaves the system once heated. In indirect systems, a heat exchanger is used so that the working fluid need not be potable water and can have antifreeze characteristics. Further, the temperature of the working fluid affects use: above 50°C for residential water and heating, and below 50°C for pool heating, an important niche market in the United States. Several configurations exist with different trade-offs between upfront costs, maintenance costs, reliability, and efficiency.

The global installed capacity of solar water and space heating in 2009 was about 257 million square meters (m^2), which corresponds to about 180 GW of thermal power or GW_{th} (REN21, 2010). This is approximately a 21 percent increase from 2008. According to the International Energy Agency (IEA), the total global solar thermal collector capacity in 2008 was about 217 m^2, which corresponds to about 152 GW_{th} using the IEA's conversion factor of 0.7 kW_{th}/m^2 (Werner and Mauthner, 2010). Of this total, about 54 percent was accounted for by evacuated tube collectors, about 33 percent by flat-plate collectors, about 12 percent by unglazed plastic collectors, and the remainder by air collectors.

The major markets for solar water heater installations are in China (58 percent of the global total), Europe (19 percent), and the United States and Canada (10 percent). Other large markets include Australia, Israel, Jordan, Japan, and New Zealand. The application of solar heat and the type of collector used vary greatly by country. In China, Japan, and Europe, flat-plate and evacuated tube collectors are dominant and the solar heat is primarily used to prepare hot water and to provide space heating. In the United States, Canada, and Australia, unglazed plastic absorbers are predominantly used for swimming pool heating.

2 Solar Water Heating in the United States: California Boom and Bust

The central episode of SWH experience in the United States occurred as a boom and subsequent bust in California in the 1980s. The boom was driven by federal and state subsidies as well as expectations of high energy prices during the late 1970s and early 1980s (Hirst, Carney, and O'Neal, 1979). The crash that occurred after 1984 was due directly to the expiration of tax credits for new installations, and was perhaps reinforced by the high-profile announcement that solar water heaters had been removed from the White House.

2.1 U.S. and California Policies

A government report from the late 2000s made a retrospective analysis of the policies that led to the solar water heater boom and bust in the 1980s (Taylor et al., 2007). That study found more than a dozen policy programs that had the potential to influence the rates of invention and diffusion of SWH. Ratings elicited from industry, government, and academic experts on which policies had the most important effects are shown in Figure 8.1. A rating of five represents "most important." Four policies stand out. Most important, federal tax credits provided from 1978–85 reduced the cost of installation by up to 25 percent, subject to the availability of tax liabilities against which to use the credits. In California, state tax credits effectively provided a 15 percent reduction in costs in addition to the federal credits. Another important policy was the requirement that installers be licensed, although deficiencies with stringency and enforcement of the licensing requirement became important later. Note that experts rated the U.S. Department of Energy solar hot water R&D program as substantially less influential than these other policies.

Finally, one of the highest-rated programs was Hawaii's program to provide incentives for SWH, though at much lower levels than in the 1970s. Among the most salient features of Hawaii's program was its requirement that all new installations receiving a subsidy be inspected a year after installation to ensure proper operation. Inspections cost on the order of $40/system and provide strong incentives to avoid the poor installations and subsequent non-operation that characterized perhaps half of the systems installed in the 1980s.

2.2 The Early SWH Industry

This set of policies resulted in a dramatic scaling up of the industry from 1975–85 and an even more rapid subsequent crash (Figure 8.2). Annual production was more than 1 million square meters of panels from 1982–85. In 1986, production fell by an order of magnitude and has stayed below even those levels since.

The industry structure followed a broadly similar pattern albeit with some key differences. The explosion of new firms producing solar water heaters increased from a couple of dozen in 1974 to more than 300 three years later (Figure 8.3). Consolidation and exit kept the number of firms relatively constant during the years of increasing and large production. Once production declined in 1986, hundreds of firms left the industry and for the next two decades the industry was left with ten to twenty firms, many of them sustained by building solar water heaters for the niche market of swimming pools.

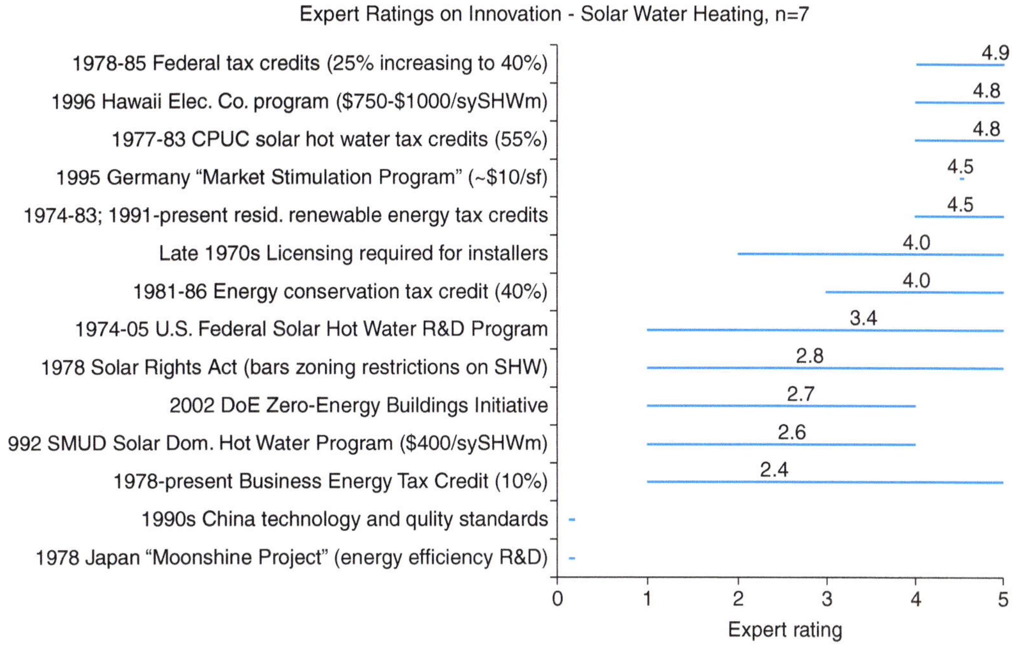

Figure 8.1. Elicited ratings of SWH policy influence on innovation (Taylor et al., 2007). *Notes*: CPUC = California Public Utilities Commission; "resid" = residential; DoE = Department of Energy; SMUD = Sacramento Municipal Utility District. Expert rating (x-axis) scale from 1-5, with 5 = most important according to policy's impact on innovation, and 0 = not familiar with policy.

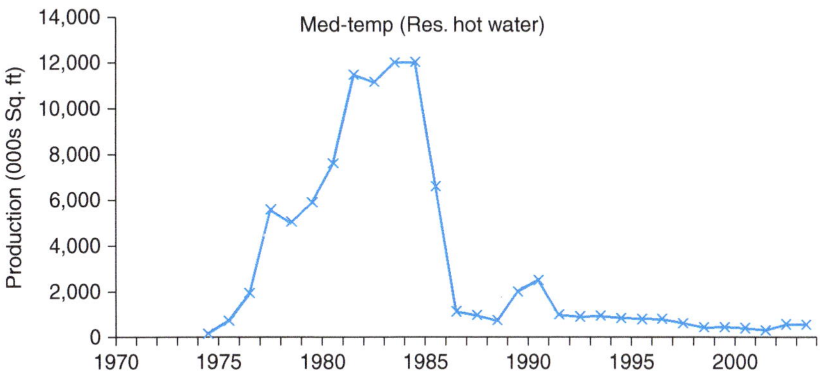

Figure 8.2. Production of solar water heaters in the United States (1974–2003).

One is left with a mixed impression of the firms in the industry. While many contributed to improvements in the quality of the equipment, such as designing systems that would not leak or freeze, many had a much shorter time horizon, such that performance and long-term reliability were less important than quantity installed. These so-called "solar profiteers" likely had the dominant effect on how the industry was perceived (Taylor, 2008).

With this much churn of firms and personnel, it seems likely that there was a high level of knowledge depreciation in the solar hot water industry, particularly with respect to the technical improvements discussed later in this chapter. Fewer than 300 people were employed in the industry from the late 1980s until the late 2000s (EIA,

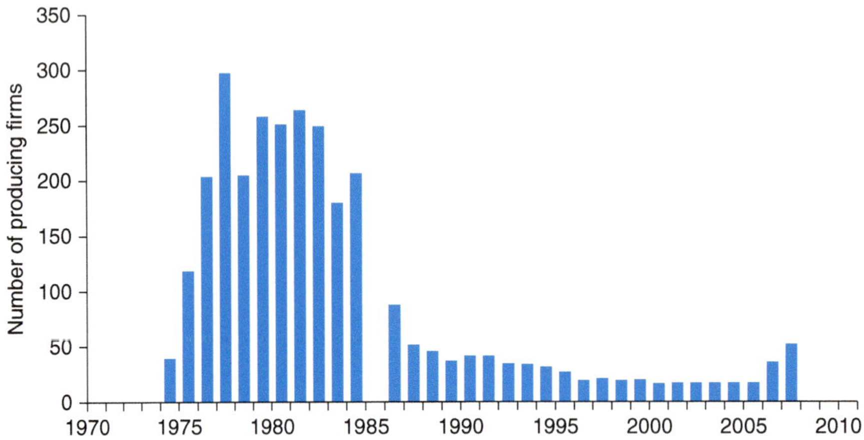

Figure 8.3. Number of firms producing solar water heaters in the United States (1974–2007).

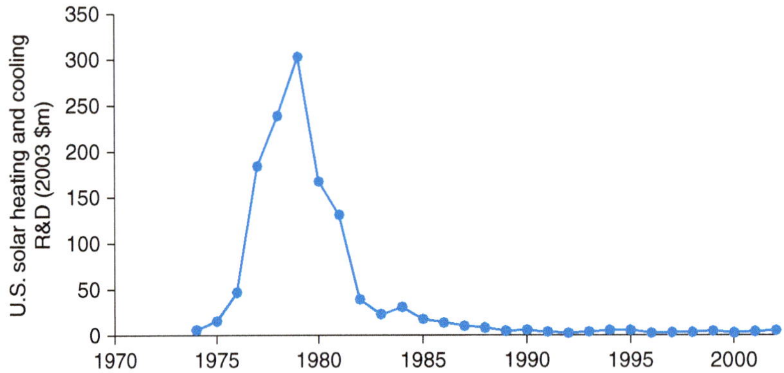

Figure 8.4. U.S. federal R&D for solar heating and cooling.

2005). One compensating factor to this loss of tacit knowledge was the persistence of an annual conference devoted to solar energy that included large participation from the SWH community. The American Solar Energy Society held occasional conferences beginning in 1955 and then annually beginning in 1976. An analysis of the conference proceedings finds that eleven hundred papers on SWH were presented at the conference from 1955 to 2004, and that papers increased by a factor of five in 1975 and remained at that level until 1980, after which they declined steadily (Taylor et al., 2007). Academics and government laboratories authored three-quarters of the papers – firms and electric utilities wrote only a quarter. Actor networks appear to be dominated by researchers rather than installers. Notably, papers on SWH declined as the industry expanded to its highest levels of production; this negative correlation between expanding markets and creative efforts has also been observed in wind power (Nemet, 2009). Similarly, the solar hot water R&D program peaked before the market expanded and by 1982 had practically disappeared (Figure 8.4).

The composition of the authors as well as the timing of the decline in conference papers suggests that different types of knowledge depreciated at different rates. The knowledge that was codified and preserved was much more likely to consist of technical efforts to improve systems, and less likely to capture learning by doing in installing and operating systems in real-world commercial environments. The tacit knowledge of installers was likely lost as many left the industry to pursue employment in other areas after what industry survivors referred to as the "tragedy of 1985" (Taylor, 2008). For SWH in the United States, knowledge depreciation is much more of a concern for system integration and use than for the components themselves.

2.3 Technological Change

Four main sources of technical improvement in SWH existed in the 1980s and 1990s (Taylor et al., 2007):

- Selective coatings: This technology originated in Israel in the 1960s and was developed by the U.S. Department of Energy in the 1970s. Coatings were applied to collector tubes that blocked the emission of infrared radiation that typically escaped from the system as the working fluid was heated. These coatings increased the amount of heat absorbed by 20 percent.
- Polymer-based systems: The switch from using glass and steel to plastic components has the potential to reduce system costs by a factor of three. Polymer materials degrade after lengthy exposure to ultraviolet rays, so have only recently begun to be used commercially despite having been in development for several decades. The use of degradation-resistant materials such as polybutelene and polysulfone has improved performance.
- External heat exchangers: Designing heat exchangers so they are not integrated with the water tank allows the use of standard hot water tanks for storage. The use of these mature and mass-produced systems reduces storage costs by half.
- Absorber plates: Improvements in coatings and use of new materials in absorber plates have enabled the shift to thinner systems that require less material.

While the quality of the technology has improved steadily, the cost of the technology has remained relatively stable over three decades. Figure 8.5 shows that the cost per watt of energy output in the mid-2000s was close to the level of the late 1970s. In this case, the quality improvements were focused on reliability, which is not captured in the energy produced at peak solar insolation. Rather, the cost per watt has been influenced by increases in the price of materials, especially steel and glass.

Frequency of filing applications for patents provides another perspective on the patterns of technological change in SWH. Figure 8.6 shows numbers of SWH patents by the year in which they were filed, again showing a boom and bust pattern (Taylor et al., 2007). The timing of the peak, 1977, precedes the peak market year by five years. The 1970s was a time of experimentation in solar water heaters during which many designs were tried. (This is similar to the case of wind power as described in another case study examined in this book.) Relative to other energy technologies, SWH patent holders are more likely to be individuals (rather than firms) and from the United States, rather than foreign. By the 1980s, when the market took off, many of the design questions were settled and patenting subsequently declined. The types of improvements that occurred during the period of market growth were subtle changes in avoiding leaks, and increasing reliability – many of these were not patentable or perhaps were best protected as trade secrets.

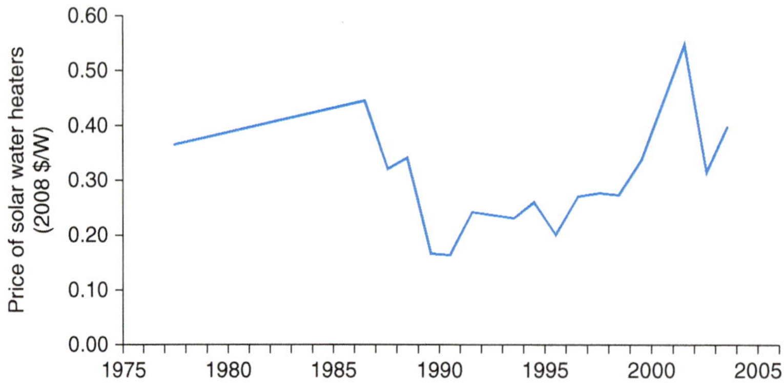

Figure 8.5. Price of U.S. solar water heaters.

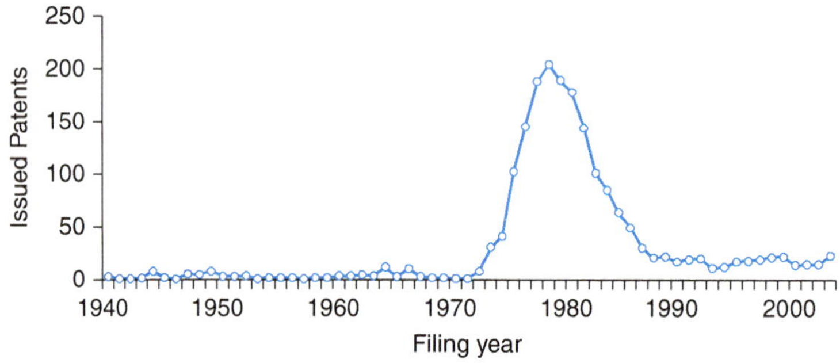

Figure 8.6. U.S. solar water heater patents.

The factors that supported the emergence of the SWH industry in the United States in the late 1970s – a shared perception of an energy crisis, expectation of high energy prices, technological promise, and strong government support – had all disappeared or diminished by the mid-1980s. In 1986, the SWH panels installed on the White House in 1977 were removed. Interviews with experts in the industry suggest that rampant abuse of generous subsidies in the 1980s had led to poorly installed systems (Taylor et al., 2007). The fallout of the end of the boom led to a shrinking of the industry from $1b/year in 1982 to $30m/year in 2002. The number of companies involved in the industry diminished from a few hundred to a few dozen. A consequence has been the loss of people from the industry and of their experience of how to build and install systems, as well as how not to. Subsequently almost all growth in the industry has taken place in China and to a lesser extent in Europe.

3 China: From Steady Growth to Dominant Market

Currently, China is the world's largest market for solar water heaters. By the end of 2008, China had installed about 125 million m^2 of solar water heating capacity, about 60 percent of the global total (Werner and Mauthner, 2010). One study estimates that China accounted for more than 80 percent of global solar water and space heating capacity installed in 2009 (REN21, 2010). This study also estimates that the newly added capacity in 2009 increased by almost 34 percent compared with the new capacity addition in 2008. By the end of 2008, China had more than thirteen

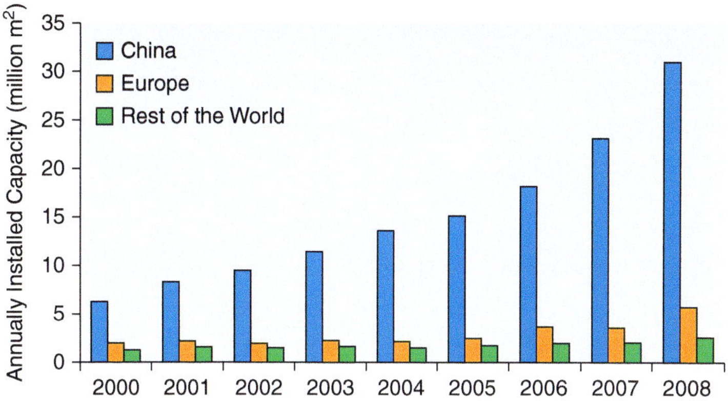

Figure 8.7. Newly installed capacity of flat-plate and evacuated tube collectors from 2000 to 2008 (Werner and Mauthner, 2010).
Notes: Europe includes EU 27, Albania, Macedonia, Norway, Switzerland, and Turkey.

hundred SWH manufacturers operating in China (REN21, 2009). It is important to note that these growth rates are not just a recent phenomenon. The SWH market has grown steadily in China since the early 1990s, with an average increase of 49 percent per year for the period 2000–2009 (Figure 8.7). Urban areas accounted for almost 90 percent of the installations (Junfeng and Runqing, 2005). Around 30 percent of installations are concentrated in large cities and about 60 percent in suburban areas and small cities (Wallace and Wang, 2006).

More than half of the increase in the newly installed capacity in 2008 and 2009 was due to the central government's program called "home appliances going to the countryside" (REN21, 2010). In addition to domestic installations, China is positioning itself as a global leader in production as well as technology advancement of vacuum-tube solar heating systems. Chinese products dominate more than 90 percent of the global vacuum-tube systems market (REN21, 2009). This is an incredible feat considering that China only recently began exporting solar hot water systems to other developing countries in Asia, Africa, and South America. Chinese systems have also made their way to Europe through joint ventures with European companies.

Sustained high growth in China's solar hot water market for more than a decade has primarily been driven by government initiatives and by increasing living standards of the Chinese people resulting from rapid economic development (REN21, 2009). China was the first country with long-term national targets for solar hot water capacity: 150 million m^2 by 2010 and 300 million m^2 by 2020 (REN21, 2010). The Ministry of Construction issued specifications for building-integrated SWH systems in 2005 (REN21, 2009). As a result, building design and construction in many urban areas in China requires solar hot water integration. Some cities in China mandate solar hot water in all new residential buildings and in new construction and renovation of hotels and commercial buildings. The rapid expansion of SWH systems has reduced costs significantly and as a result, the price of solar heaters is competitive with electric and gas water heaters.

The main challenge facing the Chinese solar hot water industry is the lack of quality control and product standardization (REN21, 2010). Addressing these issues will help China to maintain its dominance in the global SWH market. Another major issue facing China is that the policies governing solar water heater installation are too vague and nonstandardized (Han, Mol, and Lu, 2010). Many provinces in China have economic subsidies on the installation of solar water heaters, but there is no uniform subsidy structure. Subsidies differ between different cities and provinces. National installers therefore face a complex regulatory and incentive environment, which makes it difficult to standardize business models and installation approaches. Other technical issues need to be addressed as well. As an example, the water tanks and pipes installed outdoors can crack during winter.

4 Europe: Active and Diverse Policies

In Europe, the SWH market grew at an annual average of about 8 percent from 2000 to 2009 (Figure 8.7). The heat generated by these solar thermal technologies is primarily used for domestic hot water and space heating. Recently, a small fraction of the solar heat is also being used for cooling purposes, industrial process heat, desalination, and swimming pools. The cumulative SWH capacity in Europe in 2009 was about 30 million m^2 (ESTIF, 2010; REN21, 2010). Figure 8.7 indicates that the newly installed capacity increased consistently from 2000 to 2008, though decreasing from about 5 million m^2 in 2008 to about 4 million m^2 in 2009 as a result of the recent economic downturn. However, new capacity added in 2009 was more than any other previous years except for 2008.

According to the European Solar Thermal Industry Federation (ESTIF), the SWH market in Europe can broadly be divided into three zones based on cumulative installed capacity. The first zone comprises Germany, the second zone comprises Italy, Spain, Austria, France, and Greece, and the third zone comprises the remaining EU countries and Switzerland. The share of the SWH market in zone 1, 2, and 3 was 38 percent, 39 percent, and 23 percent respectively in 2009. Figure 8.8 shows the SWH capacity in operation per capita in major European markets. On a per capita basis, Cyprus is not only the leader in Europe, but in the entire world. Austria and Germany are the other top two installers on a per capita basis in Europe after Cyprus. According to ESTIF, for many years almost 70 percent of SWH installations in Europe have occurred in Germany, Austria, and Greece (ESTIF, 2007).

Public policies conducive to growth have played a critical role in the development and high growth rates of the SWH market in Europe. Other related drivers include increasing prices of conventional energy and concern with supply security and climate change (ESTIF, 2007). Financial incentive schemes in the form of direct grants, tax reductions, and loans at reduced rates have been very important in various markets. For example, Germany implemented a program called the Market Rebate Program for Renewable Energy (MAP) that provided direct grants to eligible solar thermal installers based on the size of the system installed (Hack, 2006). In addition to this financial incentive scheme, Germany also benefited from campaigns conducted at the federal and local levels to raise awareness of the benefits of solar power. Similarly, Greece implemented a program that offered investment cost deductions from personal income tax from the late 1970s to the early 1990s (Hack, 2006). This program helped the solar thermal market reach a critical mass in Greece and as a result the market is still flourishing even after the termination of the program (ESTIF, 2007). This stands in sharp contrast to the U.S. experience. Austria has

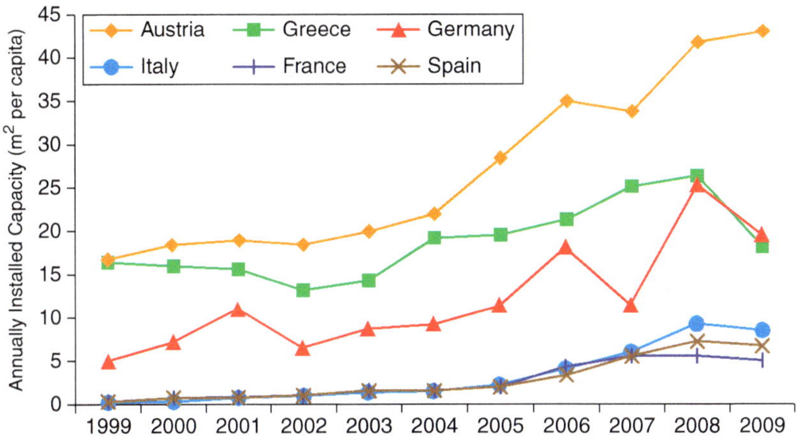

Figure 8.8. Newly installed capacity per capita in major European markets from 1999 to 2009 (ESTIF, 2010; Werner and Mauthner, 2010).

also benefited from public support programs in the form of financial incentives, professional building codes, R&D funding, and awareness campaigns.

Although Europe has seen a healthy growth rate in SWH markets, this growth is driven by only a handful of nations as noted earlier. France, Italy, and Spain have experienced healthy growth rates in the past few years, but their market size per capita is still relatively small (Figure 8.8). Europe can increase its solar capacity significantly by following the examples of Germany, Austria, Greece, and Cyprus, which have shown the promise of policy instruments including: implementing national-level as well as EU-level solar thermal targets; implementing campaigns to raise public awareness; professional development and training; setting new standards for buildings and encouraging building-integrated SWH systems; implementing targeted financial incentives schemes; implementing best practices identified in member countries as well as other countries; and encouraging emerging technologies and applications.

5 Conclusion

The discussion presented in this case study indicates the important role government policy has played in shaping the solar water heater market in the United States, China, and Europe. The policy inconsistency in California led to stagnation in the United States. Sustained and innovative policy support has contributed to a steadily growing solar market in Europe, and to an even greater extent in China. The lack of quality control and product standardization was an issue in California and continues to be in China. Implementation of mechanisms to enhance quality control has proven an effective way to improve reliability. A good example is Hawaii's program to inspect an installed system in order for purchasers to receive a tax credit. The discussion also indicated that the commercialization of SWH in the United States might have been too rapid compared to China and Europe where growth has been steady over an extended period of time.

The ensuing two decades of stagnant demand in the United States offer perhaps the most important insight from this case. The technology that was adopted very rapidly in the late 1970s was widely perceived as a failure. Stories abound of installations causing thousands of dollars of damage due to leaking roofs and fly-by-night installers. Perhaps 50 percent of systems were no longer functioning within five years. This large and prominent early failure of the technology has proven extremely difficult to overcome. The technology was no longer trusted, even though the quality of the technology improved steadily over the 1990s and 2000s (e.g., by avoiding freezing of the working fluid and so removing the issue of leaks). Moreover, policy innovations have found effective ways to improve reliability at a modest cost. The inspection program in Hawaii provides a way to create incentives for performance rather than just capital investment.

In many ways it is easier to learn from successes than from failures. The sustained support in Europe and China is the central reason for the success in growing those markets. Still, the failure of the California experience is important because of the shadow it cast over the U.S. market for decades. The experience of the solar water heaters in the United States raises the questions: Did the incentive program in the 1980s destroy the U.S. industry for twenty years? Was the technology commercialized too early? Too rapidly? Was this an example of relying too much on installers to learn by doing? The rapid removal of government and nongovernment incentives in the mid-1980s had adverse effects on the industry. But it is also possible that the combination of blunt policy instruments that were prone to abuse and the very rapid expansion of the industry in the late 1970s and early 1980s led to the proliferation of fraudulent installers and low-quality installations that became the most lasting legacy of the U.S. experience with solar water heating.

FURTHER READING

For a comprehensive policy history of solar water heating, see Taylor and colleagues (2007) for the United States and Han and colleagues (2010) for China.

REFERENCES

EIA, 2005. *International Energy Outlook 2005*. Washington, DC: U.S. Dept. of Energy, Energy Information Administration.

ESTIF, 2007. *Solar Thermal Action Plan for Europe*. Brussels: European Solar Thermal Industry Federation.

ESTIF, 2010. *Solar Thermal Markets in Europe: Trends and Market Statistics 2009*. Brussels: European Solar Thermal Industry Federation.

Hack, S., 2006. International Experiences with the Promotion of Solar Water Heaters (SWH) on Household-Level. Prepared for Deutsche Gesellschaft für Technische Zusammenarbeit (GTZ) GmbH.

Han, J., Mol, A. P. J., & Lu, Y., 2010. Solar Water Heaters in China: A New Day Dawning. *Energy Policy*, 38(1): 383–91.

Hirst, E., Carney, J., & O'Neal, D., 1979. Alternative Technologies for US Residential Water Heating: Energy Savings and Economic Benefits. *Energy Policy*, 7(4): 307–20.

Junfeng, L. & Runqing, H., 2005. Solar Thermal in China: Overview and Perspectives of the Chinese Solar Thermal Market. *Refocus*, 6(5): 25–27.

Nemet, G. F., 2009. Demand-Pull, Technology-Push, and Government-Led Incentives for Non-Incremental Technical Change. *Research Policy*, 38(5): 700–709.

REN21, 2009. *Recommendations for Improving the Effectiveness of Renewable Energy Policies in China*. Paris: REN21.

REN21, 2010. *Renewables 2010 – Global Status Report*. Paris: REN21.
Taylor, M., 2008. Beyond Technology-Push and Demand-Pull: Lessons from California's Solar Policy. *Energy Economics*, 30(6): 2829–54.
Taylor, M., Nemet, G., Colvin, M., Begley, L., Wadia, C., & Dillavou, T., 2007. *Government Actions and Innovation in Clean Energy Technologies: The Cases of Photovoltaic Cells, Solar Thermal Electric Power, and Solar Water Heating, (CEC)-500–2007–012*. Sacramento: California Energy Commission.
Wallace, W. & Wang, Z., 2006. Solar Energy in China: Development Trends for Solar Water Heaters and Photovoltaics in the Urban Environment. *Bulletin of Science, Technology & Society*, 26(2): 135–40.
Werner, W. & Mauthner, F., 2010. *Solar Heat Worldwide 2008*. IEA Solar Heating & Cooling Programme.

ACKNOWLEDGMENTS

Thanks to Bickey Rimal and Peter Braden for excellent research assistance.

9 Heat Pumps: A Comparative Assessment of Innovation and Diffusion Policies in Sweden and Switzerland

Bernadett Kiss, Lena Neij, and Martin Jakob

Authors' Summary

The development and introduction of heat pumps provides an interesting illustration of policy influence and effectiveness in relation to energy technology innovation. Heat pumps have been supported by several countries since the 1970s as a strategy to improve energy efficiency, support energy security, reduce environmental degradation, and combat climate change. Sweden and Switzerland have been essential to the development and commercialisation of heat pumps in Europe. In both countries, numerous policy incentives have lined the path of technology and market development. Early policy initiatives were poorly coordinated but supported technology development, entrepreneurial experimentation, knowledge development, and the involvement of important actors in networks and organisations. The market collapse in the mid 1980s could have resulted in a total failure – but did not. The research programmes continued in the 1980s, and a new set of stakeholders formed – publicly and privately funded researchers, authorities, and institutions – and provided an important platform for further development. In the 1990s and 2000s, Sweden and Switzerland introduced more coordinated and strategic policy incentives for the development of heat pumps. The approaches were flexible and adjusted over time. The policy interventions in both countries supported learning, successful development and diffusion processes, and cost reductions. This assessment of innovation and diffusion policies for heat pump systems can be used to generalise some insights for energy technology innovation policy.

Editors' Guide

Relationship of Case Study to Energy Technology Innovation System Framework

In both Sweden and Switzerland, diverse networks of innovation system actors formed in the initial market formation stage. In Switzerland, for example, this occurred through conferences, training, and testing activities supported by public policy. These *actor networks* proved critical to sustain the technologies during the period of declining energy prices through the 1980s. They provided the context for *knowledge feedback and exchange*, for example between developers, installers, and users of the heat pump systems. Of particular importance were the testing and certification requirements to ensure credible *performance* of the technology given the technical difficulties experienced early on. Knowledge generated through these evaluations was commonly *codified* and shared, for example in articles and seminars. This in turn supported *knowledge spillovers* among innovation system actors. The application of preexisting knowledge of well drilling in Sweden provides another example of *knowledge spillovers* that supported the development of the heat pump market.

These innovation system processes relating to *knowledge* and *actors* were further strengthened in the second phase of the heat pump market development through the 1990s. Sweden's procurement programme specifically tried to bridge the gap between manufacturers and end users; Switzerland's market development programme included a strategic position in a federal agency to coordinate innovation and diffusion activities among private and public actors. These network-strengthening and coordination efforts helped build up *shared expectations* around the market potential for the technology, in turn facilitating a continued *mobilisation of resources* as the market grew. And crucially, throughout both the Swedish and Swiss experiences, *market formation* activities including subsidies and consumer awareness campaigns were complemented by *knowledge-generation* activities such as R&D in a good example of a systemic approach to heat pump innovation. One measure of the outcome of these technology-push and market-pull incentives was the characteristic effects of *learning* and *economies of scale* on the cost of heat pumps as cumulative experience grew and the technology became mass produced.

Insights for Energy Technology Innovation Policy

- Strategic, long-term, and continuous support for energy technology innovations is critical. First attempts to introduce heat pumps in Sweden and Switzerland largely failed, and continuous support was needed to overcome technical and market shortcomings.
- The combination of policy instruments may have to change and the government's approach should be flexible. Initial uncoordinated interventions to support entrepreneurial testing should develop into credible, stable, and transparent strategies that enable industry to make long-term investments.
- Policy interventions need to consider both the development of the technology and its emerging market. R&D is a necessary but not sufficient part of a systemic policy approach. Testing and certification processes to support technical quality and build credibility and legitimacy provide a link between R&D initiatives and stable market development.
- The support of networking to improve strategic interactions between actors helps to ensure knowledge feedback and spillovers.

1 Introduction

This case study presents the results and the policy programmes applied to introduce and commercialise heat pumps, focussing on ground source heat pumps for residential space and hot water heating. Heat pumps are of interest because they have been supported by several countries since the 1970s as a strategy to improve energy efficiency, support energy security, reduce environmental degradation, and combat climate change (IEA, 2008). Two countries essential for the development and commercialisation of heat pumps in Europe are Sweden and Switzerland. Since the early 1970s, the number of installed heat pumps in Sweden and Switzerland has been rapidly growing, and as a result these two countries have the highest number of ground source heat pumps per capita and per land area respectively (EGEC, 2009). An important underlying reason may be that the two countries have had energy systems with similar characteristics, particularly a substantial supply of electricity for heating in the 1970s and 1980s. Heat pumps were therefore seen as energy efficiency measures rather than a means of substituting electricity for fossil fuels at the point of use.

Heat pump systems offer an energy-efficient solution as they use "free energy" (the solar energy stored in the earth) to provide heating, cooling, and hot water for homes. A heat pump moves heat from a low-temperature heat source (i.e., bedrock, surface soil, water, and outdoor air) to a higher-temperature heat sink (i.e., the indoor space). Heat pumps are usually characterised by their heat sources (air, water, and ground) and/or by the mediums between which they transfer heat (air-to-air, air-to-water, water-to-air, water-to-water). Ground source heat pumps (GSHPs) are a variant of either water-to-air or water-to-water heat pumps. Depending on the available land areas and the soil/rock types, GSHPs can be installed horizontally (soil), vertically (bedrock), or in a pond/lake. The most common heat pumps are motor driven (electrically powered or gas fuelled). The ratio of thermal energy gained to electric power used is called the Coefficient of Performance (COP) and is in the order of 2.5–5. So if electrical space heating is replaced by a heat pump, considerable electricity will be saved. If a heat pump replaces an alternative heating system, savings will depend on the configuration of that system, including its conversion efficiency and distribution losses.

2 Market Development of Ground Source Heat Pumps

2.1 Market Growth

The heat pump market took off during the 1980s following the oil crisis in the late 1970s (see Figure 9.1 and Figure 9.2). In Sweden, Swedish heat pump companies have sold more than nine hundred thousand heat pumps since the early 1980s (SVEP, 2009). As of 2010, approximately 98 percent of the heat pumps sold serve the residential market, with most of the sales being small heat pumps (<20kW) for single-family houses. Reports show that ground source heat pumps add value to properties (Boverket, 2008; Energimyndigheten, 2009). Until the mid 2000s, ground source heat pumps dominated market sales, constituting on average 45 percent of heat pumps sold each year with annual growth rates between 1993 and 2006 exceeding 30 percent (SVEP, 2009). Since 2000, the market share of air-to-air heat pumps has

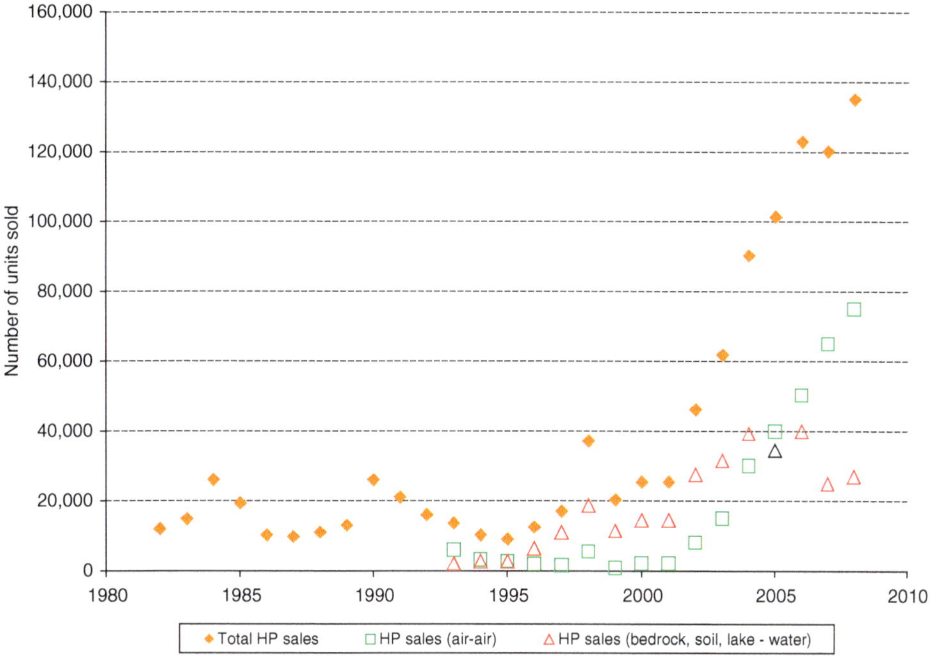

Figure 9.1. Annual sales statistics of heat pumps in Sweden, 1980–2008, based on data from the Swedish Heat Pump Association (Nowacki, 2007; SVEP, 2007).
Notes: HP = heat pumps. Ground source heat pumps (installed in single-family houses) provided 3.9 percent of the total demand for space and water heating in 1990. This had risen to around 10 percent of total demand in 2008 (Energimyndigheten, 2008; Nowacki, 2007).

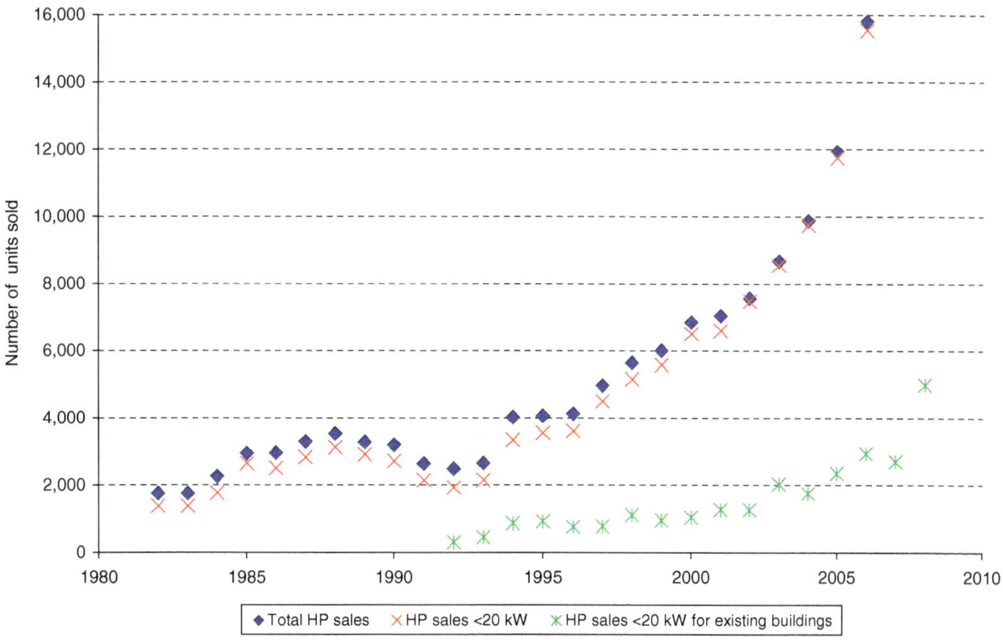

Figure 9.2. Annual sales statistics of heat pumps in Switzerland, 1982–2006 (all types of heat pumps) based on data from the Swiss Heat Pump Promotion Group (Basics, 2002; FWS, 2009).
Notes: HP = heat pumps. Ground source heat pumps provided 2.4 percent of the total demand for space and water heating in 2000 in the residential sector. This had risen to around 5.4 percent of total demand in 2008 (BFE, 2010).

increased rapidly, and in 2008 air-to-air heat pumps comprised more than 60 percent of total sales (see Figure 9.1). Exports have also represented a significant share of sales. According to interviews with manufacturers, in the mid 2000s, approximately 40 percent to 50 percent of total Swedish production was exported.

In Switzerland, more than one hundred eighty thousand heat pump units have been sold since the 1980s (FWS, 2009). Most of the sales have been small heat pumps (<20kW) predominantly installed in new single-family houses. The sales of ground source heat pumps increased rapidly in the 1990s, and by the mid 2000s the market share of heat pumps among space heating and domestic hot water heating systems in new homes was around 75 percent (Rognon, 2006). In 2009, more than twenty thousand five hundred ground source heat pump systems were sold, which shows an increase of almost 20 percent per year since 1993 (FWS, 2009).

2.2 Policy Programmes

2.2.1 Early Support

To support technology and market development, both Sweden and Switzerland introduced broad-ranging policy programmes that emphasised R&D and the development of heat pumps based on available components (see Table 9.1 and Table 9.2). In Sweden, R&D programmes were complemented with subsidies, favourable loans, trainings, and information campaigns. Long-standing Swedish experience in water well drilling also supported the early development of the emerging market as it was applicable to borehole drilling used in the installation of vertical ground source heat pumps. In Switzerland, early support focussed on conferences, training, and testing activities. The policy incentives attracted many new heat pump manufacturers and installers to the market, and by the early 1980s, the number of heat pump installations increased rapidly. However, in the mid 1980s the price of oil was reduced and in Sweden, government subsidies for domestic heat pumps were terminated. As a result, the demand for heat pumps decreased significantly and the market collapsed in both countries. In 1984 there were about 130 heat pump companies (manufacturers, retailers, and installers) in Sweden, most of which were small and worked locally. In 1986, only a few companies remained (Florin, 1987; SPK, 1986; Törnell, 2007). Moreover, several types of heat pump installed in the early 1980s were registered as technically malfunctioning (Lindeberg, 1984). To support a more stable market development, new policy instruments were introduced in the early 1990s. The foundation for further trust was based on extended research activities focussing on aspects such as quality improvement, new compressor types, optimisation of heat pump systems, the reduction of refrigerant volume, and computer simulations, as well as replacement of "freons" by HFC and natural refrigerants. These research activities were then complemented with instruments to provide more stable market development.

2.2.2 Renewed Support: Sweden

In 1993, Sweden launched a technology procurement programme for ground source heat pumps. The objective of the programme was to stimulate the development and commercialisation of innovative and high-quality heat pumps. The programme aimed to bridge the gap between buyers and manufacturers, helping buyers get products better suited for their needs, and helping manufacturers reduce risks associated with developing these products. The technology procurement programme worked from the end-use side. A specification outlining the requirements of advanced heat pumps was developed by a group of dedicated purchasers and specialists in cooperation with the Swedish Agency for Economic and Regional Growth (NUTEK). The specification required a heat pump that was 30 percent more efficient and 30 percent cheaper than the existing models on the market, met high quality and reliability standards, and did not allow products using CFCs or HCFCs (stratospheric ozone-depleting chemicals). A competition was announced in which manufacturers were invited to enter prototypes with features that met the requirements of the specification. The buyers' group guaranteed that at least two thousand units would be purchased of the winning model. To ensure credibility, third-party testing of the prototypes and the whole heating system was performed. (Initial difficulties in replacing CFCs / HCFCs led to the launch of a new research programme in the mid 1990s.)

To support a stable market development of improved heat pumps, the procurement programme was combined with additional policy incentives, such as investment subsidies, information campaigns, and evaluations. In all, 25 percent of the procurement budget was earmarked for the evaluation of heat pump installations through a test and certification programme (free tests for participants and certification of the quality and technical performance of the heat pumps). Evaluation results were presented in articles and seminars. A further 50 percent of the procurement budget was allocated to information activities, including information campaigns, brochures, and articles (Miljöstyrningsrådet, 2008). The procurement programme boosted demand for heat pumps with sales doubling between 1995 and 1996. Between 1996 and 2006 the number of installations of bedrock, soil, and lake heat pumps increased at an average of 35 percent per year (SVEP, 2007). The increase in installations of heat pumps in the late 1990s and early 2000s was further supported by additional government subsidy programmes (1998–99, 2001–2003, and 2006–10). In 2005, a quality label for heat pumps was introduced (P-label) as well as a standard on the installation of the geothermal system (Normbrunn-97), including requirements for the borehole, equipment, and competence of the drillers that was also assured by a voluntary certification scheme and new certification courses for installers.

2.2.3 Renewed Support: Switzerland

In Switzerland, a ten-year programme called Energy 2000 was launched in 1990 by the Swiss Federal Office of Energy, followed in 2000 by the Swiss Energy programme. These programmes aimed at increasing the use of renewable energy and improving energy efficiency. The use of heat pumps was considered one of the promising pillars to meet quite ambitious goals. In addition to the public financed initiatives of the early 1990s, several measures financed by the private sector were launched. To a large extent, the public and private incentives were coordinated and complemented each other (Rognon, 2006, 2008). Policy measures were also adjusted over time to support continuous technical, institutional, and economic development. Moreover, several strategic and organisational measures were taken to coordinate the initiatives to support heat pumps. At the Swiss Federal Office of Energy, a position was created to coordinate research, development, technology transfer, and market development of heat pumps (Rognon, 2006). In 1993, the Swiss Heat Pump Promotion Group (FWS)

Table 9.1. Policy instruments and major events related to heat pump development in Sweden, 1974–2008.

Phase	Policy instruments and major events
1974–89 Increasing awareness of energy and environment issues	1974: First seminar on heat pumps for researchers, authorities, builders, and real estate owners (BFR, Vattenfall, NUTEK, SP) 1973–74: International Energy Agency's heat pump group 1975: First energy research programme (government support) 1970s: Industry and public funding for demonstration projects (Vattenfall, AGA Thermia, BFR, etc.) 1977–85: Investment subsidies and loans for households (1977–79; 1979–80; 1980–81; 1981–83; 1983–84; 1984–85) 1979–85: Energy Prototype and Demonstration Programme: energy guidance training and dissemination of information (municipalities) 1980s: Start of testing activities (SP and industry) 1987: CFC phase out, Montreal Protocol (government commitment)
1990–2010 The success story	1990– Continuous research programmes (government support) 1990–95: Technology procurement programme (NUTEK) including investment subsidies and information campaigns late 1990s: First quality label (P-label) for GSHPs (SP) 1998: Introduction of energy advisory offices (municipalities) 1998–99: Subsidies for residential houses to convert from electric heating systems to other energy sources (1998–99; 2001–2003; 2006–10) 2006: First Swan eco-label for heat pumps (criteria development from 1998 with the involvement of multiple stakeholders)

Notes: BFR: Swedish Building Research (Byggforskningsrådet), after 2001 the Swedish Research Council for Environment, Agricultural Sciences and Spatial Planning (Formas) was formed and included BFR NUTEK: Swedish National Board for Industrial and Technical Development (from 1998, the Swedish Energy Agency).
SP: Technical Research Institute of Sweden (Sveriges Tekniska Forskningsinstitut).
Sources: Åstrand, Nordqvist, and Khan (2005); Energimyndigheten (2002a, 2002b, 2007a, 2007b); Neij and Öfverholm (2001); Nilsson, Åhmans, Nordqvist (2005); SFS1997:635 (1997); Nowacki (2007); Svensson and Havsjö (2006); SVEP (2007); Törnell (2007).

was constituted to coordinate market activities. This group involved heat pump producers, distributors, installers, leading electricity utilities, sector and cantonal authorities, and a professional marketing company (Zogg, 2008).

To further support technology development and quality control in the 1990s, a test centre was created in 1993 in Winterthur-Töss. After some scepticism and reluctance, heat pump companies agreed to test their products, which increased confidence in the technology. In 1993–95, a public subsidy programme was launched for investing in heat pumps in existing buildings. In relative terms, sales increased significantly (by a factor of two), but after the subsidy period sales declined again. Although in absolute terms, the subsidy programme did not have a tremendous impact (about 800 units were sold per year), it built up trust in the heat pump technology and was seen as a catalyser for further market development (Rognon, 2006). In 1996, marketing activities were reinforced and the first heat pump exhibition took place. In 1997, some utilities launched subsidy programmes and special electricity tariffs, and professional education and training were also improved. In 1998, the Swiss Retrofit

Table 9.2. Policy instruments and major events related to heat pumps in Switzerland, 1973–2008.

Phase		Policy instruments and major events
1973 – 89 Awareness	Increasing awareness of energy and environment issues	1974: First guidelines of the Swiss Association for Refrigeration 1980: First conference on heat pump technology in Switzerland 1980s: First heat pump testing facility (EPFL) 1981/82: Start of heat pump system field testing (NEFF and SFOE) 1983: Meeting on simplification of approval procedure (SFOE and authorities)
1990 – 2010 The success story	1990–92 First steps	1990: Launch of Energy 2000 (SFOE) 1992: Heat pump promotion programme (Energy 2000)
	1993–95 Bundling of activities	1993–95: Subsidy for heat pumps in existing buildings 1993–96: Handbooks for better heat pump installations 1995: FAWA – heat pump systems, field testing (SFOE)
	1996–2000 Consolidation	1996: First heat pump exhibition (trade fair for the general public) 1997–98: Subsidies supported by some electricity utilities 1997: Standards including that max. 80 percent of heat and hot water of new buildings could be covered by nonrenewable energies (Canton of Zurich) 1998: Heat pump retrofit programme and competition (R&D and subsidies) 1998: Creation of heat pump quality label DACH (Germany, Austria, Switzerland)
	2001–2005 Industrialisation	2000: Launch of Swiss Energy (follow-up programme of Energy 2000) 2001: DACH label for drilling companies
	2006–2010 Path to independence	2006: Regular three-day training programme for installers

Notes: AWP: Swiss Working Committee of Manufacturers and Distributors (Arbeitsgemeinschaft Wärmepumpen).
EPFL: Swiss Federal Institute of Technology in Lausanne (Ecole Polytechnique Fédéral de Lausanne).
FAWA: field testing of heat pump systems (Feldanalyse Wärmepumpen)
FWS: Swiss Heat Pump Promotion Group (Fördergemeinschaft Wärmepumpen Schweiz).
NEFF: private national energy research fund, sponsored by oil, gas, and electricity utilities (Nationaler Energieforschungs Fonds)
SFOE: Swiss Federal Office of Energy.
WPZ: Heat Pump Test Centre in Winterthur-Töss.
Sources: Rognon (2006, 2008); Zogg (2008).

Heat Pump Competition Programme was launched to develop high-efficiency and competitive heat pumps for heating domestic hot water. This programme brought together manufacturers, universities, and other actors and was important for creating a common understanding of technology and market development.

In 1997, the canton of Zurich legally restricted the share of nonrenewable energies for heating and hot water purposes in new buildings to 80 percent of the allowed useful energy demand per m^2. Most of the other Swiss cantons subsequently implemented this requirement. The requirement could be satisfied by additional insulation, by using wood or solar energy, or by using heat pumps. As electrically driven heat pumps were a cost-effective option, this legal measure represented a strong policy incentive for the deployment of heat pumps. In 1997, to support energy efficiency and renewable energy in residential buildings, the concept of the "Minergie house" was introduced. Minergie is a voluntary standard whose requirements can also be met by using heat pumps. To support product quality, in 1998 the DACH label was introduced for heat pumps. In 2001, the first DACH labels were also given to drilling companies.

3 Evaluation of Innovation and Diffusion Policies

Over time, numerous policy incentives have lined the path of heat pump technology and market development in Sweden and Switzerland. In both countries, the early policy initiatives were poorly coordinated but supported technology development, knowledge development, the involvement of important actors, and market formation. The market collapse in the mid 1980s could have resulted in a total failure – but did not. The research programmes, continuing in the 1980s, and the set of stakeholders and networks formed in the early years, that is, public and private funded researchers, authorities, and institutions, provided an important platform for further development. The influence of informal "advocacy coalitions" (Sabatier and Jenkins-Smith, 1993) supported further development of heat pumps and new policy initiatives (Nilsson, Åhmans, and Nordqvist, 2005).

In the 1990s and the 2000s, both Sweden and Switzerland introduced more strategic, coordinated, and flexible policy incentives for the development of heat pumps. The focus was on knowledge development, networking, and market formation, but also on quality control, credibility, and legitimacy. R&D, testing, and certification were developed in close cooperation with additional strategic market initiatives. Subsidies were introduced as a strategic incentive and catalyst in the 1990s in both Sweden and Switzerland; the role of later subsidies in Sweden is, however, more uncertain. Both countries had a focus on networking, which encouraged important processes of learning. International networking through IEA research may have also played an important role for international learning and spillovers.

3.1 Technology Developments

Both in Sweden and Switzerland, continuous government and private R&D support was essential for the development of heat pumps. The R&D programmes supported technical development throughout the boom in heat pump sales in the early 1980s, the market collapse in the mid 1980s, and the reinforcement and consolidation of the market in the 1990s and 2000s. Moreover, R&D support extended to the infrastructure in terms of boreholes and drilling processes. In Switzerland, early R&D support in borehole modelling and computer simulation in combination with prior experience in water drilling was essential for heat pump development. In Sweden, less R&D was needed and the development relied on existing experience in water drilling. Both countries developed a leading international position in drilling and boreholes in the early years.

Another key initiative for supporting technology development was the development of test facilities in the 1970s (in Switzerland) and 1980s (in Sweden). In the 1990s the test facilities were considered a vital element for providing reliable and high-quality heat pumps. The requirements of testing provided essential quality control and support of the 1990s subsidy programmes. Consequently, the poor reputation of heat pumps from the 1980s could be overcome and credibility reestablished. Quality labels were introduced in Switzerland in 1998 and in Sweden in 2005. Quality labels were then also given to drilling companies and drilling installations, reflecting the competence of drillers. In Switzerland, the test centre in Winterthur-Töss measured an increase of 20 percent to 36 percent in heat pumps' performance (COP) and energy efficiency between 1992 and 2002 (WPZ, 2001). In Sweden, the Technical Research Institute of Sweden measured 13 percent to 22 percent improvement in heat pumps' performance (COP) between 1995 and 2005 (SVEP, 2007).

3.2 Market Developments and the Involvement of Actors

The main driver of the early market development of heat pumps in the 1970s and 1980s was high oil prices. In both countries, additional policy instruments accentuated this market demand. In Sweden, the focus was on demonstration programmes and subsidies; the subsidies were, however, fragmented and uncertain. In Switzerland, incentives for market development also addressed networking. In the beginning, market actors involved in both the sale and production of heat pumps were small and worked locally. As the market grew in the 1980s the number of actors increased; relevant actors included heat pump manufacturers, retailers, driller and installation suppliers, research organisations, authorities, certifying bodies, and test institutes. The collapse of the heat pump market in the mid 1980s, as a result of cheap heating oil (and a subsidy withdrawal in Sweden), reduced the number of heat pump manufacturers and retailers as well as maintenance personnel in both markets. This had severe consequences not only on the sales of new heat pumps but also on the maintenance of the installed ones.

In the early 1990s, as a result of increasing concerns about environmental pollution as well as strong lobbying from the advocacy coalition, both Sweden and Switzerland decided to strengthen the heat pump market. In Sweden, a well-coordinated market transformation programme was launched based on a technology procurement programme in combination with test and certification programmes, subsidies, and massive information activities. The procurement not only provided high-quality technology and substantial market support but also essential interactions among actors. In Switzerland, strategic research activities and subsidies as well as education, training, and networking were introduced. As in Sweden, the subsidies were to have a catalytic effect on the market, but were also limited in time. The establishment of organisations for coordination and networking was also important. The programmes in both countries should not be seen in terms of individual policy instruments but rather as strategic and coordinated programmes to reignite the market. As a result of these programmes the market for heat pumps took off again. Industrial production ramped up and the number of manufacturers stabilised. In Sweden, "one-stop-shop contractors" formed – taking responsibility for the entire heat pump system (design, installation, and start-up).

In the 2000s, the support of heat pumps continued in both Sweden and Switzerland. In Sweden, market development was, again, supported through subsidies. However,

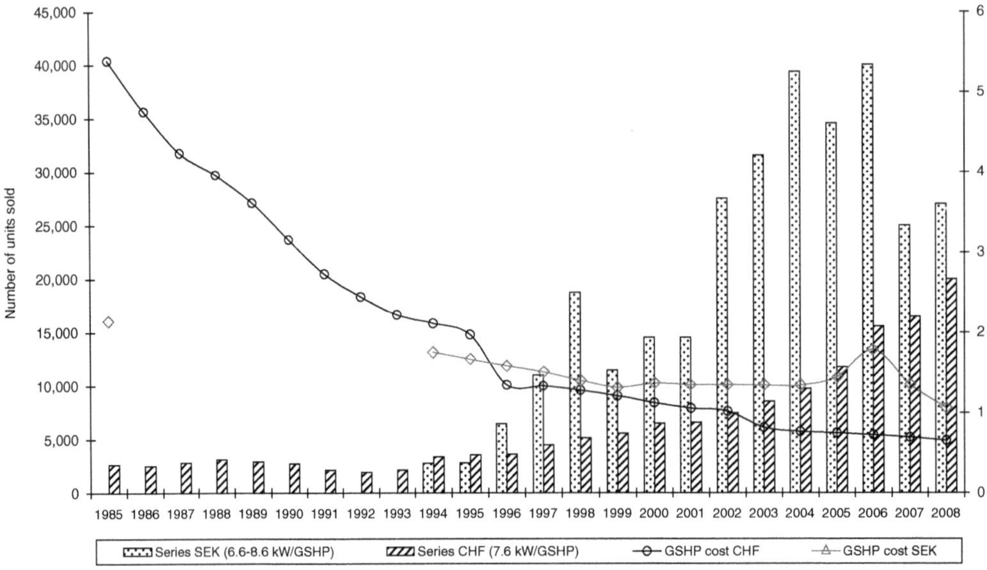

Figure 9.3. Number and cost of heat pumps installed in Sweden and Switzerland.
Notes: Costs include the cost of heat pump units and exclude costs related to installation and drilling. In Sweden, the sales figures are based on data from the Swedish Heat Pump Association (SVEP, 2007); cost figures in Swedish krona (SEK) are based on data from the Swedish Energy Agency (Energimyndigheten, 1999, 2002b, 2004, 2006) and the Swedish Consumer Agency (Konsumentverket, 1986, 2004), as well as online test magazines (Lagergren, 1995, 1999) and company product sheets. In Switzerland, sales figures are based on data from Basics (2002) and FWS (2009); cost figures in Swiss francs (CHF) are from FWS statistics. The base year for the cost index is 2008; the base currency is SEK. Prices of Swiss heat pumps in CHF are converted into SEK using annual average market exchange rates (www.oanda.com). In 2008, the average price of a ground source heat pump unit (6.6–8.6 kW) in Sweden amounted to around 53,000 SEK (8,140 US$). In Switzerland, comparable prices for heat pump units of 7.6 kW were around 7,800 CHF (47,000 SEK or 7,230 US$).

uncertainties regarding the duration and magnitude of these subsidies undermined manufacturers' long-term investments in technology development. As a result, between 2000 and 2003, 26 percent of the ground source heat pumps installed in Sweden were reported to have imperfections (Snaar, 2005). In Switzerland, the market development of heat pumps in the 2000s was driven by law (maximum share of nonrenewable energy for heating) and by voluntary standards ("Minergie" houses). The approaches in Sweden and Switzerland, involving subsidies as well as legal requirements and voluntary standards, effectively supported strong market growth in the late 1990s and 2000s. However, the subsidies applied in Sweden over the years have contributed to large government expenditures in comparison with the use of mandatory and voluntary standards to direct industry expenditure in Switzerland.

3.3 Cost Developments

The policy programmes and their effect on increased production and sales of heat pumps have provided opportunities for cost reductions over time. In Switzerland, cost reduction is reflected in the decreasing consumer prices of ground source heat

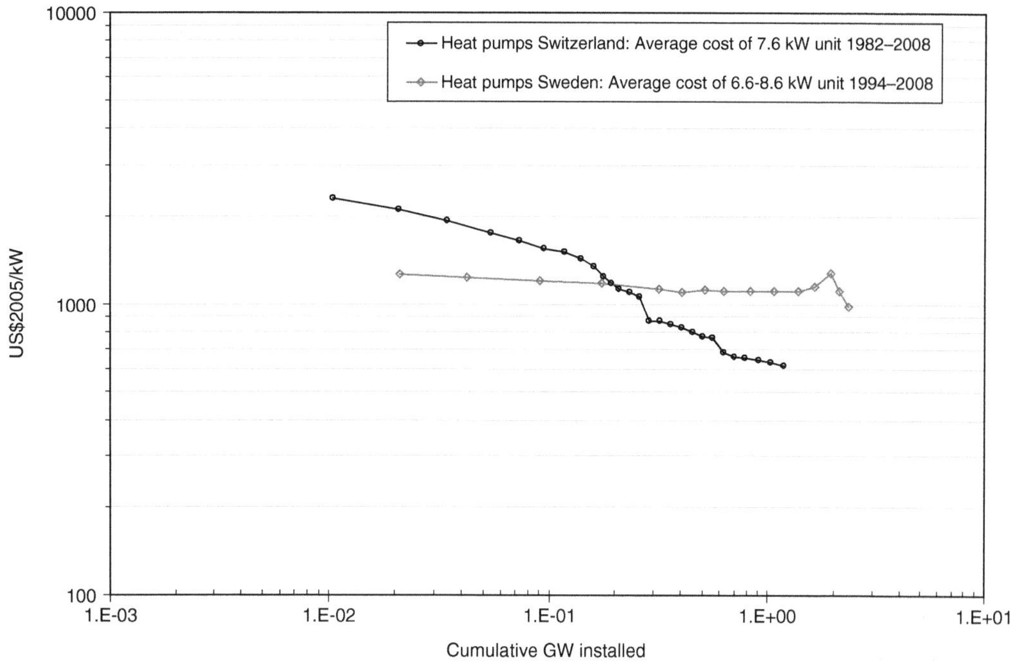

Figure 9.4. Cost of heat pumps as a function of cumulative experience: Learning curves.
Note: Log-scale x and y axes.

pumps for new single-family houses (see Figure 9.3). During the past three decades, costs have reduced by more than a factor of two. In the early 1980s, the total cost of heat pump systems was almost twice as high as for fossil fuel heating systems; currently heat pumps are cost competitive. Important drivers of cost reduction have been economies of scale (mass production), not only for heat pumps as such, but also for borehole drilling (for instance, capital costs of drilling systems could be distributed to more clients), and continuous technology improvements. Moreover, parts of the heat pumps are now imported at a lower cost than when produced in Switzerland.

In Sweden, as in Switzerland, consumer prices of ground source heat pumps have reduced over time; however, the trend differs in the two countries. In Sweden, the consumer price merely slightly decreased between 1985 and 1995; despite several drivers of cost reduction (i.e., incremental technology improvements, standardisation, and industrialisation of production) it then remained fairly stable until the mid 2000s. As Figure 9.3 shows, the cost reduction has been stronger in Switzerland over the years, however, the consumer price level has been lower in Sweden until the beginning of the 2000s. The reason for the constant consumer prices in Sweden could include limited competition of contractors and fragmented and uncertain subsidies; however this has not been verified by interviews carried out as part of this study. The price increase in 2006 can clearly be related to the introduction of a subsidy in an already saturated market. Another explanation may be that the components of the Swedish heat pumps are all manufactured in Sweden, whereas manufacturers in Switzerland import most of the components and claim this is a major source of cost reductions.

The cost reductions over time observed in Figure 9.3 are shown in Figure 9.4 as a learning curve that expresses capital or investment costs as a function of cumulative deployment. The characteristic declining cost profile is clearly seen, as is the contingency of the rate of decline and stabilisation on market conditions (e.g., Swiss and Swedish differences) and technological maturity.

4 Insights for Innovation and Diffusion Policies

Assessment of the policy incentives applied in Sweden and Switzerland to support the development of heat pumps and their emerging market can be used to illustrate some important characteristics of policy learning.

First of all, this case study shows the need for long-term and continuous support for energy technology innovation. The first attempts to influence the introduction of a new technology may fail, thus continuous support is needed to overcome initial shortcomings. The technology life cycle, from innovation to widespread diffusion, takes time. In the case of heat pumps, early government support was introduced in the 1970s but not until the 2000s was a major market increase seen and production industrialised. Over time, the combination of policy instruments may have to change and the approach used by the government needs to be flexible. Initially, policy may allow and support entrepreneurial testing, but this should be developed into stable strategies over time to allow industry to make long-term investments in standardised products.

Secondly, policy interventions need to consider both the development of the technology and its emerging market and actors. In other words, R&D is necessary but not sufficient. Market formation also requires policy incentives that support learning processes related to the use of the products. Moreover, various support for building and strengthening actor networks is essential for improving strategic integration and learning to ensure feedback and spillover effects.

Thirdly, testing and certification processes are needed not only to support technical quality but also credibility and legitimacy. R&D initiatives as well as subsidies require testing and certification to support a stable market development. Because of an emphasis on quality assurance in the early 1990s, the market of heat pumps started to grow again after having collapsed in the mid 1980s. By establishing a system for quality assurance, both Sweden and Switzerland created reliable products and a high level of public acceptance.

FURTHER READING

For an international review of the technical development of heat pumps and the Swiss examples and contribution to this development, see Zogg (2008). Rognon (2006, 2008) provides a good insight into the role of the state in the promotion of heat pumps on the market. For an overview of the Swedish heat pump market development, see Törnell (2007).

REFERENCES

Åstrand, K., Nordqvist, J., & Khan, J., 2005. Stakeholder Participation in the Policy Process: What are the Effects on the Implementation of Policy Instruments? In Sophie Attali, K. T.

(ed.) *Energy Savings – What Works & Who Delivers?* – ECEEE 2005 Summer Study, May 30–June 4, Mandelieu La Napoule, France. eceee, 53–62.

Basics, 2002. *Neue Elektro-Wärmepumpenstatistik – Dokumentation der Ergänzungsarbeiten* [*New Heat Pump Statistics*]. Bern: Basics AG on behalf of Swiss Office of Energy (SFOE).

BFE, 2010. *Analyse des schweizerischen Energieverbrauchs 2000–2009 nach Verwendungszwecken* [*Analysis of Switzerland's Energy Consumption 2000–2009*]. Bern: Bundesamt für Energi.

Boverket, 2008. *Alternativa uppvärmningsformer i befintliga permanentbebodda småhus* [*Alternative Heating Methods in the Existing Small Family-House Stock*]. Karlskrona: The National Board of Building, Housing and Planning.

EGEC, 2009. *Geothermal Heat Pumps – Ground Source Heat Pumps*. Brussels: European Geothermal Energy Council.

Energimyndigheten, 1999. *Villavärmepumpar* [*Heat Pumps for Single Family Houses*]. Stockholm: Swedish Energy Agency.

Energimyndigheten, 2002a. *Värme i Sverige* [*Heating in Sweden*]. Eskilstuna: Swedish Energy Agency.

Energimyndigheten, 2002b. *Villavärmepumpar – Energimyndighetens sammanställning över värmepumpar för småhus* [*Heat Pumps for Single Family Houses*]. Eskilstuna: Swedish Energy Agency.

Energimyndigheten, 2004. *Villavärmepumpar – Energimyndighetens sammanställning över värmepumpar för småhus* [*Heat Pumps for Single Family Houses*]. Eskilstuna: Swedish Energy Agency.

Energimyndigheten, 2006. *Villavärmepumpar – Energimyndighetens sammanställning av värmepumpar för småhus* [*Heat Pumps for Single Family Houses*]. Eskilstuna: Swedish Energy Agency.

Energimyndigheten, 2007a. *Effektivare energianvändning* [*More Efficient Energy Use*]. Eskilstuna: Swedish Energy Agency.

Energimyndigheten, 2007b. *Hushåll och energibeteende* [*Household and Energy Behaviour*]. Eskilstuna: Swedish Energy Agency.

Energimyndigheten, 2008. *Energiindikatorer 2008 – Uppföljning av Sveriges energipolitiska mål – Tema: Förnybar energi* [*Energy Indicators – Follow-Up on the Goals of the Swedish Energy Policy – Renewables*]. Eskilstuna: Swedish Energy Agency.

Energimyndigheten, 2009. *Energistatistik för småhus, flerbostadshus och lokaler 2008* [*Energy Statistics for Dwellings and Non-Residential Premises 2008*]. Eskilstuna: Swedish Energy Agency.

Florin, L., 1987. *Värmepumpar. Aktuella förändringar och framtidsutsikter* [*Heat Pumps. Actual Changes and Future Perspectives*]. Stockholm: Swedish Energy Agency.

FWS, 2009. *FWS Heat Pump Statistics* [Online]. FWS, Fördergemeinschaft Wärmepumpen Schweiz. http://www.fws.ch/zahlen_04.html. June 11.

IEA, 2008. *Energy Technology Perspectives 2008 – Scenarios and Strategies to 2050*, Paris: Organization for Economic Cooperation & Development.

Konsumentverket, 1986. *Värme i småhus – en handbok om renovering och byte av värmeanläggningar.* [*Heating in Single Family Houses – Manual on Refurbishment and Changing Heating Devices*]. The Swedish Consumer Agency.

Konsumentverket, 2004. *Värme i småhus – marknadsöversikt* [*Heating in Single Family Houses – Market Overview*]. Karlstad: Swedish Consumer Agency.

Lagergren, M., 1995. Billig i drift dyr att installera [Cheap to Run Expensive to Install]. *Råd & Rön.*

1999. Pumpen som spar pengar [The Pump that Saves Money]. *Råd & Rön.*

Lindeberg, P., 1984. *Forskning kring värmepumpar. Teknikutveckling och systemuppbyggnad med Byggforskningsrådets stöd – en översikt* [*Research on Heat Pumps*]. Stockholm: BFR, Byggforskningsrådet, Statens råd för byggnadsforskning.

Miljöstyrningsrådet, 2008. *Teknikupphandling – Verktyg för att främja innovationer och miljöteknik* [*Technology Procurement – Tool to Promote Innovation and Environmental Technology*]. Stockholm: The Swedish Environmental Management Council.

Neij, L. & Öfverholm, E., 2001. Strategies for Improving Energy Efficiency. In Silveira, S. (ed.) *Building Sustainable Energy Systems – Swedish Experience*. Stockholm: Svensk Byggtjänst AB, 301–33.

Nilsson, L. J., Åhmans, M., & Nordqvist, J., 2005. *Cygnet or Ugly Duckling – What Makes the Difference? A Tale of Heat-Pump Market Developments in Sweden*. In Energy Savings – What Works & Who Delivers? ECEEE 2005 Summer Study. Mandelieu La Napoule, France.

Nowacki, J. E., 2007. *Heat Pumps in Energy Statistics – Suggestions*. Lidingö: Nowab.

Rognon, F., 2006. Förderung der Wärmepumpe in der Schweiz mit marktkonformen Instrumenten [Support of Heat Pump in Switzerland]. 4. Wärmepumpen-Symposium, October 25, Berlin.

2008. *The Role of the State and the Economy in the Promotion of Heat Pumps on the Market*. 9th International IEA Heat Pump Conference, May 20–22, Zürich, Switzerland.

Sabatier, P. A. & Jenkins-Smith, H., 1993. *Policy Change and Learning: An Advocacy Coalition Approach*. Boulder, CO: Westview Press.

SFS1997:635, 1997. *Förordning om statligt bidrag till vissa investeringar för att minska elanvändningen i bostäder och vissa lokaler* [*Ordinance on Governmental Subsidy to Specific Investments to Reduce Electricity Use in Residential Houses and Given Locals*] [Online]. http://www.notisum.se/rnp/SLS/LAG/19970635.HTM. June 11.

Snaar, J., 2005. *Många fel på nyinstallerade bergvärmepumpar* [*Failures of Newly Installed Heat Pumps*] [Online]. Folksam. http://media.folksam.se/2005/06/19/sifoundersokning-bekraftar-manga-fel-pa-nyinstallerade-bergvarmepumpar/. August 28.

SPK, 1986. *Värmepumpar en konkurrensbeskrivning* [*Heat Pumps – Description of Competition*]. Stockholm; Statens pris- och kartellnämnd (now: Swedish Competition Authority).

Svensson, K. & Havsjö, N., 2006. *Nordiska Bergvärmemarknaden – Marknadsundersökning* [*The Nordic Bedrock Heat Pump Market – Marketanalysis*]. Ängelholm: Ängelholms Högre Utbildningar.

SVEP, 2007. *Small-Scale Ground Source Heat Pumps on the Swedish Market* [Online]. Swedish Heat Pump Association. November 20.

SVEP, 2009. *Ytterligare ett starkt år för värmepumpbranchen in Sverige* [*Another Strong Year for the Heat Pump Industry in Sweden*]. Swedish Heat Pump Association.

Törnell, M., 2007. *Heat Pump Development in Sweden – Success Story*. Eskilstuna: Energimyndigheten, Swedish Energy Agency.

WPZ, 2001. COP Values, Wärmepumpen-Testzentrum Töss. WPZ Bulletin.

Zogg, M., 2008. *History of Heat Pumps – Swiss Contributions and International Milestones*. SFOE, Swiss Federal Office of Energy and DETEC, Department of Environment, Transport, Energy and Communications.

10 Sources and Consequences of Knowledge Depreciation

Arnulf Grubler and Gregory F. Nemet

Authors' Summary

Technological knowledge, like all knowledge, can be learned or accumulated, but it can also be lost or "unlearned." In other words, knowledge capital depreciates. This case study considers the different sources of knowledge depreciation including reasons knowledge is lost (e.g., staff turnover) as well as why knowledge becomes obsolete (e.g., rapid innovation). Estimates of typical knowledge depreciation rates are reviewed, and the limited examples related to energy technologies are discussed in more detail. Knowledge depreciation rates reach 100 percent per year in service industries, characterised by high staff turnover. In the energy technology field, knowledge depreciation rates of between 10 percent per year (wind turbines) and 30 percent per year (solar PVs) have been identified. Illustrative calculations based on public energy R&D statistics of IEA countries show the implications of knowledge depreciation for two groups of energy technology innovations: nuclear power and energy efficiency. A conclusion is that a stable, gradually rising trajectory of policy support is as important, if not more important for mitigating knowledge depreciation than the absolute level of policy support if characterised by "boom and bust" cycles. With knowledge depreciation, continuous knowledge "recharge" becomes critical, for example, through stable R&D efforts and sustained market formation incentives.

Editors' Guide

Relationship of Case Study to Energy Technology Innovation System Framework

The tendency for technology and innovation studies to focus on success stories and their reasons has long been observed. This case study covers in depth an often ignored innovation system process on the negative side of the knowledge dimension: *knowledge depreciation*. Based on the available data, the depreciation, turnover, obsolescence, and loss of accumulated knowledge is found to be broadly relevant for energy technologies. Yet it is typically sidelined by the preferred emphasis on *knowledge generation* (R&D) and *learning* as sources of technological improvement. This one-sided and unduly positive story precludes a comprehensive treatment of the *knowledge* dimension within the energy technology innovation system.

Knowledge accumulated by different means is differentially exposed to the risk of depreciation. Knowledge gained through *learning by doing* appears particularly susceptible to rapid depreciation given its reliance on tacit knowledge acquired and held by specific workers (who can retire, move, or leave). This is starkly illustrated by the estimated three-month half-life of new knowledge gained from experience in the early boom years of the Californian wind industry.

Knowledge stocks for different technologies are also differentially exposed to depreciation. End-use technologies with short lifetimes in consumer goods markets with rapidly evolving *adopter preferences* are more likely to face technological or economic obsolescence. Yet far more innovation system *actors* are involved in the development and use of these granular or small unit scale technologies. This dispersion of the knowledge stock may help insulate end-use technologies from the types of depreciation observed in large-scale concentrated energy supply technologies like nuclear power.

Knowledge depreciation thus illustrates the interactions between *knowledge*, *actors and institutions*, and *resources*. Innovation system *actors* ranging from large organisations down to managers or individual employees are ultimately the holders of many forms of applied knowledge (which can not be easily codified). *Institutions* within the innovation system are also integral to the maintenance and sharing of that knowledge, through informal *exchanges and interactions* within networks, or through formal *codification*, documentation, and archiving activities. These actors and institutions are important determinants of knowledge depreciation rates. In contexts that can lead to high rates of knowledge depreciation (e.g., high staff turnover, rapid innovation), stability in knowledge-generation activities and associated institutions that enable knowledge exchange, sharing, and codification become ever more important.

Insights for Energy Technology Innovation Policy

- Conditions of rapid knowledge depreciation make stable knowledge-generation activities and associated resource inputs (e.g., R&D funding) key. The pattern of resource mobilisation over time is as, if not more, important than the absolute levels of resources mobilised. Continuous knowledge recharge through sustained R&D investments is needed to counter depreciation.
- Stable, gradually rising levels of policy support outperform larger-scale "boom and bust" cycles that can rapidly depreciate accumulated knowledge stocks even under otherwise high levels of support. As examples, estimates of the R&D knowledge stock for IEA countries suggest that nuclear has suffered substantially more from knowledge depreciation than energy efficiency, characterised by much lower, but more stable R&D expenditures.

- Overambitious efforts to scale up technologies or grow technology industries by pushing rapid knowledge generation are associated with many innovation risks, of which rapid knowledge depreciation is one.
- Codification of knowledge generated, particularly during the early growth years of an industry, can help preserve knowledge stocks during slower or downturn years. Careful codification is rarely a priority during boom years of an infant industry (with patent applications being an exception), yet can (and should) be written into policy support mechanisms if these play a role in market formation. The companion case study of solar thermal electricity in the United States provides a useful example.
- Next to codification, human capital management to prevent rapid staff turnover is key for minimising volatility-driven knowledge depreciation. Innovation policy can therefore help sustain knowledge stocks indirectly through training, skills development, exchange and interaction activities, and, most of all, through a stable and sustained policy framework in which market actors have confidence.

1 Introduction

Not unlike physical capital, knowledge capital depreciates. Technological knowledge, like all knowledge, can thus be accumulated (learned) but equally lost (unlearned). The underlying reasons for, as well as the rates of, knowledge depreciation hold important implications for innovation policy.

The relevance of knowledge depreciation for the knowledge basis of innovation is straightforward. And yet it is often ignored in analysis that unduly focuses on knowledge creation rather than on an integrated perspective that also includes the negative side of the knowledge equation. The topic of knowledge depreciation is thus underresearched, with no single review paper available in the literature. The explicit treatment of energy technologies within the context of knowledge depreciation is even more patchy and hence motivates this case study, which reviews a (very limited) research field.

Knowledge depreciation can be conceptualised as depending on two main variables. The first is the degree of innovation-driven technological obsolescence. The second dimension is the rate by which (otherwise unchanged) technological knowledge depreciates because of turnover of the holders of that knowledge.

The rapidly changing information and communication technology (ICT) field provides an important example of innovation-driven technological obsolescence. ICT also illustrates the need to differentiate two types of technological obsolescence: technical and economic. Consumers discard older vintages of ICTs (e.g., cell phones, computers) not because they are broken (technical obsolescence) but rather because newer, better-performing models (with rapidly falling prices) are a preferred alternative (rapid innovation-driven economic obsolescence). Changeover times in laptop computers are typically in the range of three years, and for cell phones often less than one year in some markets (e.g., younger consumers in OECD countries).

The second variable influencing knowledge depreciation relates to staff turnover within organisations. Staff members are the holders of the knowledge stock, and a high rate of staff turnover (for whatever reason) is associated with knowledge depreciation, particularly if these staff members hold key positions with respect to knowledge management.

Knowledge depreciation rates thus depend on the twin influences of the rate of innovation and the degree of policy and human capital volatility.

Knowledge depreciation is known to particularly affect settings in which knowledge remains largely tacit (i.e., uncodified, "hands-on" experience), residing in individuals (e.g., staff) or organisational entities (e.g., management) that need to be acquired again in case of staff turnover or "stop and go" production schedules (e.g., Argote, Beckman, and Epple, 1990). A second type of depreciation occurs as old knowledge becomes obsolete. Knowledge can depreciate because of an insufficient "recharge" of knowledge (Evenson, 2002) in cases where innovation proceeds rapidly such that old technological knowledge is no longer relevant for updated processes/techniques but new learning cannot proceed quickly enough (e.g., because of financial constraints). Both dimensions of knowledge depreciation are of particular concern in energy technology innovation systems when rapid rates of innovations coincide with erratic funding and policy support (see the case studies on solar water heaters, solar thermal electricity, and synfuels).

Another (even if atheoretical and empirically largely unfounded) distinction often advanced in the energy technology innovation literature is the differentiation between "learning by searching" (e.g., R&D) and "learning by doing" (e.g., accumulated production experience or market deployment). Kouvaritakis, Soria, and Isoard (2000) provide an example of such a model formulation. The assumed perfect substitutability of the two sources of learning, as well as formidable statistical challenges due to multi-collinearity and unknown lags, are major critiques of so-called two-factor learning curve models (as formulated, e.g., by Miketa and Schrattenholzer, 2004).

Knowledge depreciation affects the knowledge gained from R&D as well as the experience gained through production and market deployment. Recent work on knowledge depreciation in decisions made by U.S. wind turbine operators shows the high rates of depreciation in a learning by doing context (Nemet, 2012). This is discussed further later in this chapter, as is the case of nuclear power in the countries of the International Energy Agency (IEA). There is evidence of R&D knowledge depreciation of nuclear power, but also of nuclear knowledge depreciation in the ability to construct nuclear power plants as in the recent disappointing experiences at the two construction sites of the European Pressurised Water Reactor (EPR). When the first EPR, the AREVA/Siemens Olkiluoto-3 project in Finland, went at least three years behind schedule and 50 percent over budget, AREVA could and did blame this on its foreign partners, but no such explanation was plausible for the identical Flamanville-3 EPR built by and for French institutions in France. When after a year of construction the project was a year late and 20 percent over budget, suspicions arose that AREVA's last order was perhaps so long ago that critical design and construction skills atrophied (even in routine tasks such as quality assurance and documentation for concrete) (Grubler, 2010).

2 Knowledge Depreciation through Policy and Human Capital Volatility

Knowledge gained from experience may be inherently more susceptible to depreciation than other types of knowledge. Scientific knowledge goes through a process of peer review and dissemination; research and development activities are carefully monitored and often intentionally organised as experiments. In contrast, tacit knowledge may be particularly vulnerable to depreciation. This implies that learning by doing is prone to especially high rates of depreciation because that experience in production is less likely to be codified than in research and development activities.

In a study of the (temporary) wind power industry boom in California in the 1980s and 1990s, Nemet (2012) found evidence of learning by doing, both in the operation of existing wind projects and in the installation of new ones. However, this learning was subject both to knowledge depreciation and diminishing returns to experience. Knowledge gained from experience appears to have depreciated rapidly, with the half-life of new knowledge less than three months. Lessons about how to run wind farms were gained as a side effect of producing electricity. Careful codification of the outcomes of trials was unlikely to have been a priority in the context of an investment boom for an infant industry. Bankruptcies were frequent and employee attrition was high. Even if important knowledge was gained in the process of installing and operating wind turbines, it was dispersed and much of it was not available for future projects.

The rate of depreciation reported by Nemet (2012) for the Californian wind industry is very similar to that estimated by Argote and colleagues (1990) for shipbuilding, though it is slower than the depreciation rates in other industries reported in Darr, Argote, and Epple (1995), and faster than those cited in Benkard (2000) and Thompson (2007). The management literature has long recognised that in service industries (e.g., pizza franchises), organisational and production knowledge can be lost quickly, especially under high rates of staff turnover (that can be up to 300 percent per year). Argote and colleagues (1990), Argote (1996), and Darr and colleagues (1995) provide reviews and report knowledge depreciation rates from 25 percent to 50 percent *per month* in service industries. Such high rates of knowledge depreciation basically imply that after a year only zero percent to 5 percent of the original knowledge of an organisation remains. Kim and Seo (2009) arrive at similar high depreciation rates of some 26 percent per month in their analysis of Liberty ships manufacture during WWII (even if earlier studies on the same case, e.g. Thompson (2007) report much lower rates of 4 percent to 6 percent per month, highlighting the large uncertainty and challenges involved in estimating knowledge and its depreciation rates).

A classical case discussed in the literature of technological learning is the case of the Lockheed L1011 Tri-Star aircraft (Argote and Epple, 1990). Other studies in the aircraft industry suggest a significant reduction in manufacturing costs or "learning by doing" as more production experience (output volumes) is accumulated. The Lockheed Tri-Star aircraft was an exception to that rule. When production resumed after an extended production halt, manufacturing costs were much higher and also did not decline subsequently, reflecting experience/knowledge lost or "unlearning" that could no longer be recovered. The reason for this knowledge depreciation was basically the same as for pizza franchises: during the production halt the entire staff of the manufacturing plant was fired (including managers who, according to Mishina

(1999), are the main locus of organisational learning and knowledge in aircraft manufacturing). Benkard (2000) reports corresponding knowledge depreciation rates of typically 40 percent *per year* in aircraft manufacturing. The wind industry in the study by Nemet (2012) had similar challenges due to the extreme seasonality in the California wind resources, which relied on thermal gradients between cool marine air and hot inland areas. Very little production occurred from September to March each year, although maintenance activities during the off-season helped retain a base level of knowledge generation processes.

Next to knowledge codification and preservation, human capital management is key to minimise volatility-driven knowledge depreciation. Similarly, the design of policy instruments to create incentives for learning should account not only for the existence of spillovers from learning, but also for the persistence of that knowledge.

Boone, Ganeshan, and Hicks (2008), in their study of knowledge depreciation of professional services, conclude that the extremely low rates of knowledge depreciation found in engineering design firms is explained by comprehensive knowledge documentation (earlier CAD designs are documented and kept for subsequent use) and a stable business, as well as low staff turnover rates (only 3 percent per year), particularly among senior engineers, which is markedly different compared to typical service industries such as food retailing. As such, the study provides valuable lessons for improved knowledge management in energy technology and other innovation systems highlighting in particular the importance of documentation, codification, and preservation of knowledge as well as the need for a minimum degree of continuity in senior staff that is the "living memory" of organisations. Institutions practising comprehensive reassignment (job rotation) policies can frequently lack any institutional memory of earlier corporate strategies. At a renewable energy investment strategy workshop attended by one of the authors, a group of managers was entirely unaware that the company had previously invested into solar PV and biofuels but had since sold these activities.

Similarly, the case study on solar thermal electricity discusses the importance of national laboratory studies of the performance and improvements of the 350 MWs of solar energy generating station (SEGS) plants in California in the 1980s. These careful studies preserved much of the knowledge about how to design, build, and operate large-scale solar power plants. The resulting reports were important in offsetting knowledge depreciation during the fifteen-year period in which no new commercial plants were built, the U.S. R&D programme was almost completely eliminated, and the original builder of the plants filed for bankruptcy, selling off the plants to multiple investors.

3 Knowledge Depreciation through Technological Obsolescence

Estimates of innovation-driven knowledge depreciation for entire industries are extremely limited. Hall (2007) provides one of the few comprehensive efforts to estimate knowledge depreciation across various industry sectors relating the market value of firms to patent data to estimate the R&D knowledge depreciation in six U.S. industry sectors. The study found knowledge depreciation to vary significantly over time and across industries, with median R&D knowledge depreciation rates of between 15 percent *per year* (drugs and instruments) up to 36 percent *per year* (electrical). An alternative model formulation using a production function approach did not yield any statistical significant results of knowledge depreciation.

Watanabe and colleagues (2002) provide one of the few estimates of knowledge depreciation for energy technologies. By constructing a knowledge stock model for the Japanese PV industry that includes both R&D by firms and knowledge spillovers from other firms (measured via patent citations), they estimate a mean PV knowledge depreciation rate of some 30 percent per year. This implies that without continuous recharge (R&D) an existing technology knowledge stock is reduced to some 25 percent of the original value after five years and to less than 5 percent after ten years. Nemet (2009) provides an illustration for the U.S. wind turbine industry by analysing citations to a set of "highly cited" (i.e., valuable) wind energy patents. He finds that 40 percent of all (cumulative) citations occur during the first five years, after which citations decline to basically zero after twenty-five years. While lags in knowledge diffusion account for the gradual increase in citations in the first five years, the declining trend in patent citations after year five reflects their decreasing significance and can be used as a proxy for knowledge depreciation, which corresponds to a rate of approximately 10 percent per year after the fifth year. The lower number compared to the Watanabe and colleagues (2002) study can be explained by technology differences (PVs being on a much more dynamic technology trajectory compared to wind turbines), but also by the fact that the Nemet sample focuses on "highly cited" patents, that is, only a subset of all patents that are the most successful ones whose knowledge base can be expected to depreciate at a slower rate than the industrywide knowledge stock, which was the subject of the Watanabe study.

It's important to note that the use of patent citation data helps distinguish between the effects of technological obsolescence and employee turnover. Patents require a detailed and fully public invention disclosure. This codification, including diagrams and detailed schematics, preserves knowledge that would otherwise be lost through employee attrition. The depreciation rates we see in these studies essentially isolate the effect of technological obsolescence. As a result, they produce lower estimates of depreciation rates than studies that include the effects of both attrition and obsolescence.

4 Synthesis of Knowledge Depreciation Rates

The available literature suggests typical knowledge depreciation rates of between 10 percent to 40 percent per year in industries comparable to energy, that is, industries in which innovation and R&D play a significant role. (Human capital volatility remains an important driver independent from R&D intensiveness.) Given such high rates of obsolescence of technological knowledge, continuous "knowledge recharge" and maintenance becomes extremely important. To date, no study has attempted to assess the relative contributions of human capital volatility and innovation-driven technological obsolescence to knowledge depreciation. Nonetheless, Figure 10.1 provides a qualitative ranking to the rates of knowledge obsolescence reviewed earlier along the two dimensions of knowledge depreciation.

It is quite intuitive that knowledge depreciation rates tend to be highest in cases when rapid technological obsolescence combines with high staff turnover and/or

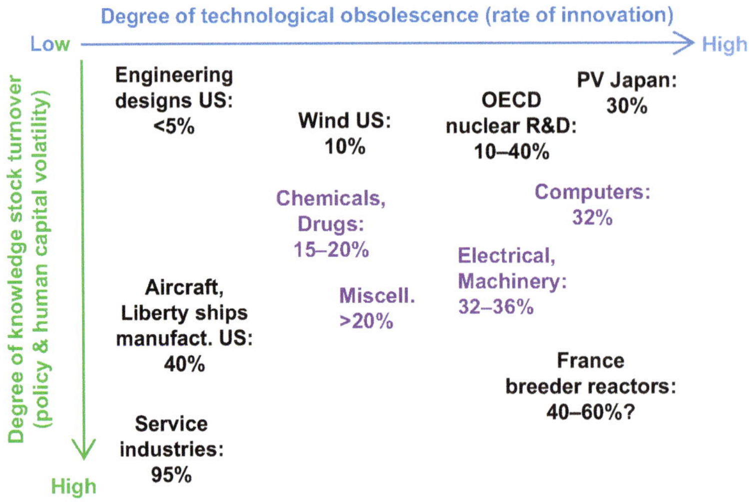

Figure 10.1. Typical knowledge depreciation rates (% per year).
Notes: U.S. industries shown in magenta (Hall, 2007); selected energy technologies and activities in manufacturing and services shown in black (see text for sources); rates for breeder reactors in France estimated by the authors (see text for explanation).

erratic policy support. As an illustration consider the case of an innovation failure: the French plutonium fuelled fast breeder reactor programme whose culmination aimed at an early up-scaling of reactor size to commercial scale in the form of the 1.2 GW Superphénix reactor in Creys-Malville, France. After completion in 1985, and a stuttering performance record until reaching full capacity in 1996, the basic design flaws of using a liquid metal (highly corrosive sodium) as coolant provided for the ultimate demise of this technological innovation. Superphénix was permanently shut down in 1998. French public sector breeder reactor R&D peaked in the years prior to Superphénix's start-up and declined precipitously along with the disappointing reactor performance and accumulating technical problems. Private sector R&D into breeder reactors was insignificant.

Breeder reactor R&D expenditures declined between 40 percent and 60 percent per year during the period from 1985 to 1989/1990 (IEA, 2008). It is likely that the staff attrition in R&D engineers and in operating engineers (after the shutdown of the reactor) reached similarly high levels (explaining our speculative estimated range of knowledge depreciation rates presented in Figure 10.1). In such extreme cases, high-tech energy technologies and industries might not be all that different in terms of knowledge depreciation than more mundane human endeavours such as making pizzas.

5 R&D Spending: "How" Is at Least as Important as "How Much"

To illustrate the importance of R&D knowledge depreciation, we draw on the IEA public sector R&D expenditure statistics using annual R&D expenditures of all IEA countries over the period from 1974 to 2007 for nuclear and energy efficiency (IEA, 2008). The two examples represent contrasting patterns of R&D expenditure: high but erratic (nuclear) versus low but stable (efficiency). We then compute the remaining knowledge stock over time using a range of knowledge depreciation rates

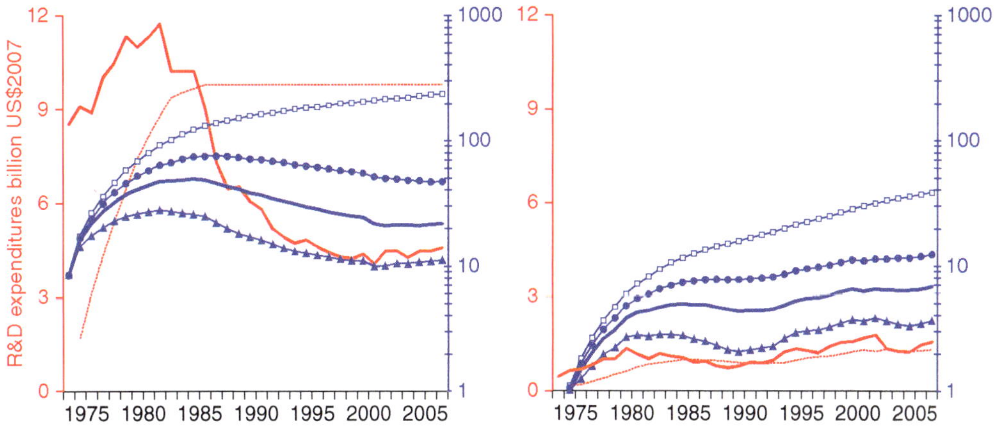

Figure 10.2. R&D expenditures in IEA countries and knowledge stocks under different depreciation rates for nuclear (left panel) and energy efficiency (right panel).
Notes: R&D expenditures ($2007) in IEA countries shown on left axis in red (solid lines show actual R&D, dashed lines show R&D required to maintain knowledge stock at 20 percent depreciation rate); knowledge stock from actual R&D expenditure shown on right axis in blue at zero percent, 10 percent, 20 percent, and 40 percent knowledge depreciation rates per year.

from zero percent, 10 percent, 20 percent, up to 40 percent per year. The knowledge stock estimates shown in Figure 10.2 describe how much of the original knowledge (cumulative R&D) remains relevant at any given year. In the case of a zero percent per year depreciation rate, the knowledge stock is simply the cumulative R&D until any particular year. In the other cases of greater than zero percent depreciation rates, annual expenditures are discounted by the corresponding depreciation rate and then summed up to estimate the knowledge stock. Downward sloping curves represent a decline in technological capability (remaining knowledge stock). A simplifying assumption underlying these calculations, which is frequently employed in the literature (e.g., Watanabe et al., 2002), is that a linear relationship exists between R&D investment and knowledge creation. Neither increasing nor decreasing returns to R&D are assumed to exist. This restricting assumption is easier to criticise than to disprove empirically.

As the distinction between the left and right panels of Figure 10.2 illustrates for nuclear and energy efficiency respectively, two variables define a technology's knowledge stock: the knowledge depreciation rate and *the pattern of knowledge generation over time* (here: R&D expenditures). Erratic expenditures, as in the case of nuclear, are particularly prone to knowledge depreciation. Taking a mean knowledge depreciation rate of 20 percent per year, only 9 percent of the original knowledge remained in 2007 (22 billion US$2007 out of a total cumulative nuclear R&D of 236 billion US$2007). Conversely, in the case of energy efficiency R&D with the same assumed knowledge depreciation rate, 18 percent of the original knowledge remained (7 billion US$ out of cumulative R&D expenditures of some 38 billion US$). This was due to a more consistent R&D expenditure pattern allowing continuous knowledge recharge.

This finding of a much higher portion of knowledge retention in stable R&D expenditures is robust against the large variation in the rate of knowledge depreciation

Table 10.1. Current (2007) and cumulative (1974–2007) R&D expenditures of IEA countries (Billion US$2007) with estimate of remaining knowledge stock.

	Current R&D		Cumulative R&D		Remaining knowledge capital stock	
	(2007)	(2007)	(1974–2007)			
	10^9 US$ 2007	%	10^9 US$ 2007	%	US$2007	%
Energy efficiency	1.6	13.0	38	8.9	7	12.9
Fossil fuels	1.4	11.3	55	12.8	6	11.6
Renewables	1.5	12.3	37	8.7	6	10.6
Nuclear fission	3.7	30.6	194	45.1	18	33.3
Nuclear fusion	0.9	7.3	42	9.7	4	7.9
Others	3.1	25.4	64	14.8	13	23.7
Total	12.0	100.0	431	100.0	53	100.0

Notes: Assumes an average knowledge depreciation rate of 20 percent applied to annualized expenditure profiles.
Source: IEA (2008).

included in the studies discussed previously. After a peak in nuclear R&D in 1982 at some 12 billion US$, R&D expenditures declined precipitously resulting in a substantial reduction of the nuclear knowledge stock. In fact, maintaining the knowledge stock would have required annual R&D expenditures to the tune of 10 billion US$ (assuming a mean knowledge depreciation rate of 20 percent per year), that is, twice above the actual R&D expenditures. This "R&D replacement" funding shortfall limited the continued knowledge recharge and consequently led to an additional knowledge depreciation beyond the "natural" rate of knowledge obsolescence. R&D into energy efficiency, while substantially below nuclear, was much less erratic and thus enabled continued knowledge recharge. Conversely, one could have achieved the same $22 billion nuclear knowledge stock in 2007 with a more stable investment pattern, similar to that of efficiency. In this illustrative calculation we assume a starting nuclear R&D budget of 1.4 billion US$ (actual: 8.5 billion US$) in 1974 growing continuously by 4 percent per year to 2007 (similar to the average growth in energy efficiency R&D) and a mean knowledge depreciation rate of 20 percent. This would have required only 100 billion US$ R&D expenditure, or 42 percent of the $236 billion that was actually spent over thirty-three years, saving $136 billion.

Table 10.1 provides a further illustration, again assuming a mean knowledge depreciation rate of 20 percent per year. Considering the entirety of energy R&D expenditures of IEA countries (IEA, 2008) suggests that only the equivalent of 53 billion US$ out of the cumulative R&D spending of 431 billion US$ over the period from 1974 to 2007 may still constitute useful technology knowledge at present, that is, some 12 percent, with the remaining investments lost to knowledge depreciation. Technologies such as nuclear that have suffered a particularly erratic R&D funding profile of "boom and bust" are affected much more by knowledge depreciation than technologies that have received less but more stable funding. This is seen in Table 10.1

by the corresponding lower percentage for nuclear compared to efficiency in the knowledge stock variable compared to cumulative expenditures.

In other words, for knowledge management, *a continuous and regular pathway of funding is as important as the absolute level of funding* with "boom and bust" cycles particularly detrimental in maintaining and recharging technological innovation knowledge.

6 Discussion

This case study's survey of the literature on knowledge depreciation in the field of energy technologies identified large gaps in available knowledge. There are intricate measurement issues with a complex unobservable and a lack of comprehensive technology-specific data on innovation inputs and outputs. Distinctions between depreciation via obsolescence and depreciation via human capital attrition provide a promising avenue for research. Another important research area will be to examine the impacts of different technological and organisational configurations on knowledge codification, and how public disclosure can help minimise knowledge depreciation. Distributed knowledge and technology cooperation networks could hold the promise of overcoming conventional patterns of knowledge depreciation. Technology, including open source, Internet-based information systems, could play an important role.

It is also important to establish empirically whether rates of knowledge depreciation are linked to the very nature of technologies themselves. In the case of knowledge accumulation this indeed seems to be the case. "Granular" small unit scale technologies appear to have twice as high rates of knowledge accumulation or learning than large scale "lumpy" technologies that by their very nature only allow experimentation at the scale of dozens to hundreds of projects compared to "granular" technologies produced and experimented with the scale of millions to billions (Wilson et al., 2012). More granular technologies, in part because they are dispersed among a large number of technological pathways with heterogeneous incentives for innovation, may also be less susceptible overall to the boom and bust cycles of prices, markets, and high rates of knowledge depreciation in the energy system.

Knowledge depreciation is an important and pervasive yet underrecognised phenomenon of technological innovation. Knowledge depreciation rates of 10 percent to 40 percent per year seem common. In some situations characterised by, for example, a lack of codification, boom and bust innovation incentives, and high rates of innovation, knowledge depreciation rates are even higher. Depreciating knowledge has been observed in careful studies of a wide variety of industries, including several energy technologies. It occurs for knowledge acquired through R&D as well as knowledge acquired through experience.

These empirical findings and research challenges hold important implications for technology policy in terms of stability of support for knowledge accumulation as well as incentives for knowledge preservation and sharing. To mitigate knowledge depreciation, public policy support for R&D and market formation incentives should be made contingent on public disclosure of technological knowledge, from scientific papers, technical reports, patent applications, all the way to the documentation of actual market technology performance of subsidised technologies. Public disclosure of information is an essential first step for mitigating depreciation particularly associated with tacit technological knowledge by facilitating documentation, archiving, and knowledge sharing.

More generally, knowledge depreciation needs to be more explicitly recognised and managed in technology policies. The life cycle of technological innovation from early experimentation to successful market deployment spans many decades (three decades being a rough rule of thumb). Recognising this lengthy process cautions against overambitious efforts to "scale up" innovations too quickly (e.g., the Superphénix breeder reactor) and definitively provides a cautionary tale against erratic policy signals and financial innovation support. When the pervasiveness of knowledge depreciation is taken into account, concerns about the inertia of entrenched government technology programmes (Cohen and Noll, 1991) are balanced by concerns about the losses associated with volatile programming. "Small and stable" as opposed to "big but boom and bust" is an important policy implication this case study on knowledge depreciation can provide to policy makers.

FURTHER READING

Argote and Epple (1990) is a widely cited study in the field of knowledge depreciation, particularly in manufacturing and service industries. Nemet (2012) provides a recent application of thinking on knowledge depreciation to the early Californian wind industry.

REFERENCES

Argote, L., 1996. Organizational Learning Curves: Persistence, Transfer and Turnover. *International Journal of Technology Management*, 11(7–8): 759–69.

Argote, L., Beckman, S. L., & Epple, D., 1990. The Persistence and Transfer of Learning in Industrial Settings. *Management Science*, 36(2): 140–54.

Argote, L. & Epple, D., 1990. Learning Curves in Manufacturing. *Science*, 247(4945): 920–24.

Benkard, C. L., 2000. Learning and Forgetting: The Dynamics of Aircraft Production. *American Economic Review*, 90(4): 1034–54.

Boone, T., Ganeshan, R., & Hicks, R. L., 2008. Learning and Knowledge Depreciation in Professional Services. *Management Science*, 54(7): 1231–36.

Cohen, L. R. & Noll, R. G., 1991. *The Technology Pork Barrel*, Washington, DC: Brookings.

Darr, E. D., Argote, L., & Epple, D., 1995. The Acquisition, Transfer, and Depreciation of Knowledge in Service Organizations: Productivity in Franchises. *Management Science*, 41(11): 1750–62.

Evenson, R. E., 2002. Induced Adaptive Invention/Innovation and Productivity Convergence in Developing Countries. In Grubler, A., Nakicenovic, N., & Nordhaus, W. D. (eds.) *Technological Change and the Environment*. Washington, DC: Resources for the Future Press, 61–96.

Grubler, A., 2010. The Costs of the French Nuclear Scale-Up: A Case of Negative Learning by Doing. *Energy Policy*, 38(9): 5174–88.

Hall, B. H., 2007. Measuring the Rates of Return to R&D: The Depreciation Problem. *NBER Working Paper Series, Working Paper 13473*. Cambridge, MA: National Bureau of Economic Research.

IEA, 2008. *Energy Technology R&D Statistics*. Online Edition. International Energy Agency (IEA).

Kim, I. & Seo, H. L., 2009. Depreciation and Transfer of Knowledge: An Empirical Exploration of a Shipbuilding Process. *International Journal of Production Research*, 47(7): 1857–1876.

Kouvaritakis, N., Soria, A., & Isoard, S., 2000. Modelling Energy Technology Dynamics: Methodology for Adaptive Expectations Models with Learning by Doing and Learning by Searching. *International Journal of Global Energy Issues*, 14(1): 104–15.

Miketa, A. & Schrattenholzer, L., 2004. Experiments with a Methodology to Model the Role of R&D Expenditures in Energy Technology Learning Processes; First Results. *Energy Policy*, 32(15): 1679–92.

Mishina, K., 1999. *Learning by New Experiences: Revisiting the Flying Fortress Learning Curve.* Chicago patent application.

Nemet, G. F., 2009. Demand-Pull, Technology-Push, and Government-Led Incentives for Non-Incremental Technical Change. *Research Policy*, 38(5): 700–709.

2012. Subsidies for New Technologies and Knowledge Spillovers from Learning by Doing. *Journal of Policy Analysis and Management*, 31(3): 601–22.

Thompson, P., 2007. How Much Did the Liberty Shipbuilders Forget? *Management Science*, 53(6): 908–18.

Watanabe, C., Griffy-Brown, C., Zhu, B., & Nagamatsu, A., 2002. Inter-Firm Technology Spillover and the "Virtuous Cycle" of Photovoltaic Development in Japan. In Grubler, A., Nakicenovic, N. & Nordhaus, W. (eds.) *Technological Change and the Environment*. Washington, DC: Resources for the Future, 127–59.

Wilson, C., Grubler, A., Sims Gallagher, K., & Nemet, G. F., 2012. Marginalization of End-Use Technologies in Energy Innovation for Climate Protection. *Nature Climate Change*. 2(11): 780–88.

11 The French Pressurised Water Reactor Programme

Arnulf Grubler

Author's Summary

The case study reviews the French nuclear Pressurised Water Reactor (PWR) programme as an example of successful scaling up of a complex and capital-intensive energy technology. Starting in the early 1970s, France built fifty-eight PWRs with a total gross installed capacity of sixty-six GWe. On completion in the year 2000, they produced some 400 TWh/y of electricity, or close to 80 percent of France's electricity production. The institutional setting enabled a high degree of standardisation, external learning via the use of proven U.S. reactor designs under a Westinghouse licence, high regulatory stability, and the effective absence of public opposition. The case study then considers the economics of this successful scale-up of nuclear reactor technology, identifying a significant cost escalation in real-terms reactor construction costs, but also a remarkable stability in reactor operation costs. This is all the more noteworthy considering the need for load modulation in a system relying as heavily on base-load nuclear as is the case in France.

Editors' Guide

Relationship of Case Study to Energy Technology Innovation System Framework

The initial phase of the French nuclear Pressurised Water Reactor (PWR) programme illustrates the importance of the different dimensions of the innovation system for supporting successful innovation outcomes. The small number of innovation system *actors*, both state and industrial, were tightly networked, shared *clear expectations* around the technology's deployment, and enjoyed a marked absence of *resistance* or opposition. Clear and stable *public policy* underwrote the substantial *resources* needed for widespread deployment. A particular institutional configuration of ÉDF (Électricité de France) as both principal (utility, operator) and agent (designer, engineer) in the construction process proved successful at managing costs and risks. Standardisation and clear market demand enabled *economies of scale* through the repeat use of the same reactor design that embodied *knowledge generated* in an earlier (U.S.) market. These *knowledge spillover* benefits were obtained through licensing of Westinghouse reactors.

A salient finding from the case study is that *learning* from the experiences of commercial deployment can be negative. In the later years of the PWR programme, a shifting emphasis towards bespoke, nonstandardised technology amidst a slowdown of the nuclear build out programme and a fragmentation of industry interests restricted the continued application of earlier experiences. *Learning* is thus highly contingent on market conditions and does not always cause costs to fall. Embedded in these reasons for the observed negative learning were *diseconomies of scale* associated with a scaling up of reactors from the tried-and-tested 900 MW sizes to 1,300+MW. As with learning, scale economies are not always beneficial.

Despite the steady increase in reactors' real construction costs, their operating costs remained low and flat in France (for similar reasons relating to France's peculiar institutional and market settings). The case study therefore makes an important distinction in how innovation system processes affect *the adoption and use* of a technology.

Insights for Energy Technology Innovation Policy

- The pervasive adoption of nuclear power in France was enabled by stable public policy and market conditions and a clear alignment between policy and industry interests that helped nuclear overcome the relative disadvantage of its extremely high upfront costs. This was further helped by ÉDF's choice – as owner, construction engineer, and operator – to replicate a proven and standardised technology. This congruence of supporting factors was in many ways peculiar to France. The issues and risks associated with scaling up large-scale, complex technologies with very high ("lumpy") initial investments are substantial, particularly in deregulated markets with multiple actors and fragmented design, construction, and operational roles. In other words, the inherent characteristics of a technology can militate against successful innovation outcomes, and so present greater risks to public policy.
- Cost projections for novel technologies are an inherent element in policy analysis. Assumptions of the inevitability of future learning and cost reduction need to be firmly grounded within the historical observational space, reflecting both uncertainty and the possibility for cost escalation.

1 Introduction

The French nuclear Pressurised Water Reactor (PWR) programme is legitimately considered the most successful scaling up of a complex and capital-intensive energy technology system in the recent history of industrialised countries. Starting in the early 1970s, France built fifty-eight PWRs with a total gross installed capacity of sixty-six GWe. On completion in the year 2000, they produced some 400 TWh/y of electricity, or close to 80 percent of France's electricity production.

Successful scaling up of a new technology entails three dimensions: an increase in technology deployment that is a) *substantial* (approaching 80 percent nuclear electricity), b) *rapid* (50 GWe, or 75 percent of the total installed capacity of 66 GW went "on-grid" within the decade 1980–90), and c) *systemic* (developing the industrial capacity to manufacture PWR components, the capability of building reactors within – by international standards – astonishingly short construction times, and developing a domestic industry covering the entire nuclear fuel cycle from enrichment, fuel manufacture, and reprocessing to nuclear waste management).

On all three counts, the French nuclear PWR programme stands out as the most successful of all similar efforts worldwide. While the reasons for this success are specific to the French political/technocratic system and may not be replicable in other countries (not even in France in the new millennium), the economic dimensions, especially the costs, of this nuclear scaling up have remained shrouded in mystery for a long time. Grubler (2009, 2010) provides the first quantitative economic history of the French PWR programme demonstrating that it has exhibited significant escalation in real-term construction costs. This is in stark contrast to the anticipated cost-lowering effects of economies of scale and learning arising from ordering whole series of standardised reactor designs that were at the core of the economic rationale of the French nuclear scale-up. As such the case study illustrates the limitations of engineering cost projections as well as simplistic learning or experience curve models to describe the cost dynamics of very large-scale, complex technologies.

2 Institutional Setting as Key to Successful Scaling Up

Following Jasper's (1992) perceptive analogy from Greek mythology, the main groups of actors in a nuclear scale-up are "gods" (governments), "titans" (large industries and utilities), and "mortals" (the general public). The institutional key to success in France was the extremely limited number of institutional actors: "mortals" never played any decisive role either in the technocratic decision-making process or in hindering rapid expansion. The senior actors were extraordinarily well coordinated through the "invisible hand" of a small technocratic elite – the state engineers of the Corps d'État. In other words, "god" (the French government) and the two "titans" that really mattered – the nationalised utility Électricité de France (ÉDF), and the state nuclear R&D organisation (CEA) – acted in a well-coordinated way, overcoming inevitable rivalries and differences of opinion. They ended up with a clearly formulated vision, mobilised the necessary resources, and proved quite apt in executing this extremely large-scale and complex technology programme, summarised in Table 11.1.

Finon and Staropoli summarise the unique institutional framework as consisting of four elements: "strong political support, a state-owned electricity monopoly endowed with [substantial] engineering resources ... a highly concentrated electromechanical manufacturing industry [emerging in the scale-up process], and an influential R&D public agency [the CEA that operated under] high regulatory stability ... and efficient co-ordination resulting from long-term organizational arrangements" (2001: 179). Standardised reactor series, ordered in bulk and profiting from external learning through the use of existing U.S. reactor designs via the Westinghouse licence, complemented the unique French nuclear institutional setting.

Table 11.1. Overview of French PWR programme.

Order series	Reactor type	Reactor size, typical MWnet (mean MWgross)	Number built	Constructed between	Mean construction time in months	Mean investment costs "best guess" and (uncertainty range) 1,000FF98/kWgross	Sites
CP0	PWR Westinghouse licence	900 (927)	6	1971–79	63	4.9 (4.2–5.9)	Bugey Fessenheim
CP1	as CP0	900 (949)	18	1974–85	65	5.5 (5.0–6.0)	Blayais Dampierre Gravelines Tricastin
CP2	asCP1	900 (955)	10	1976–87	67	6.5 (6.1–7.2)	Chinon Cruas St Laurent
P4	1.3 GW PWR Westinghouse licence	1330 (1382)	8	1977–86	78	6.9 (6.5–7.1)	Flamanville Paluel St Alban
P'4	P4 "frenchyfied" Westinghouse	1300 (1366)	12	1979–93	90	8.4 (8.0–8.8)	Belleville Cattenom Golfech Nogent Penly
N4	PWR new French design	1500 (1561)	4	1984–99	126	11.0*–13.3 (10.3–14.5)	Chooz Civeaux
EPR	EPR Framatome-Siemens	1600 (1650)	1	2007–	under construction	n.a.	Flamanville

* lower range excludes Civeaux-2 reactor

Notes: Construction time measured from construction start to first grid connection.

Source: Grubler (2009).

The single, most noticed measure of success in the French nuclear scale-up is undeniably the construction times from construction start to first grid connection (see Table 11.1), which are short by international standards. While a certain increase across the various successive reactor generations built is evident from the data (IAEA, 2012), particularly for the later P'4 and especially the N4 reactor types, construction times within the entire programme remain quite remarkable. The mean construction time is 76 months compared to a mean of 108 months in the U.S. reactor sample that Koomey and Hultman (2007) analysed. About half of the French reactors – 55 percent of reactors and 47 percent of total gross capacity added – have construction times of less than seventy-two months (six years); around three quarters – 76 percent of reactors, 70 percent of gross capacity added – have construction times of less than eighty-four months (seven years), which only a third of all U.S. reactors achieved.

When discussing the importance of standardisation in reactor designs as well as short construction times as key technical success factors, special reference needs to be made to ÉDF. From ÉDF's perspective (cf. Boiteux (2009: 411–12)), the success factors are due to: a) size of the order programme; b) standardisation (series effects); c) client engineering of the construction process (i.e., by ÉDF rather than supplier Framatome); and d) rigorous quality and costs control (that in the words of Boiteux "extends to the beefsteak," that is, even to the food offered in the canteens on the construction sites).

Boiteux's emphasis on "client engineering" echoes similar findings from earlier analysis of the economics of U.S. reactors. McCabe (1996) developed a statistical model explaining reactor construction costs by differentiating various learning effects between "principals" (the utility) and "agents" (the architect-engineer/construction firm). He found that learning declined with larger dispersion between principals and agents, and also when there was cost uncertainty (inherent in the U.S. contractual arrangements for compensating architect-engineers on a "cost-plus" basis). McCabe also found that in the United States the locus of learning shifted from agents to principals (utilities). From this perspective, ÉDF – by overcoming the principal-agent dichotomy and by having the institutional capacity with its thousands of well-trained engineers to engineer and manage construction projects as a client – can be considered key in explaining the success in short construction times and moderated cost inflation, at least for the first four order series of 900-MW and the first 1,300-MW reactor units. Conversely, the gradual erosion of ÉDF's determination to standardise (caving in to proposals of numerous design changes in the wake of the "frenchifying" of the Westinghouse design in the P'4 series, and above all to the new N4 reactor design pushed by the CEA), as well as the abrupt slowdown of the expansion programme after 1981, paved the way towards a gradual demise of the French success model, as borne out in lengthened construction times and ever higher cost escalation towards the end of the programme.

3 Costs of Scaling Up

The entire nonmilitary costs of the French PWR programme amounted to some 1.6 trillion (10^{12}) FF98 (constant 1998 French francs), which translates into 230 billion

Euros2008 or 330 billion US$2008 (but not considering any cost escalation after 1998). These cost data come from official French government documents as detailed in Grubler (2009), including especially Charpin, Dessus, and Pellat (2000) and Girard, Marignac, and Tassard (2000).

Total costs are split between 810 billion FF98 capital expenditures and 833 billion FF98 operating cost expenditures. The 833 billion FF98 operating cost expenditures are roughly equally split between 402 billion FF98 operation and maintenance costs and 431 billion FF98 fuel costs. Together these translate into total, average levelized systems costs per unit of generated electricity of some 0.2 FF98/kWh for the entire programme (Grubler, 2010). This is discussed further later in this chapter.

The 810 billion FF98 capital expenditures include:

- 480 billion FF98 reactor investment costs plus interest accrued during construction;
- 170 billion FF98 investments and provisions for end-of-fuel-cycle facilities (waste disposal and reactor decommissioning);
- roughly 160 billion FF98 cumulative (public sector, civilian) nuclear R&D expenditures, considered here as knowledge capital investments.

These capital expenditures do not, however, include investments in the fuel cycle facilities, whose amortisation and finance are reflected in the fuel costs accounted as operating expenditures. Bataille and Galley (1999) summarise those fuel cycle investments at about 169 billion FF98. About half of these investments relate to the fuel cycle (enrichment, reprocessing, etc.) capacity of the French PWR programme (some 85 billion FF98) with the remainder covered by foreign clients, presumably for foreign contractual use. Also excluded are expenditures related to the unsuccessful fast breeder reactor, Superphénix. Schneider presents French estimates (presumably in current francs) of some 65 billion FF total life cycle costs of the 1.2 GW fast breeder reactor (2009: 77).

Specific costs per unit of installed capacity can also be estimated with an uncertainty range as a function of minor uncertainties in construction costs reported across various official documents, inclusion or exclusion of R&D expenditure, and differences in capacity between 65.9 GWgross or 63.1 GWnet. Average specific costs range between 10,400 to 12,800 FF98/kW installed. In US$2008, these numbers translate into a range between 2,100 to 2,600 US$2008, with lower bound values in both cases being per kW gross capacity and excluding knowledge investments, and the upper bound values being per kW net capacity and including R&D. These numbers do not include any cost escalation after 1998. They also reflect the average costs during the entire programme, although as suggested by the lengthened construction times, actual costs trended upwards during that period.

3.1 Investment Costs over Time

Although the available data do not report investments costs per individual reactor, average annual construction costs over time and inferred reactor-specific construction costs can be derived. Details of the method are reported in Grubler (2009, 2010) and consist of combining a constrained vintage-structure optimisation model using as inputs total annual construction expenditures and typical construction expenditure

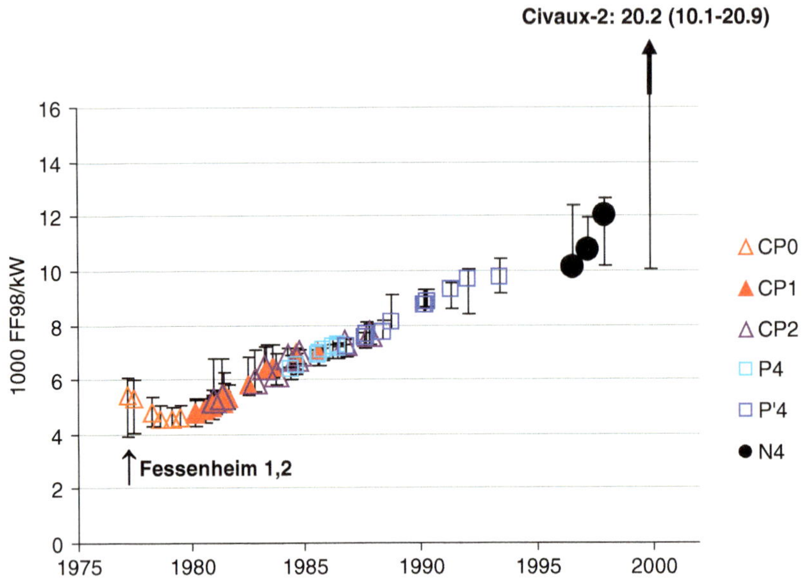

Figure 11.1. Specific investment costs of French PWRs over time and per reactor.
Notes: Costs shown on y-axis as 1,000 FF98 per kW, including both best guess model estimates (large symbols, differentiated by reactor generation) and estimate uncertainty ranges (error bars).
Source: Grubler (2010).

profiles over construction duration, as well as official construction duration data (IAEA, 2012) as constraints in the model calculations to infer average as well as reactor-specific construction costs. The results are summarised in Figure 11.1 giving "best guess" model estimates as well as estimate uncertainty ranges. Detailed numerical results are reported in Grubler, (2010).

Figure 11.1 illustrates the substantial real cost escalation of the French PWR programme. Even leaving aside the last four N4 reactors (which all exhibit significantly higher construction costs compared to all French reactors built earlier), real-term construction cost escalation across all reactors was significant as well as substantial: from some 4,000 to 10,000 FF98/kW, that is, by a factor of 2.5. Including the N4 reactors in the cost escalation calculation reveals an increase by a factor of close to three (but this still excludes the Civaux-2 reactor, which was extremely disappointing in terms of cost). This cost escalation, although significantly below that experienced in other countries, most notably the United States (cf. Koomey and Hultman, 2007), is far above what would be expected just from longer construction times. The reasons for cost escalation await further detailed research, but have been already alluded to: loss of the cost-dampening effects from standardisation, partly due to scaling up to 1,300 MW, but especially in the "frenchifying" of the tested Westinghouse design (as evidenced in the differences between the P4 and the P'4 reactor series); a certain "stretching" in the construction schedules after 1981 to maintain human and industrial intellectual capital during the significant scale back of the expansion programme as a result of built overcapacity; and above all, the unsuccessful attempt towards the end of the programme to introduce a radically new, entirely French design, the P4 reactor (the precursor to the current European Pressurised Reactor (EPR) design) that did not allow any learning spillovers in design or construction.

The reactor design changes undeniably improved certain safety features (Bataille and Birraux (2003), who compare the N4 with the EPR reactor; Thomas, (1988)). But that was never a prime motivation for the changes in design and is therefore unlikely to have been a significant factor in the cost escalation compared to the much more drastic and cost-consequential design changes aimed at improving reactor economics, higher domestic value added for the nuclear industry, and export market potentials. These endogenous non-safety drivers of design changes can be summarised simply: ever larger unit scales and more output (the interest of ÉDF), more French equipment and components (the interest of the nuclear equipment industry), and technological leadership (the interest of the CEA). In the view of the author, these endogenous drivers and the radical design changes they caused need to be analysed as primary causes for the significant real cost escalation, with the influence of improved safety features likely to be small.

3.2 Operating Costs

Contrary to the substantial cost escalation in construction costs observed in the French nuclear scale-up, operating costs have remained remarkably flat. Since 1984, operating costs have averaged some 0.13 FF98 (or 13 centimes, cFF98) per kWh produced. To put this number into perspective: operating costs equal some 18 euro2008 per MWh produced, or some 30 US$2008/MWh. This stability in operating costs, whilst not suggesting any positive learning effects in terms of cost reductions, are nonetheless remarkable considering the increasing need to operate the reactors in load-following mode (and resulting lowered load factors) in a country where base-load nuclear supplies well above three-quarters of all electricity generated.

3.3 Levelized Total Costs

It is instructive to integrate construction (capital) and operating costs over time, for example through a customary levelized costs calculation (Figure 11.2). As noted earlier, total capital expenditure includes investment costs as well as decommissioning capital provision. Figure 11.2 also shows the main components of total levelized costs: the net present value of the cumulative capital and operation and maintenance costs (always until the year reported and discounted with a 5 percent discount rate). Their sum, divided by the (discounted) cumulative electricity generated yields the average, total levelized costs of nuclear electricity (in FF98/kWh) reported in Figure 11.2. Also shown are sensitivity analyses with a zero percent and 10 percent discount rate respectively on top of the 5 percent used in the reference levelized costs calculation. The declining levelized costs should not be interpreted as evidence of some sort of learning phenomenon, being instead simply the result of the mathematical formulation of levelized costs that integrate progressively shifting shares of capital and operating costs as the nuclear scale-up

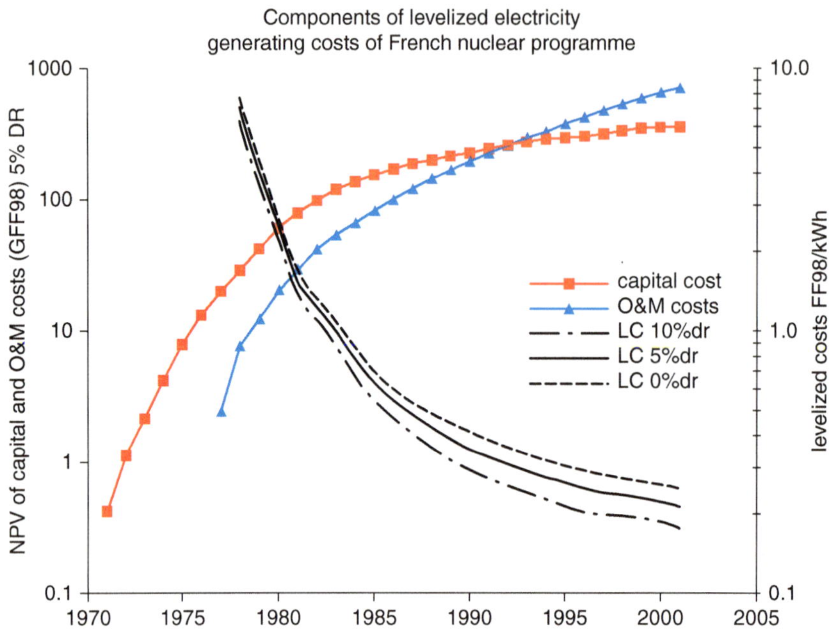

Figure 11.2. Capital and operating cost components of total average levelized costs under different discount rates.
Notes: Left y-axis shows net present value (NPV) of capital costs (construction costs plus decommissioning capital) and operation costs (in billion FF98) until year reported and discounted at 5 percent); right y-axis shows total, average levelized costs (in FF98/kWh) of French nuclear electricity at 5 percent, zero percent, and 10 percent discount rates (dr).

programme matures. And yet it allows the opposing trends between significantly escalating construction costs and stable, low operating costs to be put into perspective. With increasing programme completion, the escalating construction cost trends are increasingly less important determinants in total levelized costs, which increasingly become dominated by operating costs. Total levelized costs converge to an average value of 0.2 FF98/kWh (equivalent to 0.03 US$98 or 30 mills per kWh nuclear electricity generated), which represents an attractive economic prospect for base-load electricity generation. This remains the case even when considering the price escalation beyond 1998, which is not included in these numbers.

4 Discussion

4.1 Successful Scaling Up

The ambitious French PWR expansion programme can be considered the most successful scaling up of a complex, large-scale technology in the recent history of industrialised countries. The success in terms of rapidity of scale-up, comparatively short construction times, and pervasiveness of diffusion (close to 80 percent nuclear electricity, full fuel cycle industry) arises from a unique French institutional setting that allowed centralised decision making, regulatory stability, dedicated efforts for standardised reactor designs (which could long profit from knowledge spillovers via the Westinghouse licence), and a powerful nationalised utility, ÉDF, whose substantial in-house engineering resources enabled it to act as principal *and* agent of reactor construction simultaneously.

4.2 Cost Escalations and Prospects for Nuclear

However, the economic assessment of this scale-up yields a more differentiated picture. Despite a most favourable setting, the French PWR programme exhibited substantial real cost escalation. Specific investment costs increased by a factor of 2.5. While this increase is substantially lower than in other countries (most notably the United States), it nonetheless raises a number of fundamental issues worth considering with respect to the potential role of nuclear in a climate-constrained world.

First, while the nuclear industry is often quick to point at public opposition and regulatory uncertainty as reasons for real cost escalation, it may be more productive to start asking whether these trends are not intrinsic to the very nature of the technology itself: large-scale, lumpy investments reliant on a formidable ability to manage complexity in both construction and operation. These intrinsic characteristics of the technology limit essentially all classical mechanisms of cost improvements – standardisation, large series, and a large number of quasi-identical experiences that can lead to technological learning and ultimate cost reductions – except one: increases in unit size, that is, economies of scale. In the history of steam electricity generation, these indeed led initially to substantial cost reductions, but after the 1960s that option has failed invariably because of the corresponding increases in technological complexity.

Second, whilst reactors' real construction costs increased steadily, their operating costs remained low and flat in France, as well as for many reactors elsewhere. As a result, when looking at total levelized costs for the programme as a whole, the significant construction cost escalation becomes less and less significant once the reactor expansion programme is brought to a successful completion and continues to run at low operating costs. The implication is that nuclear's "valley of death" lies between commercial application and widespread diffusion given its inherently high investment costs and their tendency to rise beyond economically viable levels. Perhaps new institutional configurations that separate centralised reactor construction from decentralised operation should be explored, if indeed a nuclear expansion is deemed in the public interest to respond to climate concerns. ÉDF's success in combining principal and agent in the construction process could be at the core of such considerations. This logic may also suggest that competitive nuclear power is unlikely to be achieved in a private free market, which instead tends to produce the rapid innovations that now competitively challenge nuclear power.

Third, this innovation history of the French PWR programme provides valuable lessons for energy technology and climate policy analysts. Cost projections of novel technologies are an inherent element in any climate change policy analysis. This case study has reconfirmed Koomey and Hultman's (2007) conclusion that projections of the future need to be grounded much more firmly within the historical

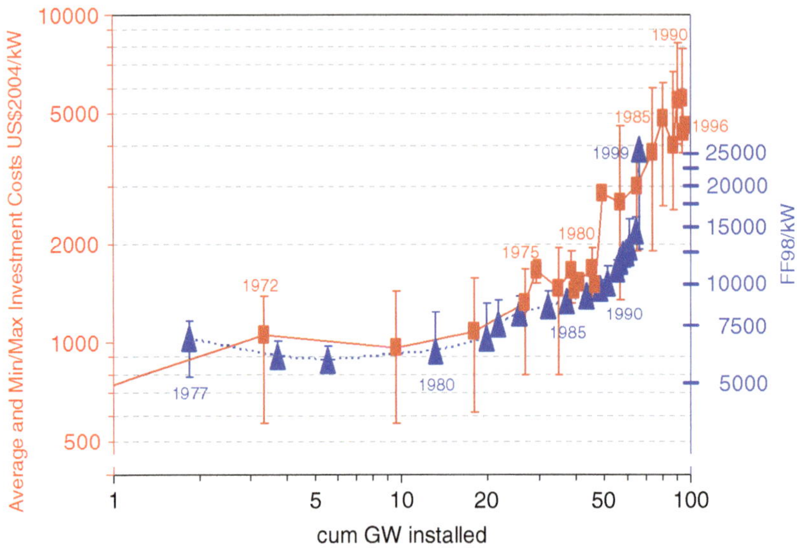

Figure 11.3. Reactor construction costs per year of completion date for the United States and France as a function of cumulative installed capacity.
Notes: Reactor costs include averages and min/max range. In both the United States and France, nuclear exhibits a pattern of "negative" learning despite a radically different institutional environment (centralised and favourable in France, and fragmented and less favourable in the United States).
Sources: U.S. data from Koomey and Hultman (2007); France data from Grubler (2009).

observational space, requiring much more careful arguments and logic in scenario design and model runs before suggesting "robust" or "optimal" climate stabilisation pathways. Again, agreeing with Koomey and Hultman (2007), detailed justification needs to be provided if assumptions differ radically from historical experience. For example, Kouvaritakis and colleagues (2000) assume substantial cost declines along a learning curve for nuclear reactors. This is counterfactual to even the most successful nuclear scale-up, and results in biased scenario modelling as well as bad policy advice.

These findings also suggest a need for in-depth sensitivity analysis across a much wider range of technological cost uncertainties. Perhaps climate policy analysis could begin by embracing in sensitivity analyses the engineering rule of thumb that large-scale, complex infrastructure construction projects tend to cost three times the original estimate. Nuclear is not the only example of a large-scale, complex technology that might be subject to this engineering rule: coal-based integrated gasification combined cycles with carbon capture and sequestration (or very large-scale solar plants in desert areas) could be other prime candidates.

4.3 Negative Learning

Lastly, the French nuclear case has also demonstrated the limits of the learning paradigm: the assumption that costs invariably decrease with accumulated technology deployment. The French example serves as a useful reminder of the limits of the generalisability of simplistic learning/experience curve models. Not only do nuclear reactors across all countries with significant programmes invariably exhibit negative learning, that is, cost *increase* rather than decline, but the pattern is also quite variable, defying approximations by simple learning-curve models, as shown in Figure 11.3 comparing the French and U.S. experiences.

In symmetry to the often evoked "learning by doing" phenomenon, there appears not only to be "forgetting by not doing" (Rosegger, 1991) but also "forgetting by doing," suggesting that technology learning possibilities are not only structured by the actors and institutional settings involved, but are also fundamental characteristics of technologies themselves. In the case of nuclear, a theoretical framework explaining this negative learning was discussed by Lovins, who referred to the underlying model as the Bupp-Derian-Komanoff-Taylor hypothesis (1986: 17–21). In essence, the model suggests that with increasing application ("doing"), the complexity of the technology inevitably increases, leading to inherent cost escalation trends that limit or reverse "learning" (cost reduction) possibilities. In other words, technology scale-up can lead to an inevitable increase in systems complexity (in the case of nuclear full fuel cycle management, load-following operation mode, and increasing safety standards as operation experience and unanticipated problems are accumulating) that translates into real cost escalation, or "negative learning" in the terminology of learning/experience curve models. Note that this is quite different from the examples of "negative learning" discussed in the traditional management literature (e.g., the case of the Lockheed Tristar aircraft referred to by Argote and Epple, 1990) where cost escalations arise from erratic (roller coaster) production scale-ups leading to organisational "forgetting-by-not-doing" (Rosegger, 1991).

The potential role of nuclear in a climate mitigation technology portfolio cannot be assessed seriously if the lessons from its most successful and intensive deployment, in France, are ignored.

5 Postscript: Ex Post Model Validation, A Comparison to the 2012 *Cour de Comptes* (CdC) Report

After the completion of the analysis in this case study, the Auditor General for the French government, the *Cour de Comptes* (CdC), released a report that presented for the first time official reactor specific cost estimates (CdC, 2012). This postscript appraises the CdC numbers and compares them to the modelled reactor specific cost estimates in this case study as an *ex post* model validation exercise.

Two tables on pages 22–23 are at the core of the CdC report. These tables present total and specific (per MW installed capacity) construction costs of 29 reactor pairs still in operation today. A critical appraisal of the data shows that a number of specific accounting conventions and non-standard cost definitions result in a significant underestimation of construction costs. Four major caveats are in order.

First, the CdC cost numbers are reported in constant 2010 Euro. The original cost numbers are converted to 2010 Euros using the French GDP deflator (CdC, 2012: 302) and the official French Franc to Euro exchange rate (6.55975 FF/Euro). The cost numbers expressed in 2010 terms are misleading as they largely refer to reactor construction costs from the 1970s and 1980s. As such, the CdC cost data are not representative and should not be interpreted as indicators of nuclear construction costs in the year 2010.

Second, reported construction costs totalling some 73 billion Euro2010 for the 29 reactor pairs (CdC, 2012: 24) exclude construction-related engineering and labour costs and pre-operating charges of some 10 billion Euro2010. Consequently they are not indicative of overnight-construction costs. (It is a whole other issue as to whether the concept of overnight-construction costs has any meaning in the CdC's ex post audit of a reactor construction programme in which construction durations ranged from 58 to 144 months).

Third, the CdC's decision to aggregate cost estimates of the 58 individual reactors built to a reduced sample size of 29 reactor pairs can arguably be considered manipulative as it masks (reduces) heterogeneity in reactor costs and so the resulting estimates of cost escalation trends.

Fourth, the CdC report arbitrarily assumes a total of 13 billion Euro2010 (CdC, 2012: 30) as interest accrued during construction. By comparison, the French national utility, EdF, estimates its interest accrued during construction to total 23 billion Euro2010 (CdC, 2012: 30, footnote 10).

Taken together, these points reveal the detailed cost numbers reported by CdC to be, at best, an absolute minimum of the actual costs of the French construction programme which are more comprehensively estimated in this case study. Taking the total construction costs for the entire PWR reactor fleet as an example, the CdC cost estimates imply a grand total of 73 billion Euros2010, compared to overnight construction costs of 83 billion, and total construction costs of 96 billion or 106 billion when including the accrued interest assumed by the CdC or by EDF respectively. By comparison, previous official French government cost estimates for the total cost of PWR construction ranged from 86 to 89 billion Euro2010 (Bataille & Galley 1999, Girard et al. 2000). The cost model set out in this case study estimated total costs of 81 to 96 billion Euro2010, with a best-guess estimate of 89 billion Euro2010 that is perfectly in line with the CdC programme cost figures.

Concerning reactor-specific cost estimates, the published CdC numbers (CdC 2012: 22–23) are the lowest and least plausible, and so here are adjusted accordingly. Total additional costs beyond the 73 billion Euro2010 reported by CdC have been allocated to individual reactor pairs in proportion to their share of total fleet construction costs for the additional 10 billion Euro2010 of engineering and labour costs, and in proportion to their share of construction costs multiplied by the corresponding reactors' construction time (in months) for the additional 13 to 23 billion Euro2010 of accrued interest during construction. Figure 11.4 compares the four cost estimates of the CdC with the three cost estimates of Grubler as presented in this case study and in Grubler (2010). The critical importance of comprehensive cost accounting as well as the significant cost escalation over time becomes apparent from both data sources.

Some analysts, including Rangel and Leveque (2012), have compared the lowest CdC numbers (73 billion) to the best-guess model estimates 89 billion reported in Grubler (2010), and reached the pre-mature conclusion of a significant overestimation of costs and resulting cost escalation of the Grubler costing model.

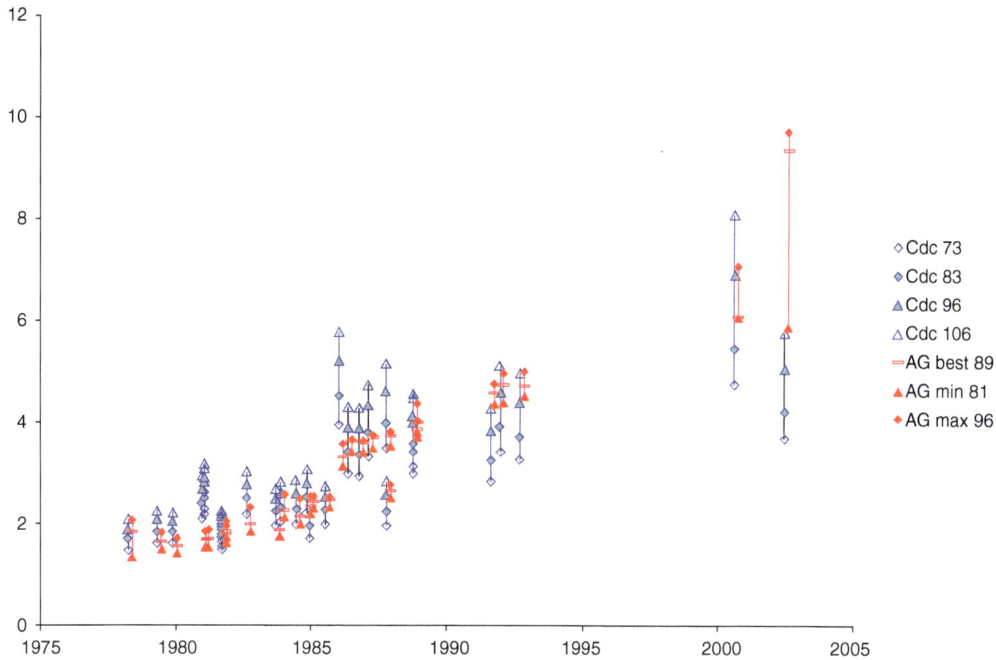

Figure 11.4. Specific construction costs in Euro2010/W of 29 reactor pairs based on the cost data presented in the CdC report versus best guess and min/max values of the original Grubler costing model versus year of reactor completion.
Notes: All costs have been adjusted to a common denominator of Euro2010 and a consistent cost accounting framework. Comparable costing system boundaries and estimates of the two references are marked with full symbols.

However, this conclusion is not supported by an apples-to-apples comparison of the two references based on comparable data sets that yield identical total programme construction costs as shown in Figure 11.5. Such a comparison in fact shows good agreement. The Grubler costing model explains between 67% to 79% of the cost variance in the total construction costs of the 29 reactor pairs presented in the CdC report. The cost escalation estimates can also be brought into a reasonable range of agreement. The ratio of cheapest to most expensive reactor pairs in the CdC costs range from a factor 2 up to a factor of 2.2 to 2.4, noting that the pairwise aggregation of the CdC data masks higher cost variance and hence cost escalation. (The low end of the range uses the CdC's original estimate excluding construction engineering and labour costs, and considers the Tricastin 3,4 versus the Chooz 1,2 reactor costs as reported in CdC 2012: 22–23; the high end of the range uses the CdC costs adjusted to include construction engineering and labour, plus the 13 to 23 billion Euro2010 accrued interest during construction). This cost escalation estimated from the specific CdC reactor costs is in good agreement with the original conclusion in this case study of a "minimum cost escalation estimate for the French PWR program of a factor 2.4 with an uncertainty range of a factor between 2.2 and 2.6" (Grubler, 2010: 5182).

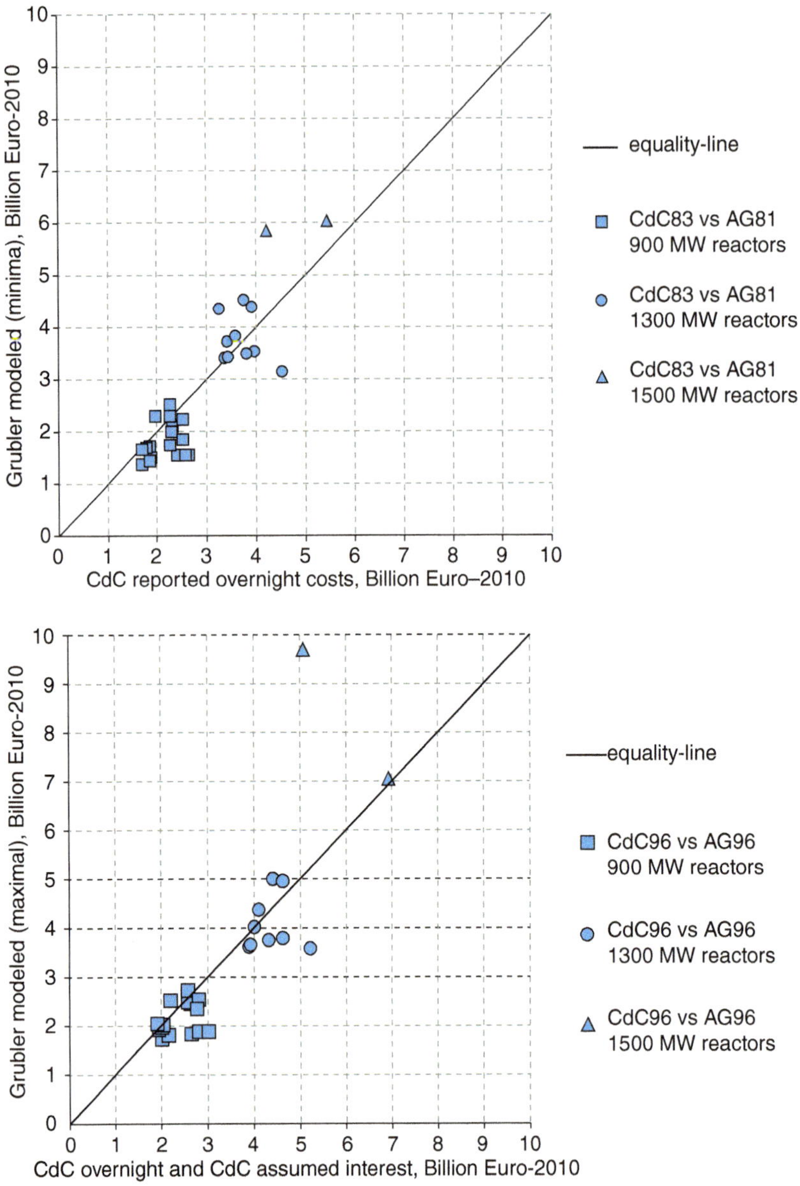

Figure 11.5. Total reactor construction costs (in billion Euro2010) for 29 reactor pairs for two comparable cost accounting frameworks from CdC (2012) versus Grubler (2010).
Notes: Top panel shows CdC overnight costs versus Grubler costing model minima (with total programme cost of 83 and 81 billion Euro2010 respectively); bottom panel shows CdC overnight costs plus CdC assumed interest versus Grubler costing model maxima (with total programme cost of 96 billion Euro2010 each). Correlation coefficients are 79% (top panel) and 67% (bottom panel) respectively.

FURTHER READING

Further details on the innovation histories of nuclear power in France can be found in Grubler (2009, 2010) and in the United States in Koomey and Hultman (2007). For additional empirical material and analysis of the French case, see Finon and Staropoli (2001), and for the French speakers, see also Boiteux (2009); Girard and colleagues (2000).

REFERENCES

Argote, L. & Epple, D., 1990. Learning Curves in Manufacturing. *Science*, 247(4945): 920–24.
Bataille, C. & Galley, R., 1999. *Rapport sur l'Aval du Cycle Nucléaire, Tome II: Les coûts de production de l'électricité*. Paris: Office Parlementaire d'Évaluation des Choix Scientifiques et Technologiques, Assemblée Nationale.
Bataille, M. M. & Birraux, C., 2003. *La Durée de Vie des Centrales Nucléaires et les Nouveaux Types de Réactuers.*, Paris: Office Parlementaire d'Évaluation des Choix Scientifiques et Technologiques, Assemblée Nationale.
Boiteux, M., 2009. Le Programme Électro-Nucléaire : EDF et ses Choix Industriels (testimony May 11, 2004). *État et Énergie: XIXe-XXe Siècle*. Paris: Comité pour l'Histoire Économique et Financière de la France, Ministère de l'Économie de l'Industrie et de l'Emplois CdC, 2012. *Thematic Public Report: The costs of the nuclear power sector*. Paris: Cour de Comptes (CdC). Available from: www.ccomptes.fr/content/download/43709/697228/version/2/file/thematic_public_report_costs_nuclear_+power_sector_012012.pdf.
Charpin, J.-M., Dessus, B., & Pellat, R., 2000. *Étude Économique Prospective de la Filière Électrique Nucléaire : Rapport au Premier Ministre*. Paris.
Finon, D. & Staropoli, C., 2001. Institutional and Technological Co-evaluation in the French Electronuclear Industry. *Industry and Innovation*, 8(2): 179–99.
Girard, P., Marignac, Y., & Tassard, J., 2000. *Le Parc Nucléaire Actuel*. Paris.
Grubler, A., 2009. *An Assessment of the Costs of the French Nuclear PWR Program 1970–2000*. Laxenburg, Austria: International institute for Applied Systems Analysis.
2010. The Costs of the French Nuclear Scale-Up: A Case of Negative Learning by Doing. *Energy Policy*, 38(9): 5174–88.
IAEA, 2012. The Database on Nuclear Power Reactors. International Atomic Energy Agency (IAEA) Power Reactor Information System (PRIS). http://www.iaea.org/programmes/a2/.
Jasper, J. M., 1992. Gods, Titans and Mortals: Patterns of State Involvement in Nuclear Development. *Energy Policy*, 20(7): 653–59.
Koomey, J. & Hultman, N. E., 2007. A Reactor-Level Analysis of Busbar Costs for US Nuclear Plants, 1970–2005. *Energy Policy*, 35(11): 5630–42.
Kouvaritakis, N., Soria, A., Isoard, S., & Thonet, C., 2000. Endogenous Learning in World Post-Kyoto Scenarios: Application of the POLES Model under Adaptive Expectations. *International Journal of Global Energy Issues*, 14(1): 222–48.
Lovins, A. B., 1986. The Origins of the Nuclear Fiasco. In Byrne, J. & Rich, D. (eds.) *The Politics of Energy Research and Development*. New Brunswick, NJ: Transaction Books, 7–34.
McCabe, M. J., 1996. Principals, Agents, and the Learning Curve: The Case of Steam-Electric Power Plant Design and Construction. *The Journal of Industrial Economics*, XLIV(4): 357–75.
Rangel, L.E., and Leveque, F., 2012. Revisiting the cost escalation curse of nuclear power: New lessons from the French experience. Working Paper 12-ME-08. Paris: CERNA, MINES ParisTech. Available from: http://www.cerna.ensmp.fr/images/CostEscalationPaper.pdf
Rosegger, G., 1991. Diffusion through Interfirm Cooperation, a Case Study. In Nakićenović, N. & Grubler, A. (eds.) *Diffusion of Technologies and Social Behavior*. Berlin, New York: Springer-Verlag.
Schneider, M., 2009. Fast Breeder Reactors in France. *Science & Global Security*, 17(1): 36–53.
Thomas, S. D., 1988. Chapter 8: France. *The Realities of Nuclear Power: International Economic and Regulatory Experience*. Cambridge University Press.

PART IV

ADOPTION AND USE IN THE ENERGY TECHNOLOGY INNOVATION SYSTEM

12 Technological Improvements in Solar Thermal Electricity in the United States and the Role of Public Policy

Gregory F. Nemet

Author's Summary

Solar thermal electricity (STE) technology has improved through a combination of learning by doing and R&D investments. Even though R&D levels for this technology were small relative to other energy technology programs, R&D played an important role in making full use of the knowledge gained through experience in manufacturing, installation, and operation of STE facilities. R&D helped codify and document the often tacit knowledge that accrued to operators through experience. Most of this knowledge about technical change in STE comes from the 350 MW (mostly parabolic trough systems) deployed in California during the 1980s. As a result of codification, this knowledge could be shared across firms and even across countries. It also preserved at least some of the value of the knowledge over time – especially during more than a decade of stagnation in the 1990s, when essentially no large STE plants were built worldwide. The rebirth of the STE industry during the 2000s indicates that at least some of this knowledge accumulated in the 1980s informs current designs and operation. Since the mid-2000s, a new round of installations has begun, primarily in Spain and the southwestern United States. These installations encompass three types of STE systems: troughs, concentrators, and dishes. By the end of 2011, worldwide STE capacity exceeded 1.5 GW, and additional GW were close to breaking ground. Aggressive renewable electricity obligations and feed-in tariffs have driven this resurgence in STE installations in both California and Spain. Complementary policies, such as loan guarantees and tax credits, have stimulated investment as well. A distinguishing feature of STE, relative to other renewables, is that STE requires big, risky investments. Early investors may have to "eat between one and three US$200 million plants" before improvements enable profitable operation. A policy implication is that the scale of technology requires different types of incentives from smaller-scale renewables such as solar PV.

Editors' Guide

Relationship of Case Study to Energy Technology Innovation System Framework

Central to the history of solar thermal electricity (STE) are the interactions between two innovation processes relating to knowledge: the *generation* of knowledge through R&D, and the *learning* processes associated with deployment experience. Both processes were essential: interactions and *feedbacks* between them explained the substantial decline over time in operating costs of STE systems. More *resources mobilized* in the form of R&D investments could not have substituted for the applied experience or learning by doing associated with installing and operating STE facilities in market settings.

This interrelationship between R&D and learning is mirrored in the importance of both *technology-push* and *market-pull* drivers of STE developments. The outcomes of early R&D investments and learning processes came to fruition in early markets formed by a combination of investment subsidies, tax credits, and guaranteed returns (through power purchase contracts), which were all necessary to attract investors.

An important and distinctive characteristic of the R&D investments in STE was the *codification* of knowledge generated. Cost improvement efforts and their outcomes were formally documented and made publicly available. During the initial growth years, this codified knowledge facilitated the exchange and *spillover* of knowledge from pioneer firms in the Californian market to later adopters in the United States and abroad. The technology testing facility constructed with public resources also proved a key *institution* in the innovation system for *exchanging* knowledge. Critically, codification also helped sustain the knowledge base during the subsequent lean years so that STEs were positioned to take advantage of the recent renewed interest and market incentives. Codification thus worked to offset the *depreciation* of knowledge following the collapse of the dominant market actor and the moving on of its staff.

Insights for Energy Technology Innovation Policy

- The relatively large unit scale of STE technology required large, lumpy (nondivisible) upfront investments. Firms knew they would need to absorb losses on the first few plants – a classic "valley of death" problem in the innovation life cycle between successful demonstration and successful diffusion. In the STE case, this was overcome by the financial sector interest in alternative energy in the early 1980s (following the 1970s oil shocks). But early plants needed both guaranteed tariffs (market-pull) and capital cost subsidies (technology-push) to mobilize these financial investments.
- Compared to granular, modular, distributed technologies that are amenable to policy encouraging experimentation and diversity, novel technologies like STE that need scale economies to lower costs pose major financial and innovation

risks and require concerted action throughout the innovation system to lower costs, improve performance, and underwrite market demand.

- Technically, the 1980s California (SEGS) plants have been successful. They steadily improved in performance and cost, are still providing power two decades after they were installed, and provide the basis for newly designed plants in Spain and California. The failure of the company that built them was due to falling energy prices and consequent changes in policy rather than technical problems. This illustrates the substantial market risks associated with policy intermittency, particularly when a legacy of failure can hang over the entire industry.
- The codification of knowledge in the form of publicly available, nonproprietary documentation enables knowledge to be shared and exchanged as a technology matures and diffuses. Codification is also essential to preserve knowledge learned through market downturns and periods of industry turnover and shakeouts. Codification should be made a quid pro quo of public R&D investment or market formation support via subsidies, feed-in tariffs, or tax credits.

1 Introduction

1.1 STE Technologies

Solar thermal electricity (STE) technologies use mirrors to concentrate solar radiation onto an absorber that converts the sunlight to heat and then transfers that heat to a working fluid used to power a Rankine or Brayton cycle to generate electricity. In some cases Stirling engines are used instead. Three prominent designs are commercially available:

1. *Parabolic trough systems* heat a working fluid in a pipe running parallel to the orientation of the troughs. Solar radiation heats the working fluid, typically oil, to approximately 400°C. Heat exchangers produce superheated steam that runs a Rankine cycle turbine.
2. In *tower systems*, an array of mirrors, called heliostats, concentrates sunlight on the working fluid at a single point, elevated on a tower. Temperatures are substantially higher, close to 600°C. The working fluid, typically a molten nitrate salt, can be run through a heat exchanger to power a Rankine cycle, or can be stored for later use. An important characteristic of tower systems is the possibility of storing heat for up to twelve hours. Storage is enabled by the centralization of heat at a single point, minimizing losses in pipes, as well as by the higher initial temperatures, which allow the use of working fluids that retain heat well.
3. Typically on a smaller scale (10s of kW), *parabolic dish systems* use a mirrored dish to concentrate sunlight on a focal point to run a Stirling engine. Temperatures reach 750°C.

An important difference with photovoltaic or PV-based solar power is that these concentrating STE systems require direct sunlight – cloudless skies and low humidity – because of the need to focus solar rays. This constraint limits their use to dry climates. In contrast, a fourth type of STE design, *solar updraft towers*, uses a skirt and chimney system to capture the energy from rising heat. Projects have been proposed on the scale of 100s of MW in China, the United States, and Spain, but to date only a few pilot plants at sub-MW scale have been built.

1.2 Knowledge Generation and Technological Improvements in STE

Most of our historical knowledge about technical change in STE comes from the 350 MW of plants (mostly parabolic trough systems) deployed in California in the 1980s. Since the mid-2000s, a new round of installations has begun, encompassing all three types of STE systems, mainly in Spain and in the southwestern United States. By the end of 2011, 1.5 GW had been installed and multiple GWs of project proposals were at various stages of development.

Solar thermal electricity (STE) technology has improved through a combination of learning by doing and research and development (R&D) investments. Even though R&D levels for STE have been small relative to those of other energy technology programs, R&D programs have nevertheless played an important role in making full use of the knowledge gained through experience in manufacturing, installation, and operation of STE facilities. R&D helped codify and document what was often "tacit" knowledge that accrued to operators through experience. As a result this knowledge could be shared across firms and even across countries. Codification also preserved at least some of the value of the knowledge over time, especially during more than a decade of stagnation in the 1990s when essentially no large plants were built worldwide. The rebirth of the STE industry during the 2000s indicates that at least some of the knowledge accumulated in the 1980s informs current designs and operation of STE plants.

2 Public Policies Relevant to Solar Thermal Electricity

Public policy has played a crucial role in stimulating the deployment of both the STE systems built in California in the 1980s and those built in the southwestern United States and Spain in the 2000s. The Californian plants in the 1980s were built primarily in response to two key policies. The first was the Public Utility Regulatory Policy Act (PURPA) of 1978, which mandated that electric utilities purchase power from "qualifying facilities," small independent providers of wholesale power. The second took the form of generous procurement contracts that the California Public Utilities Commission ruled utilities had to offer to the qualifying facilities (Price and Carpenter, 1999). One set of contracts that were available from 1983–86 were known as "Standard Offer Contract #4," which guaranteed prices for ten years at rates equivalent to $140/MWh in present-day dollars (Nemet, 2009). A third set of policies supporting STE deployment included state and federal tax credits that offset a portion of the plants' capital costs. Investments were further incentivized by the expectations of most analysts in the first half of the 1980s for high and rising electricity prices over the next decade. This included official government forecasts.

An important implementation detail of the Californian policy regime in the early 1980s was the requirement, under PURPA, that qualifying facilities not exceed thirty megawatts (MW). The rationale for this limit was to encourage a diverse set of small-scale generation technologies, whose scale had attributes that made them unique and attractive alternatives to the much larger scales used by conventional fossil fuel and nuclear generation. This limit proved awkward for STE technologies, which were initially built at the limit size, but needed to be substantially larger to reduce costs further. The size limit was temporarily raised to 80MW in 1987, which led to larger plants of exactly that scale. However, the legislation included substantial uncertainty and regulator discretion as to the future size limits. This uncertainty meant designs for much larger plants were never developed into serious commercial proposals.

The first STE pilot plant was a 10MW central receiver built in 1982 called Solar One. It had a capacity factor of less than 10 percent and was dismantled in the early 1990s (Mead and Denning, 1991). The commercialization of STE in the 1980s centered on a single company, Luz. It raised $1.2 billion from Wall Street banks to build nine parabolic trough STE plants, called Solar Electric Generating Stations (SEGS), between 1984 and 1991 (Philibert, 2004). The company went bankrupt in 1991 shortly before its ten-year procurement contracts (Standard Offer Contracts #4) were to expire. The new rates the company would have received in the early 1990s would have been much lower than those contracts and lower than the high rates anticipated in the early 1980s. Subsidies for new projects had ended by then as well. Afterward, the plants were sold to other investors and Luz employees dispersed, with many going to NASA to work on the space shuttle program. With them went much of the industry's experience and tacit knowledge about what configurations and operation activities worked best. Probably the best example of a compensating mechanism for this loss of knowledge was the extensive codification of testing and evaluation results conducted by Sandia National Laboratory and its contractors. This is discussed further later in this chapter. The SEGS plants themselves continued to produce electricity and improve, including the addition of natural gas use so that they could generate at night. They were still in operation as of mid-2012.

In the 2000s, two other types of policies drove deployment. In Spain, the renewable energy law guaranteed high prices for solar thermal electricity. In California, a Renewable Energy Portfolio Standard mandated that utilities source 20 percent of their retail sales from renewable energy by 2010 and 33 percent by 2020 (CPUC, 2007). Like in the early 1980s, expectations in the mid-2000s were that natural gas prices and so electricity would stay high and likely rise in future years. But natural gas prices fell dramatically in the last years of the decade, which provided one reason for much weaker support for STE deployment, including a substantial weakening of the STE subsidies in Spain.

3 Technological Improvements in Solar Thermal Electricity

The nine Solar Electricity Generation Station (SEGS) plants built in California in the 1980s became more efficient, cheaper to build, and substantially cheaper to operate and maintain over time. Taylor and colleagues (2007) identify three specific improvements that were most important in improving the STE plants:

1. *Mirrors*: also referred to as "heliostats," they are the largest component of capital cost and have fallen in cost over time (see Figure 12.1). Key improvements in heliostat development over time have included better optical transmission of sunlight to the receiver, lighter weight, and greater resistance to wind and storms

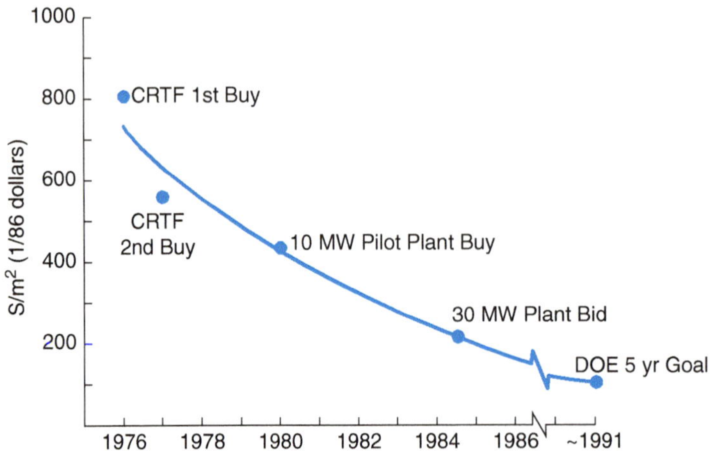

Figure 12.1. Cost of heliostats over time. Notes: CRTF = Central receiver test facility. Costs shown in 1986 US$.
Source: Grosskreutz (1996).

(Lotker, 1991). Reliable heliostats were the most successful outcome of the U.S. R&D program.

2. *Configuration*: arrangement of the solar arrays and transmission of working fluid so that sunlight absorption can be maximized while heating losses are minimized has improved the efficiency of electrical output (Kearney and Price, 2005). Better configuration was the result of learning by doing, including both reconfigurations within plants and redesigns of subsequent plants in the series of nine SEGS plants.
3. *Reheating of steam*: efficiency of the Rankine cycle has been improved by reheating the steam exiting the turbine using the working fluid. This was likely a spillover benefit from other thermal power plants, but may have been aided by R&D and learning by doing to transfer these applications to the lower temperatures in STE, rather than fossil combustion.

One can observe changes in subsequent designs for central receiver plants. For example, Figure 12.1 shows a factor of five decline in the costs of heliostats and other capital costs from the first Solar One plant in 1981 until the designs for a 200MW facility in 1987.

Changes are more dramatic in the solar trough plants. The full effect of these and other changes are summarized in Figure 12.2, which shows how eight factors interacted in reducing the costs of electricity from the SEGS plants. The arrows are meant to show the most important causal relationships. Other interactions certainly existed but are assumed to be minor. The three most important causal factors are learning by doing, R&D, and scale.

Learning by doing is captured by the cumulative amount of capacity constructed and the cumulative amount of electricity generated. Two important and observable technical improvements that are most attributable to learning by doing are the near elimination of component failures and the decrease in operation and maintenance (O&M) costs. Both of these reduced the levelized annual cost to operate the plant.

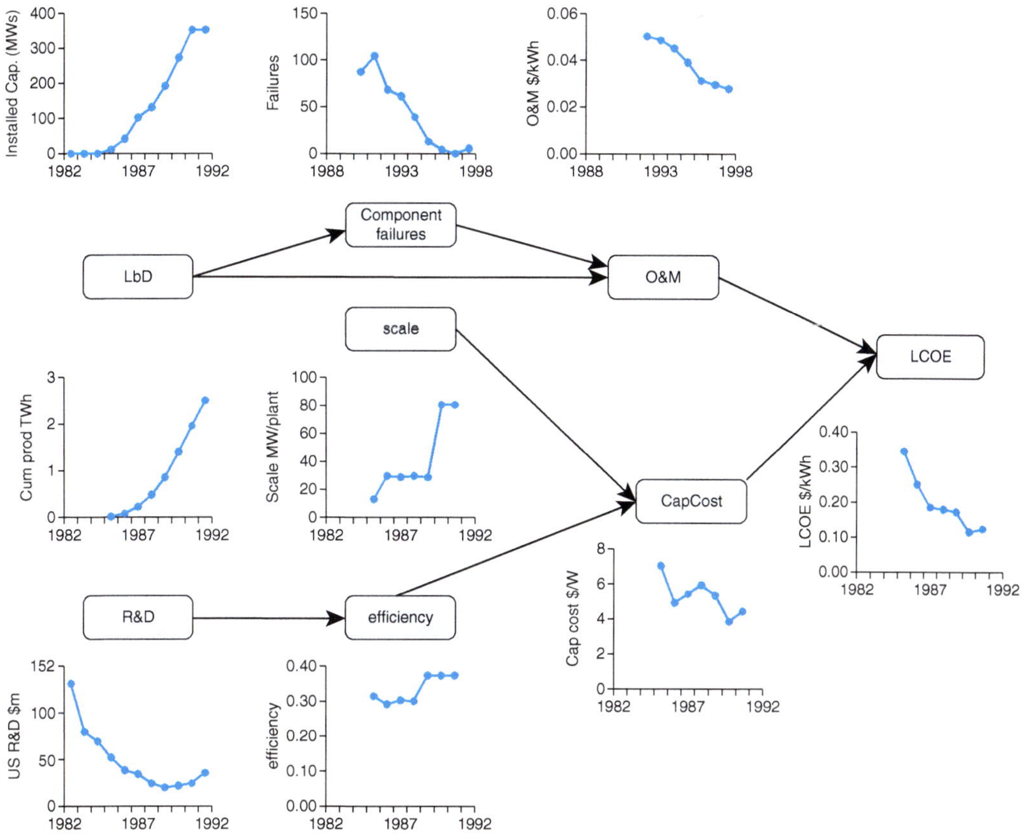

Figure 12.2. Indicators of technological change for STE plants in California.
Notes: LbD = learning by doing; O&M = operation and maintenance costs; LCOE = levelized costs of electricity production.

Federal R&D investment was highest at the beginning of the program and declined steadily over the life of the STE plants. These programs helped improve efficiency as well as capital cost.

Another important change was the up-scaling of the subsequent plants by nearly an order of magnitude in less than ten years. The pace of up-scaling was limited at least in part by policy that restricted the size of plants eligible for the generous procurement contracts.

An important observation from the data shown in Figure 12.2 is that technical improvement was not monotonic. Component failure rates increased in some new plants, as did capital costs, and even levelized cost of electricity. These short-term adverse outcomes may be attributable to design changes and experimentation.

One can also measure the amount of inventive and entrepreneurial activity associated with STE technology. This activity, as measured by counts of successful patent filings, corresponds well with the two periods of intense policy making and expectations of high energy prices. Figure 12.3 shows patents filed for STE technologies through 2005. One can see only occasional patents before a dramatic increase in the mid-1970s. A steady decline occurs after the mid-1980s with a new round of patenting activity in the mid-2000s. The most prolific STE patent filers were U.S. aerospace companies.

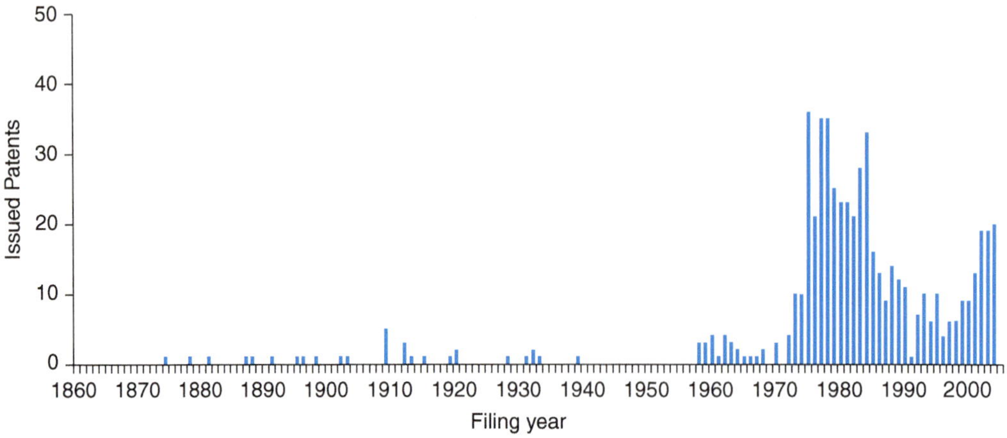

Figure 12.3. U.S. patents filed for STE technologies.
Source: Taylor and colleagues (2007).

3.1 Learning by Doing

Learning by doing was central to well-documented improvements in STE. The dominant STE firm of the 1980s, Luz, was successful in reducing the O&M costs on its plants by 40 percent (Lotker, 1991). Learning by doing at SEGS plans in California was particularly important to this improvement; it was also aided by cost-shared studies conducted by federal laboratories (Cohen, Kearney, and Kolb, 1999). Part of the performance improvements were also anecdotally attributed to having "someone looking over them," a reference to the Wall Street banks, which invested more than one billion dollars in California SEGS plants in the 1980s (Taylor et al., 2007).

3.2 R&D

The combination of PURPA and Standard Offer Contracts that guaranteed ten years of above-market prices drove one firm (Luz) to build nine commercial plants. All other STE installations until the 2000s were funded by R&D or were demonstrations (Taylor et al., 2007).

R&D investment had several specific outcomes on STE costs. Government R&D produced the heliostats, the mirrors that concentrate sunlight on focal points or axes. For example, Figure 12.1 shows the costs of heliostats over time. R&D studies also improved reliability by assessing the source of failures and damage. This led to cost savings. One example is the outcomes of R&D on central receivers. Table 12.1 lists the accomplishments of the federal R&D program on central receiver technologies (Grosskreutz, 1996). Note that R&D outcomes occur at multiple levels: improvements in the individual components, such as heliostats and receivers; experiments with larger systems, such as the use of molten salts; and finally, the construction of a testing facility, an aspect of R&D programs that has been important for other technology programs, such as wind power testing facilities at Riso, Denmark and Golden, Colorado.

Table 12.1. Outcomes of federal R&D on central receiver STE technology.

	Elements	Outcomes
Test facilities	Central Receiver Test Facility (CRTF)	Five MW_{th} facility built and successfully operated for eleven years at Sandia National Laboratories
Components / subsystems	Heliostats (glass / metal)	Three generations built and tested More than twenty-one hundred heliostats deployed
	Receivers	Three generations of water / steam receivers built and tested One sodium receiver built and tested Two generations of molten nitrate salt receivers built, tested, and operated
	Thermal storage	Oil / rock system built, tested, and operated Molten salt system built, tested, and operated
	Steam generators	Hot molten nitrate salt-to-steam system built, tested, and operated Hot sodium-to-steam heat exchanger built, tested, and operated (International Energy Agency, IEA)
Systems	Proof-of-concept experiments	0.5 MW_e sodium/steam experiment built and operated (Almeria, Spain) 750 kW_e molten salt experiment built and operated at the CRTF (Albuquerque, New Mexico)
	Pilot plant	Ten MWe water/steam plant built and operated (Barstow, California)

Source: Grosskreutz (1996).

R&D was also spent on constructing demonstration plants for central receiver stations and parabolic dish systems (the term RD&D captures demonstration as well as research and development activities). Neither plant was deployed beyond demonstration scale in the 1970–90s, although currently both of these designs are used for 100MW-scale projects.

As a second example of beneficial R&D outcomes, an important source of improvement came from a $6m 50/50 cost shared project between Sandia National Laboratory and Luz to find opportunities for cost reductions in the SEGS plants (Cohen et al., 1999). Part of the value of this work is that it was formally reported and publicly available. These aspects became crucial with the demise of Luz and the entire STE industry in 1991, as the industry lay dormant for more than a decade. As a result, the knowledge codified in these reports was not lost, as opposed to the tacit knowledge of plant operators that was almost all lost. As an example, the cost data shown in Figure 12.4 is derived from this codified knowledge.

Like other energy technologies, quality control problems were important barriers to the widespread commercialization of STE. To some extent, these issues were addressed. For example, see the declining rate of component failures in Figure 12.2. Still, quality control was an issue that technical fixes or policy remedy might have addressed. In the case of STE, quality control became an irrelevant concern with the dissolution of the industry in the 1980s. Some of the changes to improve reliability, specifically damage to heliostats from weather, were codified

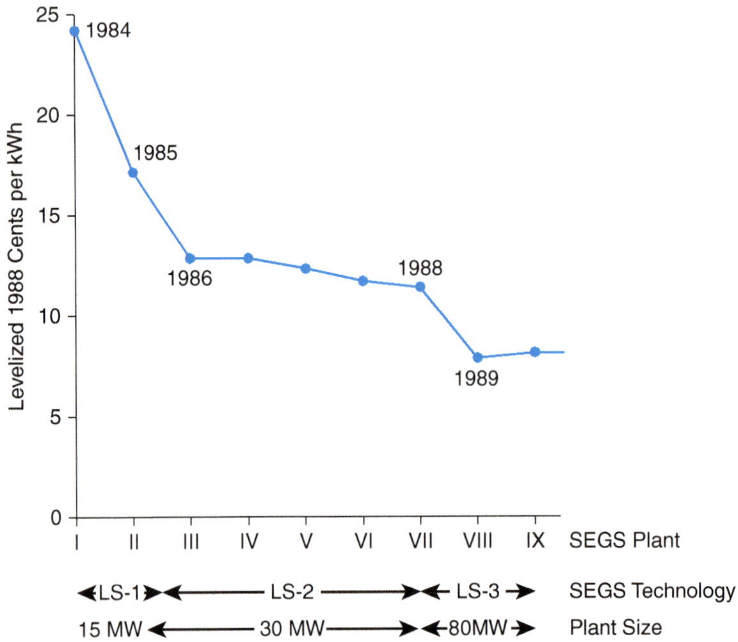

Figure 12.4. Documentation of cost reductions in STE plants.
Source: Lotker (1991).

in government reports and have the potential to inform subsequent installations. The performance of the new plants deployed in Spain and the southwestern United States over the past few years will show whether quality control continues to be a concern.

3.3 Scale

There are economies of scale in constructing larger plants partly due to: (1) relatively fixed costs for the turbine and generator units; (2) the ability to mass produce components, such as heliostats, and for the working fluid transmission system. As a result, larger STE plants are less expensive on a $/W or $/kWh basis (see Figure 12.4).

Scale itself has been important to cost reductions (see Figure 12.4). Larger scale combined with O&M improvement derived from learning by doing drove the levelized cost of STE energy down by a factor of four (Enermodal, 1999; Mariyappan and Anderson, 2001; Neij et al., 2006; Sargent-Lundy, 2003). And while the United States spent several hundred million dollars on STE R&D in the late 1970s and early 1980s, it is important to realize that the concept behind the Luz facilities originated in Israel, when an American engineer found financing to build an early pilot plant (Shinnar, 1993). A 15MW demonstration plant was subsequently constructed in California, followed by the 30MW and 80MW SEGS installations (Kearney and Price, 2005).

Plant-level scaling (15 to 30 to 80MW plants) was initially constrained by policies primarily focused on encouraging small-scale distributed generation. The upper limit of 30MW/plant was later raised to allow construction of the 80MW units. By 2012, the SEGS plants were still the largest ones operating, but five projects of more than 100MW were under construction and due to come online in 2013. These plants are intended to satisfy California's renewable portfolio standard.

4 Prospects for STE and Recent Experience in Spain and the United States

Two policy instruments have driven the resurgence in STE installations over the past ten years: California's aggressive renewables obligation and Spain's feed-in tariff for renewables. California's Renewable Portfolio Standard originally obligated utilities to generate 20 percent of their retail sales from renewables by 2020. In 2006 that 2020 target was increased to 33 percent and the original 20 percent target was met by 2011. Almost 3GW of new renewables had come online by 2012. More than 1GW of STE projects were under construction in California as of mid-2012. Spain's feed-in tariff began in 1998 but was substantially increased in 2007. At their peak, STE projects were receiving nearly 30 euro cents/kWh, guaranteed for twenty-five years. These tariffs have been reduced in the past five years, and in 2012 the feed-in tariff program was cancelled. Still, more than 500MW of STE capacity was installed and operational as of mid-2012.

An important implication for effective STE policy relates to a distinguishing feature of STE compared to other renewables. STE requires large investments approaching the magnitude of fossil plants. Early investors may therefore have to "eat between one and three $200m plants" before improvements enable profitable operation (Taylor et al., 2007). A policy implication is that the scale of technology requires different types of incentives from smaller-scale renewables such as PV. For example, the existence of capital constraints implies that STE depends more on instruments such as investment tax credits than on incentives for electricity production. While it has been essential to stimulate deployment, policy has also had adverse effects on STE: deregulation increased risk, which discouraged large-scale investments; early policies included limits on scale to encourage small producers, but these limits were detrimental to reducing STE costs.

Notwithstanding these issues, further cost improvements in STE remain possible. These include scale increases (e.g., from 100 to 200MW), the use of advanced technologies – such as molten salts as working fluids and reflective films instead of mirrors – and increases in capacity factor driven by improved O&M and by location decisions (Kutscher, 2009).

5 Summary and Conclusion

STE projects were widely viewed as failures in the 1990s. They were seen as expensive uses of public funds that benefited a few entrepreneurs and early investors and were simply unnecessary given low energy prices. But this "failure" of the California SEGS plants was due to expectations of falling energy prices rather than to technical failures. Indeed all nine plants themselves are still in operation. The reemergence of the technology in the 2000s is striking in how similar the designs are to those experimented with and deployed in the 1980s. At least some of the experience and new knowledge produced in the 1980s was not lost.

Several concluding observations can be made about the innovation histories of the STE plants built in the 1980s:

- STE technology appears to be one for which higher levels of R&D investment could not have substituted for deployment. Learning by doing was essential because of the systemic issues of operating these plants in ambient conditions for several years while producing electricity for real customers.
- The interaction between learning by doing and public R&D investment, even if small, was important for the substantial decline in operating costs observed from 1984–91.
- Part of the value of the R&D investment was the formal documentation of cost improvement efforts, which was publicly available and nonproprietary. These investments may benefit the more recent round of plant construction.
- Initial investments required were large and lumpy because of the need to produce plants at the scale of 10s of MW. Firms knew they would need to absorb losses on the first few plants in order to learn from experience that they could only glean at real commercial scale. This situation was a classic "valley of death" between demonstration and market application that, in this case, was overcome by a combination of generous subsidies and an alternative energy "bubble" on Wall Street in the early 1980s that supplied a large appetite for risk. Early plants needed guaranteed tariffs as well as capital cost subsidies.

FURTHER READING

The following references provide useful additional insights and detail into the innovation history of solar thermal electricity: Lotker (1991); Sargent-Lundy (2003); Shinnar (1993).

REFERENCES

Cohen, G. E., Kearney, D. W., & Kolb, G. J., 1999. *Final Report on the Operation and Maintenance Improvement Program for Concentrating Solar Power Plants.* Oak Ridge. TN: Sandia National Laboratories.

CPUC, 2007. *Progress of the California Renewable Portfolio Standard as Required by the Supplemental Report of the 2006 Budget Act.* San Francisco: California Public Utilities Commission.

Enermodal, 1999. *Cost Reduction Study for Solar Thermal Power Plants, Prepared by Enermodal Engineering Limited in Association with Marbek Resource Consultants Ltd.* Washington, DC: The World Bank.

Grosskreutz, J. C., 1996. High Temperature Technologies. In Larson, R. & West, R. E. (eds.) *Implementation of Solar Thermal Technology*. Cambridge, MA: MIT Press.

Kearney, D. & Price, H., 2005. *Recent Advances in Parabolic Trough Solar Power Plant Technology*. Boulder, CO: American Solar Energy Society (ASES).

Kutscher, C., 2009. Concentrating Solar Power. *Summer School on Global Sustainability*. Santa Fe, NM: Santa Fe Institute.

Lotker, M., 1991. *Barriers to Commercialization of Large-Scale Solar Electricity: Lessons Learned from the LUZ Experience*. Albuquerque, NM: Sandia National Laboratories.

Mariyappan, J. & Anderson, D., 2001. *Solar Thermal Thematic Review*. Washington, DC: The Global Environment Facility (GEF).

Mead, W. J. & Denning, M., 1991. Estimating Costs of Alternative Electric Power Sources for California. In Gilbert, R. (ed.) *Regulatory Choices: A Perspective on Developments in Energy Policy*. Berkeley: University of California Press.

Neij, L., Borup, M., Blesl, M., and Mayer-Spohn, O., 2006. Cost Development: An Analysis Based on Experience Curves. New Energy Externalities Development for Sustainability (NEEDS) Project.

Nemet, G. F., 2009. Demand-Pull, Technology-Push, and Government-Led Incentives for Non-Incremental Technical Change. *Research Policy*, 38(5): 700–709.

Philibert, C., 2004. *International Energy Technology Collaboration and Climate Change Mitigation Case Study 1: Concentrating Solar Power Technologies*. Paris: International Energy Agency.

Price, H. & Carpenter, S., 1999. *The Potential for Low-Cost Concentrating Solar Power Systems*. Oak Ridge, TN: National Renewable Energy Laboratory.

Sargent-Lundy, 2003. *Assessment of Parabolic Trough and Power Tower Solar Technology Cost and Performance Forecasts*. Golden, CO: National Renewable Energy Laboratory.

Shinnar, R., 1993. The Rise and Fall of Luz. *Chemtech*, 23: 50–53.

Taylor, M., Nemet, G., Colvin, M., Begley, L., Wadia, C., & Dillavou, T., 2007. *Government Actions and Innovation in Clean Energy Technologies: The Cases of Photovoltaic Cells, Solar Thermal Electric Power, and Solar Water Heating, CEC-500–2007–012*. Sacramento: California Energy Commission.

13 Automobile Fuel Efficiency Standards

Gregory F. Nemet

Author's Summary

Increasing fuel efficiency standards have affected innovation in automobiles by accelerating the adoption of new fuel, engine, and drive train technologies, and by shifting the use of improved efficiency between fuel consumption and performance.

In 1975, the U.S. government passed the Federal Automotive Fuel Efficiency Standards, which specified mandatory levels of miles per gallon of gasoline consumed for new vehicles averaged across each manufacturer's vehicle fleet (Greene, 1998). The required Corporate Average Fuel Economy (CAFE) standard steadily increased from 18 miles per gallon in 1978 to 27.5 miles per gallon in 1985. The standard was effective in meeting its goals – actual fuel efficiency has only ever fallen below the government requirement by small amounts for short periods. In fact, actual fuel efficiency has almost always exceeded the mandatory level. This overcompliance, combined with the simultaneous rise in the price of gasoline during the period of escalating standards, suggests that fuel prices have played a role in motivating efforts to improve fuel efficiency; change is not attributable to the CAFE standard alone (Davis, Diegel, and Boundy, 2008). The standard for passenger cars in the United States remained the same from 1985 to 2010, while actual efficiency has improved slightly in that period. Standards also exist in Europe, China, Japan, Australia, and Canada. All of these standards are more stringent than the U.S. efficiency requirements, by nearly 100 percent in the cases of Japan and the EU.

The past four decades have generated two broad technological responses to the combination of rising standards and rising oil prices: (1) changes in the use of improved engine efficiency – in particular, whether efficiency gains are used to reduce fuel consumption, increase vehicle weight, or increase acceleration; and (2) rapid adoption in certain periods of new technologies including fuel injection, front-wheel drive, and diesel engines.

Editors' Guide

Relationship of Case Study to Energy Technology Innovation System Framework

Unlike cost, the *performance* attributes of energy technologies – particularly end-use technologies – are multidimensional and often involve trade-offs as in the case of a vehicle's efficiency and its weight or acceleration. This case study shows how *performance* improvements are shaped by innovation responses to consumer preferences and market demand but also to *public policy* – in this case, fuel efficiency standards. Public policy also has a distinctive influence on the cost profile of vehicles and vehicle technologies over time. Policy acts directly on costs through subsidies and fuel prices, but also indirectly through *shared expectations* as to the direction and rates of required technological change. These in turn guide the *mobilization of resources* by other innovation system actors, which will have an impact on costs. For example, the use of labeling, disclosure requirements, information campaigns, and differential tax schemes to help form and develop markets for fuel-efficient vehicles shows how alternative *market-pull drivers* can complement more conventional regulatory approaches.

The broader point this case study illustrates is that different *actors in the innovation system* may have different interests that interact in determining the direction and outcome of the innovation process. These outcomes are therefore complex and contingent. The successful uptake of innovations, however, can be rapid and pervasive under appropriately aligned and supportive regulatory and market conditions. Diesel engines are an example here. But these concentrated periods of improving performance in a market context are founded on often lengthy R&D, testing, and market formation activities. This again reinforces the importance of a *systemic perspective* on energy technology innovation that includes both technology and market developments.

Insights for Energy Technology Innovation Policy

- Standards are an effective means of driving innovation and technological change. Commercial technologies can deploy quite rapidly under appropriate and aligned policy and market signals.
- End-user preferences for technology's performance attributes interact with policy objectives and implementation to shape innovation outcomes.
- Policies need to balance consistency with adaptability to changing market signals and behavior. Consistency does not simply mean static policy incentives or mandates. In fact, credible long-term signals (e.g., through dynamic standard settings as exemplified in the Japanese Top-Runner efficiency program) are key for creating shared expectations and incentivizing continuous innovation and technology improvements.
- Regulatory policies also benefit from allowing market actors some flexibility in their strategies to fulfill mandated outcomes.

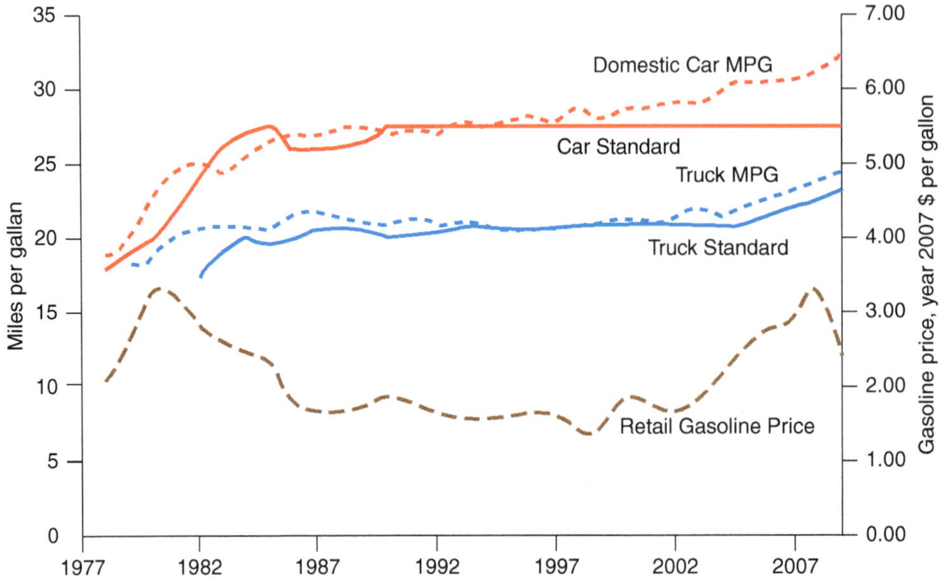

Figure 13.1. Gasoline prices, actual fuel efficiency, and mandated fuel efficiency (Anderson et al., 2010).

1 Fuel Efficiency in the United States: The CAFE Standard and Technological Improvements

Congress passed the Energy Policy and Conservation Act of 1975 in response to the Arab Oil Embargo in 1973 and in recognition of the United States' increased dependence on imported liquid fuels. The act added Title V, "Improving Automotive Efficiency," to the Motor Vehicle Information and Cost Savings Act (DOT, 2002). Title V established the mandatory corporate average fuel economy (CAFE) standards, which were based on the sales weighted average fuel economy, expressed in miles per gallon (mpg) of a manufacturer's fleet of passenger cars or light trucks. The original goal of the program was to double new car fuel economy by model year 1985. The National Highway Traffic Safety Administration (NHTSA) is responsible for establishing and amending the CAFE standards while the Environmental Protection Agency (EPA) is responsible for calculating the average fuel economy for each manufacturer. The standards for model years 1978, 1979, and 1980 were set at 18.0, 19.0, and 20.0 mpg, respectively. The Act required NHTSA to establish CAFE standards for model years 1981, 1982, 1983, and 1984 administratively and to specify the CAFE standards for model years 1985 and thereafter at 27.5 mpg (see Figure 13.1).

An important aspect of the CAFE standard is its strong enforcement capacity. The penalty for failing to meet CAFE standards is $55 per vehicle for every mile per gallon under the relevant target standard. NHTSA collected about $786 million in fines for CAFE noncompliance by the end of 2008.

The CAFE standard has affected the rate and direction of technological change in vehicles (NAS, 2002). End-use efficiency has improved almost continuously for the past thirty years. But the availability of more efficient energy conversion has been used to accomplish different ends during the period of rising fuel economy

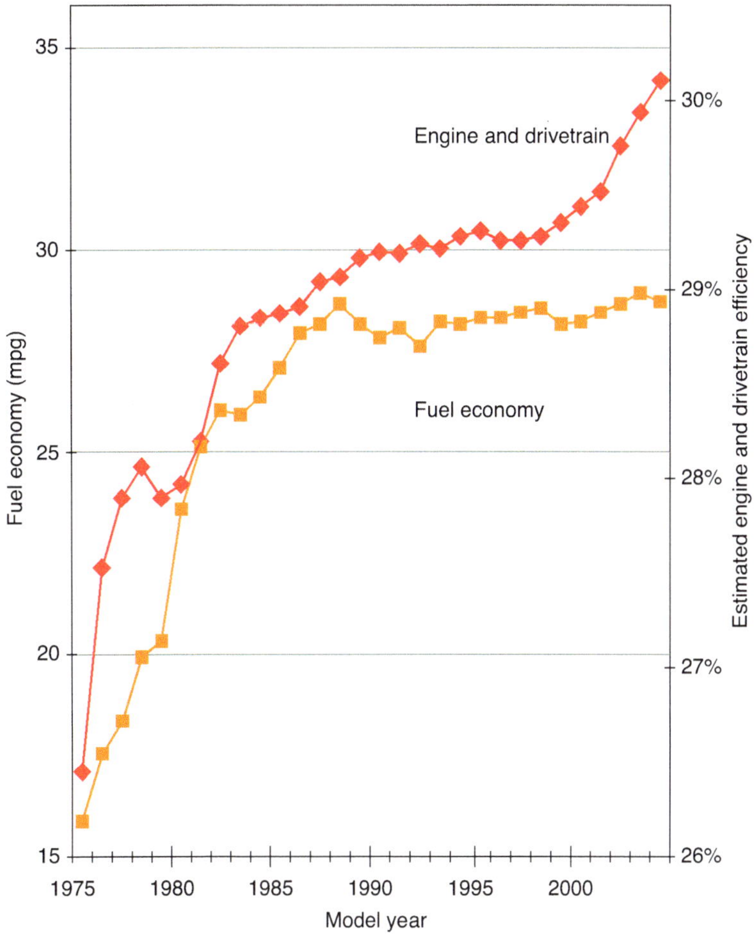

Figure 13.2. Fuel economy and efficiency of engine and drive train (Lutsey and Sperling, 2005).

standards (1975–85) compared to the period of static standards (1986–present). In the first period, efficiency improvement was directed toward improving miles traveled per gallon of gasoline consumed. After 1985, in the second period, almost all of the efficiency improvements were used to increase other vehicle attributes including acceleration, towing capacity, and size.

Energy conversion efficiency has improved in drive trains, engines, drag, and rolling resistance. As shown in Figure 13.2, drive train and engine energy conversion efficiency improved from about fifteen to forty-three ton-miles per gallon between 1975 and 2005. From 1975 to 1985, efficiency improved at the rate of 2 percent to 3 percent per year (Lutsey and Sperling, 2005). After 1985, efficiency improvement slowed to about 1 percent per year, although that rate has increased since 2000.

The continuity of this improvement – spanning the period of rising efficiency standards and then static standards – makes it difficult to attribute to CAFE. However, consideration of what end-use characteristics these efficiency gains were used for is revealing; from 1975 to 1985 vehicle weight dropped by about one-third and acceleration remained the same. After 1985, when CAFE standards stopped rising, efficiency improvements were used to power increasingly heavier vehicles,

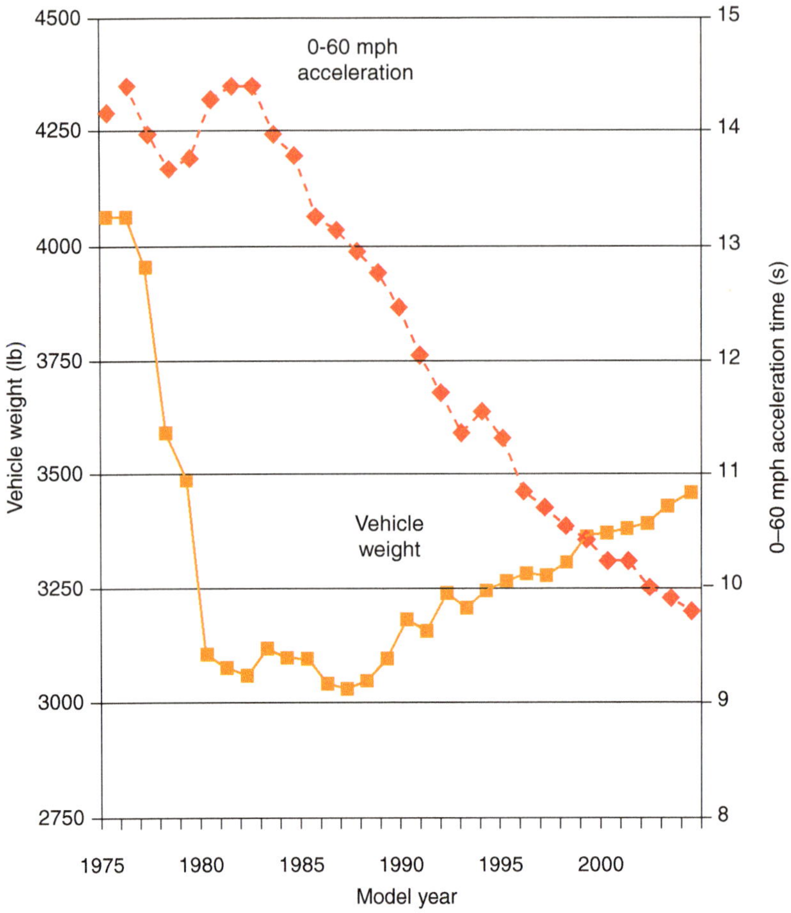

Figure 13.3. Vehicle weight and acceleration (Lutsey and Sperling, 2005).

which could accelerate considerably faster. These trends are clearly shown in Figure 13.3.

Although the use of lighter vehicles with stable acceleration characteristics accounts for about half of the miles per gallon improvement from 1975 to 1985, new technology accounts for the other half (Lutsey and Sperling, 2005). Technology adoption in passenger vehicles during and after the period of stringent CAFE standards also shows the effect on the direction of technological change. As shown in Figure 13.4, the adoption of fuel injection, front-wheel drive, and automatic transmissions with torque converter lock-up each went from installation in less than 10 percent of new vehicles in 1975 to 100 percent, 80 percent, and 60 percent respectively by 1985. Conversely, the adoption of air conditioning decreased during the same period (Figure 13.4). Seen over much longer time frames, this is a striking deviation from the long-term trend of technological change (Figure 13.5).

2 Fuel Efficiency in Europe: Voluntary Agreements and Regulation

Until the late 1990s, Europe did not have EU-wide fuel economy regulations. In 1998, the European Commission (EC) and the European automobile manufacturers

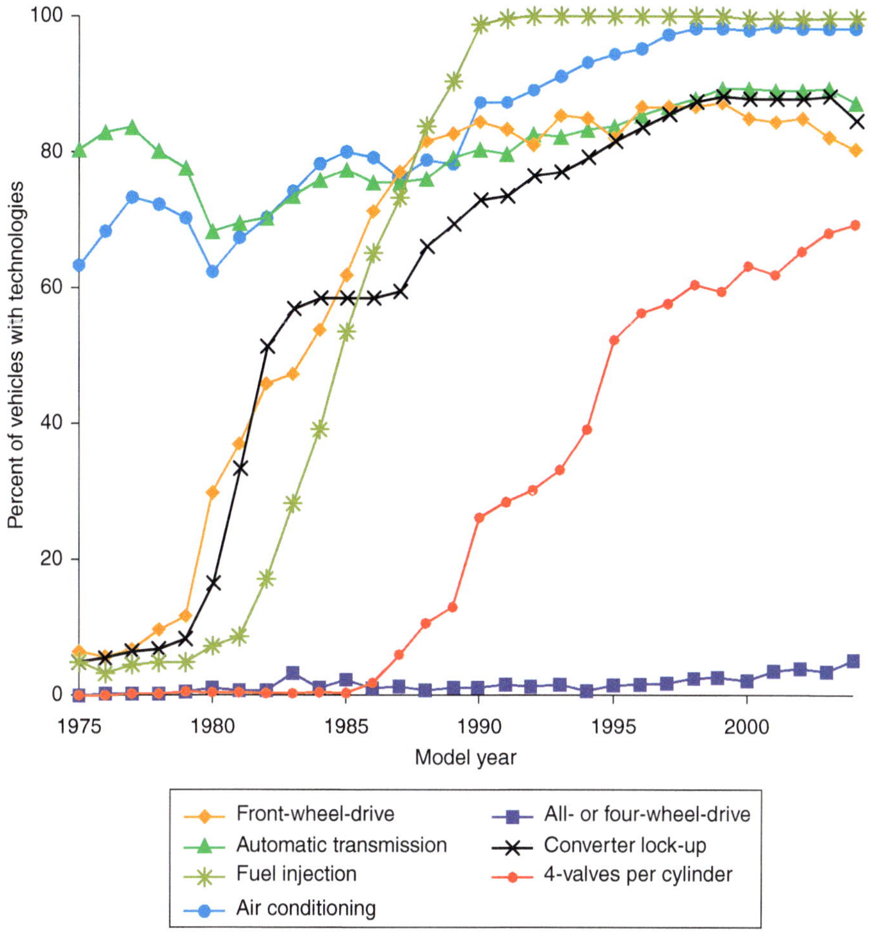

Figure 13.4. Diffusion of passenger vehicle technologies (Lutsey and Sperling, 2005).

association, Association des Constructeurs Européens d'Automobiles (ACEA), signed an agreement to voluntarily reduce the CO_2 emission rates of new passenger cars sold in the EU (An and Sauer, 2004a). The Korean and Japanese Automobile Manufacturers Associations (KAMA and JAMA, respectively) also signed similar voluntary agreements with the EC soon after. The industry committed to reducing average emission rates to 140 grams of CO_2 per km (gCO_2/km) by 2008 for ACEA and 2009 for KAMA and JAMA from the 1995 average of 187 gCO_2/km, with the possibility of extending the agreement to reduce the average emission rate to 120 gCO_2/km by 2012. The equivalence of these emission standards to fuel efficiency standards is shown in Figure 13.6.

Until 2007, the EC's strategy of reducing auto emissions was primarily based on three pillars: voluntary agreements with auto manufacturers; increased consumer awareness regarding fuel economy of cars, and market-oriented measures aimed at influencing motorists' choice. (Further details of the pre-2007 strategy can be found on the EC's Climate Action Web pages.) In December 2007, the EC adopted a proposal for reducing the average emission rates and eliminating the voluntary agreements with the industry asserting that the previous strategy had brought only limited progress toward the 120 gCO_2/km target for 2012. In December 2008, the EC

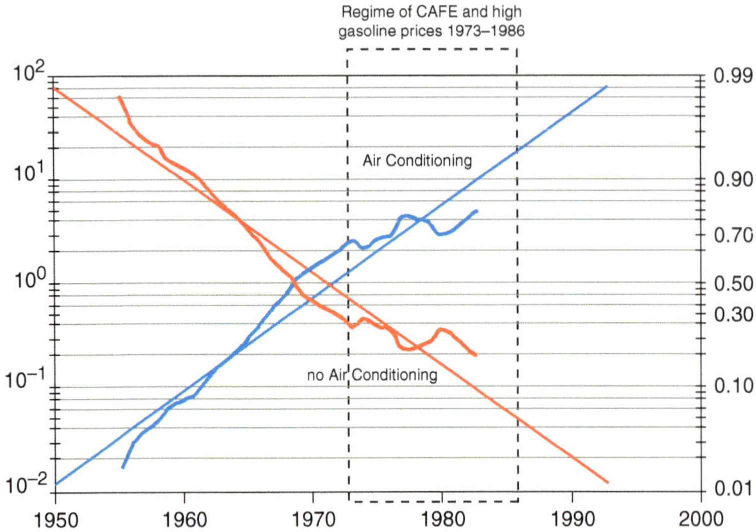

Figure 13.5. Diffusion of air conditioning in vehicles.
Notes: This figure represents substitution of one technology by another. The right-hand y-axis refers to the market share for each technology (f). The left-hand y-axis is f/(1-f), which helps define the logistic substitution curve described by Fisher and Pry.
Sources: Nakicenovic (1986); Fisher and Pry (1971).

established a new regulation to reduce the fleet average emission to 130 gCO_2/km through technology improvement related to vehicle efficiency and a further 10 gCO_2/km reduction through complementary measures (ICCT, 2010). These complementary measures can broadly be classified into two categories. The first category includes improvements in car components that have the highest impact on fuel consumption, such as efficient tires and air conditioners, tire pressure monitoring systems, and gearshift indicators. The second category includes measures to gradually reduce the carbon content of the fuel itself through increased use of biofuels and other alternative fuels (Amin, 2009; ICCT, 2010).

As indicated by Figure 13.6, along with Japan, Europe has the most stringent fuel economy standards in the world and as a result has the most fuel-efficient passenger vehicle fleet. The fuel economy standard has played a vital role in significantly reducing the average fuel consumption of passenger vehicles in Europe since 1995 (Zachariadis, 2006). Technologically, the main driver of the fuel efficiency improvement in Europe has been the rapid adoption of advanced technologies that have reduced vehicle mass and enhanced thermodynamic engine efficiency. The second driver is the widespread penetration of diesel vehicles. The share of diesel cars as a share of all new cars sold in Europe has increased significantly in the last decade – from about a quarter of new vehicles sold to over half (Figure 13.7). However, like the U.S. case, it is difficult to ascertain what portion of this diesel vehicle penetration was due to fuel economy regulation. Apart from its combustion efficiency, other factors that have contributed to the rising share of diesel vehicles in Europe are fuel tax differentiation favoring diesel in most European nations and advancements in diesel vehicle technology (Pock, 2010).

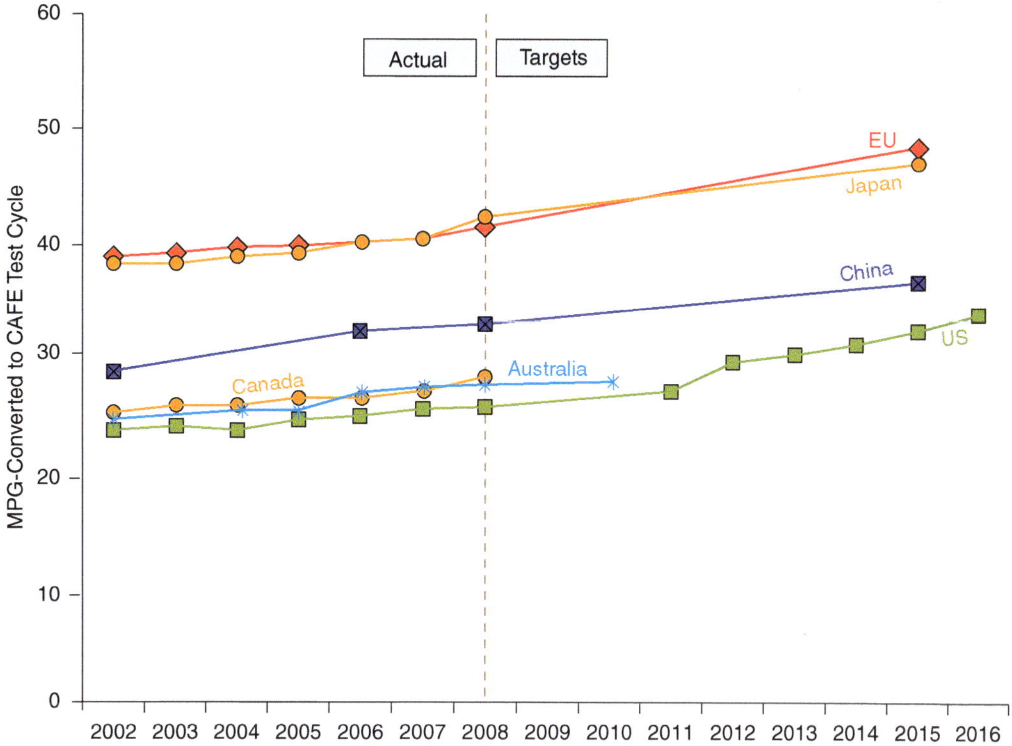

Figure 13.6. Fuel economy: historical and future targets for six countries (ICCT, 2010).

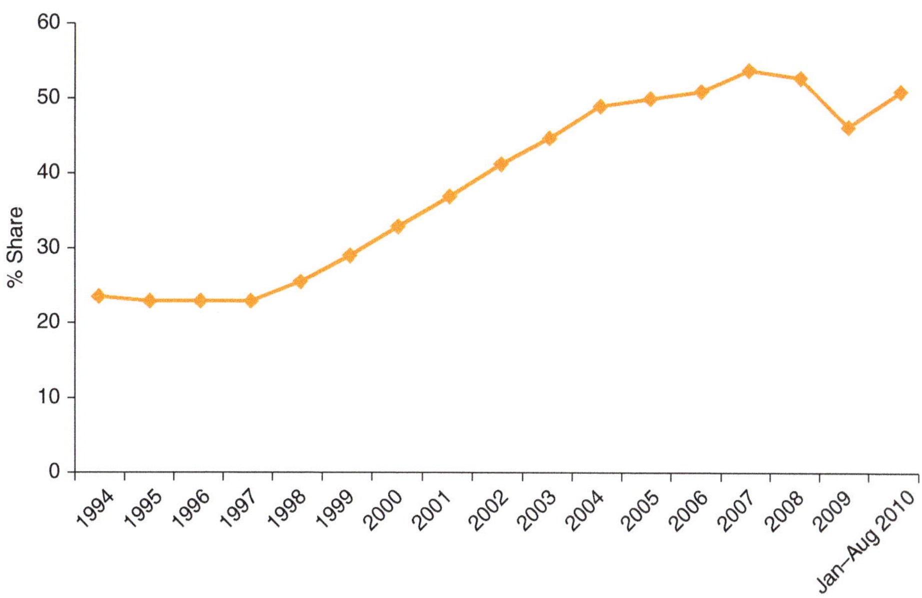

Figure 13.7. Diesels as a share of new passenger car registrations in the EU-15 (ACEA). *Note*: The EU-15 was the fifteen-member state EU from January 1995 to April 2004. Currently the EU has twenty-seven member states.

3 Fuel Efficiency in China: Rapid Improvements and Flexibility Issues

China began regulating vehicle fuel economy relatively recently. Its first national fuel economy standard for new passenger vehicles based on weight category was issued in September 2004. This standard was to be implemented in two phases, taking effect in July 2005 (Phase 1) and January 2008 (Phase 2). The main goals of this regulation were to reduce China's increasing dependence on foreign oil, to encourage foreign automakers to introduce more advanced and fuel-efficient vehicle technologies, and to encourage domestic automakers to improve their standard (Oliver et al., 2009). China is finalizing the third phase of its fuel economy standard, officially issued by the Chinese government during 2011 and expected to be fully effective by 2015 (An, Daniel, and James, 2011; Wang et al., 2010).

Phase 1 of the Chinese fuel economy standard was established for passenger vehicles based on their weight by dividing them into sixteen categories (Oliver et al., 2009). Phase 1 standards were effective from July 2005 and July 2006 for new models and existing models respectively (Wang et al., 2010). Phase 2, which tightened the standards by about 10 percent and had only fifteen weight classes, became effective on January 2008 and January 2009 for new and existing models, respectively. In addition to weight classification, vehicles were further classified as "normal structure" or "standard structure" vehicles. Vehicles that had automatic transmission or at least three rows of seats were classified as "special structure" vehicles. Vehicles in this category were subject to less stringent standards than those falling in the "normal structure" category.

The fuel economy standards have brought about some impressive outcomes in the Chinese automotive market with respect to average fuel consumption and technology advancement (see Figure 13.6). Between 2002 and 2006, normal structure vehicles and special structure vehicles reduced their average fuel consumption by 7 percent to 17 percent and 10 percent to 20 percent respectively (Oliver et al., 2009). The sales weighted corporate average fuel consumption (CAFC) for thirty-four companies that accounted for almost 90 percent of the passenger vehicle market decreased by almost 12 percent from 2002 to 2006.

The results of these standards from a technological perspective are noteworthy as well. About 800 highly inefficient domestic vehicle models were eliminated and inefficient technologies such as three-gear transmission and two-valve engines were phased out (Wang et al., 2010). The economy standards also increased penetration of vehicles with advanced technologies such as variable valves and multivalve camshafts, electronic throttle control, multigear transmission, and parts made from lighter materials.

Similar to the case of Japan, the effectiveness of the fuel economy standards in China has been complemented well by fiscal policies. As shown in Figure 13.8, China has revised the excise tax levied on automakers on multiple occasions to encourage the manufacture of smaller-engine vehicles and discourage the manufacture of larger ones (ICCT, 2007).

It is important to note that even though China's standards have been generally successful in meeting their goals, some shortcomings need to be addressed. One is the lack of flexibility. Vehicles that fail to meet the standards for their weight class are to be eliminated from the market. Other countries set standards based on average fuel economy of the entire fleet, not a particular model. This flexibility

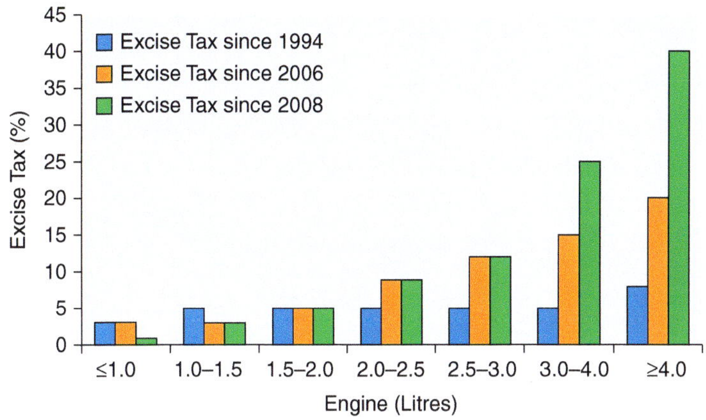

Figure 13.8. Excise tax rate in China based on the engine size of the vehicle (Zhang, 2010).

allows for lower-efficiency vehicles that may be important for some users, provided that above-standard vehicles offset these. In addition, unlike other nations, the Chinese standard does not include special provisions for alternative fuels. The Phase 3 standards are expected to be based on corporate average fuel consumption targets, which are more flexible. Again, it is important to note that the inflexibility in the first two phases resulted in the elimination of about 800 inefficient domestic vehicle models. In essence, the inflexibility achieved one of the main goals that a differential incentive-based system aims for: it gets the dirtiest vehicles off the road (Greene et al., 2005). This feature may be attractive for concerns such as local air pollution for which damages may be disproportionately caused by a small number of high-polluting sources.

4 Fuel Efficiency in other Countries

Japan, Canada, and Australia also have fuel efficiency standards for automobiles. Japan's are well in excess of the U.S. standards while Canada and Australia have standards only slightly higher than those of the United States (see Figure 13.6).

4.1 Japan

Japan established fuel economy standards for gasoline-powered and diesel-powered light-duty passenger and commercial vehicles in 1999. The weight-based Japanese standards were set using the "Top Runner" program (discussed in detail in the End-Use Efficiency (Japan) case study). Under this program, the target for each vehicle in a particular weight class is based on the performance of the most efficient vehicle in that weight class in the national market and a range of other factors (Onoda, 2008). Originally, the program placed vehicles into a total of nine weight classes. The targets for diesel and gasoline vehicles were to be met by 2005 and 2010 respectively (An and Sauer, 2004b). The initial regulation was revised in 2001 to let automakers accumulate credits in one weight class and use them in another weight class with certain limitations. In December 2006, the fuel economy targets were revised upward to be met by 2015 and the weight classes were expanded from nine to sixteen (ICCT, 2007). This upward revision of the targets resulted because most of the vehicles

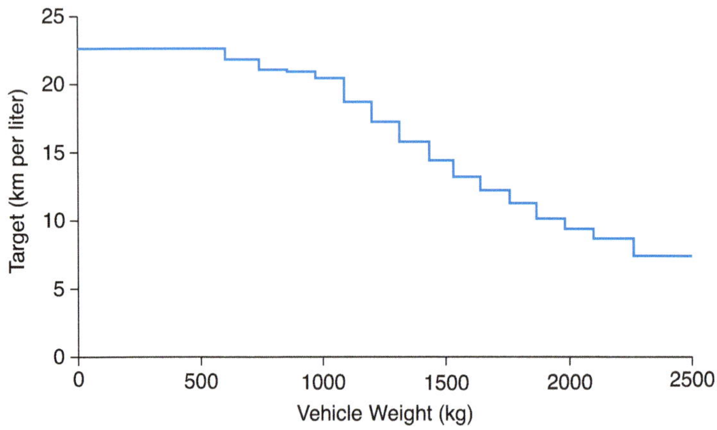

Figure 13.9. Japanese fuel economy standards by weight for 2015.
Note: 1 km/L = 2.35 mpg.

sold in Japan by 2002 already met the standards set for 2010. The 2015 targets are expected to improve the average fleet fuel economy of new passenger vehicles from 13.6 km/L (or 32.0 mpg) in 2004 to 16.8 km/L (or 39.5 mpg) in 2015. Figure 13.9 shows the Japanese fuel economy targets for 2015 for each weight class (Onoda, 2008).

As indicated by Figure 13.6, Japan has the second most stringent fuel economy standards in the world. As a result of these standards, the fuel economy performance of vehicles in Japan has been impressive. Fiscal policies in Japan complement these standards, enhancing their effectiveness. A progressive tax system based on gross vehicle weight and engine displacement during purchase and registration of vehicles encourages the purchase of lighter vehicles with smaller engines (ICCT, 2007). Additionally in 2004, Japan implemented the vehicle fuel efficiency certification program to stimulate consumer interest in fuel efficiency performance and to encourage adoption of fuel-efficient vehicles (Onoda, 2008). Under this program, vehicles are certified in one of four categories, based on whether they meet the target or levels exceeding the target by 5 percent, 10 percent, and 20 percent. Manufacturers can attach the certified stickers that indicate vehicles' fuel efficiency performance.

4.2 Canada

Canada's Company Average Fuel Consumption (CAFC) program was introduced in 1976 to reduce the fuel consumption of the new passenger vehicle fleet. This CAFC program is similar to the U.S. CAFE program. Two important differences, however, are that Canada's CAFC program does not distinguish between domestic and imported vehicles, and compliance is voluntary unlike the United States' CAFE standard (ICCT, 2007). The Canadian target has continued to match the U.S. standard each year, with the Canadian vehicle fleet consistently outperforming the U.S. vehicle fleet. This difference is primarily due to variation in tax provisions for fuel and vehicles, as well as to a variation on the ownership mix of vehicles in the two countries (An et al., 2007). For example, the ratio of passenger cars to light trucks is higher in Canada than in the United States. The CAFC program is complemented well by other measures taken by the Canadian federal and provincial governments. In 2007, the Canadian government introduced a program called Vehicle Efficiency Incentive (VEI), which contained a rebate and a tax component based on vehicle fuel efficiency (ICCT, 2007). The rebate program offers up to $2,000 for the purchase of an eligible fuel-efficient vehicle. The new excise tax, called the "Green Levy," is applied to new passenger vehicles with average fuel efficiency ratings of 18 mpg or less. In April 2010, the Canadian government issued a draft regulation to limit CO_2 emissions from passenger cars and light trucks. The regulation, which applies to models from 2011 to 2016, is expected to result in an average CO_2 emission rate of 153 gCO_2/km, a 20 percent reduction compared to the new vehicle fleet sold in Canada in 2007 (ICCT, 2010).

4.3 Australia

The Australian Federal Chamber of Automotive Industries (FCAI) has established several voluntary codes of practice for reducing fuel consumption of new passenger cars over the past several years (An and Sauer, 2004b). The first code was in effect between 1978 and 1987 and even though the industry reduced vehicle fuel consumption significantly, it still narrowly missed the voluntary target. The second code was enacted in 1996 and the FCAI members agreed to reduce average fuel consumption of passenger cars nationally to about 29 mpg by 2000. In 2003, a third voluntary agreement between the FCAI and the government was enacted. This agreement requires the industry to reduce the fleet average fuel consumption for passenger cars by 18 percent in 2010 compared to the 2002 level. The actual average fuel consumption has been slightly higher than the target under all three agreements. This is mainly due to the fact that there are no specific enforcement mechanisms or noncompliance penalties. Australia introduced a mandatory fuel consumption labeling scheme in 2001, which required a fuel consumption label to be placed on the windscreen of all new cars (Onoda, 2008). These labels are required to show the vehicle's fuel efficiency and CO_2 emissions per km.

5 Conclusion

5.1 Efficiency, Performance, and Technological Change

Automobiles are far more efficient in 2010 than they were in the 1970s – perhaps by a factor of three. However, automobile manufacturers have not used all of this efficiency improvement to reduce fuel consumption per distance traveled; they have used much of the gain in efficiency to satisfy other user preferences, notably for acceleration and for large and heavy vehicles. The allocation of efficiency improvements between reduced fuel consumption and enhanced performance characteristics has varied over time, across countries, and, of course, among vehicle types within countries and periods. Government policies have almost certainly influenced this allocation decision – both in vehicle design and in consumer choice among models. Rising fuel efficiency standards and fuel taxes have, in certain times and geographies, directed technological change toward reducing fuel consumption. However, rising affluence, consumer preferences, and low after-tax fuel prices have at times influenced technological change in the direction of improved performance.

When they are rising, and supported by taxation and rising market prices for fuels, fuel efficiency standards have also led to the adoption of lighter vehicles, as well as to the diffusion of important technologies in vehicle design. In the United States, the CAFE standards had a real effect on technological change. The improvement in miles per gallon was accomplished not only by a shift to lighter, less powerful vehicles, but also by the adoption of new energy efficient technologies. Important examples include front-wheel drive and fuel injection in the 1970s. Attribution of these changes to the regulations, rather than gasoline prices, is less clear because the two are so well correlated. The effectiveness of CAFE may actually have been more important after 1985 when it served as a fuel economy floor in the face of persistently low gasoline prices.

5.2 Outlook

The outlook for fuel efficiency and technological change in the United States is more dynamic than it has been since the 1970s; however, policy details and outcomes remain unsettled. In 2010, the National Highway Traffic Safety Administration (NHTSA) and the Environmental Protection Agency (EPA) issued a joint final rule that established a new national program to improve fuel economy and reduce greenhouse gas emissions from passenger cars and light trucks from model year 2012 to 2016. The implementation of this new rule was primarily driven by the Energy Independence and Security Act of 2007 and further administrative action in 2009 (Anderson et al., 2010). Light duty vehicles must have an estimated combined average fuel economy of 29.7 and 34.1 mpg in 2012 and 2016 respectively. These standards, although more stringent, are more flexible. Automakers can earn credits by overcomplying and these credits can be traded with other manufacturers, banked for future compliance, or transferred between car and light truck fleets. In August 2011, the White House released its plan for even tighter fuel efficiency standards, developed in conjunction with U.S. automakers and labor groups. However, these standards remain contentious and their eventual passage into legislation remains uncertain. Similarly, expectations of future fuel prices are also unclear. The steady and substantial rise in gasoline prices in the mid-2000s was followed by a period of exceptionally high price volatility, including a rapid decline with multiple years of lower prices, and a more recent rise.

The situation in Europe, China, and Japan is much different. Standards are higher and expectations for the next several years provide a much clearer view to increasingly stringent fuel efficiency standards. One important source of difference is that concern about climate change provides a sustained source of support for fuel economy measures in these countries. Conversely, in the United States, climate change ranks low in polls of social priorities, and the dominant motivation for fuel efficiency legislation is energy security, which tends to oscillate with crises and fuel prices.

Taxation of fuels and vehicles outside the United States also supports the establishment of more stringent fuel economy standards. Technological outcomes are evident. One sees the dramatically higher adoption of diesel vehicles in Europe, but also efficiency improvements to tires and air conditioners, as well as gear shift indicators and tire pressure monitoring. In China, adoption of lighter materials, variable valves, electronic throttles, and improved transmissions have helped meet the more stringent standards.

FURTHER READING

For more background see Anderson and colleagues (2010); Greene (1998); Lutsey and Sperling (2005).

REFERENCES

Amin, A. T. M. N., 2009. Reducing Emissions from Private Cars: Incentive Measures for Behavioural Change. *UNEP–Green Economy Initiative* 8.

An, F., Daniel, S., & James, S. C., 2007. International Comparison of Policies to Reduce Greenhouse Gas Emissions from Passenger Vehicles. *Driving Climate Change*. Burlington, VT: Academic Press.

An, F., Earley, R., & Green-Weiskel, L., 2011. *Global Overview on Fuel Efficiency and Motor Vehicle Emission Standards: Policy Options and Perspectives for International Cooperation*. New York: United Nations Department of Economic and Social Affairs.

An, F. & Sauer, A., 2004a. *Comparison of Passenger Vehicle Fuel Economy and Greenhouse Gas Emission Standards Around the World*. Arlington, VA: Pew Center on Global Climate Change.

2004b. *Comparison of Passenger Vehicle Fuel Economy and Greenhouse Gas Emission Standards Around the World*. Arlington, VA: Pew Center on Global Climate Change.

Anderson, S. T., Fischer, C., Parry, I., & Sallee, J. M., 2010. Automobile Fuel Economy Standards: Impacts, Efficiency, and Alternatives. *National Bureau of Economic Research Working Paper Series*, No. 16370.

Davis, S., Diegel, S., & Boundy, R., 2008. *Transportation Energy Data Book*. Oak Ridge, TN: U.S. Department of Energy.

DOT, 2002. *Effects of the Alternative Motor Fuels Act CAFE Incentives Policy*. Washington, DC: U.S. Department of Transportation (DOT); U.S. Department of Energy; U.S. Environmental Protection Agency.

Fisher, J. C. & Pry, R. H., 1971. Simple Substitution Model of Technological Change. *Technological Forecasting and Social Change*, 3(1): 75–88.

Greene, D. L., 1998. Why CAFE Worked. *Energy Policy*, 26(8): 595–613.

Greene, D. L., Patterson, P. D., Singh, M., & Li, J., 2005. Feebates, Rebates and Gas-Guzzler Taxes: A Study of Incentives for Increased Fuel Economy. *Energy Policy*, 33(6): 757–75.

ICCT, 2007. Passenger Vehicle Greenhouse Gas and Fuel Economy Standards: A Global Update. International Council on Clean Transportation.

ICCT, 2010. Global Passenger Vehicle Fuel Economy and GHG Emissions Standards: April 2010 Update. International Council on Clean Transportation.

Lutsey, N. & Sperling, D. 2005. Energy Efficiency, Fuel Economy, and Policy Implications. *Transportation Research Record: Journal of the Transportation Research Board* 1941(1): 8–17.

Nakicenovic, N., 1986. The Automobile Road to Technological Change: Diffusion of the Automobile as a Process of Technological Substitution. *Technological Forecasting and Social Change*, 29(4): 309–40.

NAS, 2002. *Effectiveness and Impact of Corporate Average Fuel Economy (CAFE) Standards*. Washington, DC: The National Academies Press.

Oliver, H. H., Gallagher, K. S., Tian, D., & Zhang, J., 2009. China's Fuel Economy Standards for Passenger Vehicles: Rationale, Policy Process, and Impacts. *Energy Policy*, 37(11): 4720–29.

Onoda, T., 2008. *Review of International Policies for Vehicle Fuel Efficiency*. International Energy Agency (IEA).

Pock, M., 2010. Gasoline Demand in Europe: New Insights. *Energy Economics*, 32: 54–62.

Wang, Z., Jin, Y., Wang, M., & Wei, W., 2010. New Fuel Consumption Standards for Chinese Passenger Vehicles and Their Effects on Reductions of Oil Use and CO2 Emissions of the Chinese Passenger Vehicle Fleet. *Energy Policy*, 38(9): 5242–50.
Zachariadis, T., 2006. On the Baseline Evolution of Automobile Fuel Economy in Europe. *Energy Policy*, 34: 1773–85.
Zhang, Z., 2010. Is it Fair to Treat China as a Christmas Tree to Hang Everybody's Complaints? Putting Its Own Energy Saving into Perspective. *Energy Economics*, 32: S47–S56.

ACKNOWLEDGMENTS

Thanks to Bickey Rimal and Peter Braden for excellent research assistance.

14 Hybrid Cars: Development and Deployment in Japan, the United States, and China

Kelly Sims Gallagher

Author's Summary

Many governments are interested in promoting hybrid electric vehicles (HEVs) and other alternatives to conventional internal combustion engines because of concerns about energy security, oil dependence, air pollution, and global climate change. HEVs achieve greater fuel efficiency than conventional vehicles, although the extent of improvement in efficiency depends greatly on the specific configuration of the HEV system.

The development of HEVs in three major markets – Japan, the United States, and China – was characterized by similar drivers of innovation. However, policy mechanisms and incentives for deployment diverged significantly. In all three countries, governments originally pushed harder for alternative automotive technologies other than hybrids, such as pure electric vehicles or hydrogen fuel cell vehicles. In some cases, private firms made research and development (R&D) choices that do not appear to have been strongly influenced by public policy but rather followed a commercial logic in exploring fuel-efficient technologies. In other cases, government policies appear to actually have turned firms away from HEVs. Once HEVs emerged in the marketplace, however, the government response was completely different in each of the three countries, especially in terms of the extent to which each government was prepared to support HEVs' transition through the early deployment phase to facilitate widespread market diffusion.

Editors' Guide

Relationship of Case Study to Energy Technology Innovation System Framework

The emphasis of this case study is on the role of *public policy* in shaping and leveraging R&D and market deployment efforts for hybrid electric vehicles (HEVs). What stands out, however, is the range of factors outside the direct purview of policy that influenced the success story of Japan relative to the slower developments in the United States and later China. Chief among these are consumer (adopter) preferences. The *adoption and use* of HEVs is clearly responsive to increased fuel prices and fiscal incentives but also environmental and energy security concerns. Adoption behavior is a function of values and beliefs as well as costs and markets. More generally, the characteristics, heterogeneity, and complexity of *adopter preferences* prove consistently important for understanding the innovation histories of end-use technologies with consumer good characteristics. The U.S. technology development program for HEVs recognized that – from the perspective of potential adopters – one attribute of *relative advantage* (fuel efficiency) should not substitute for other attributes (performance, size, safety). The later success of the larger, more powerful, second-generation HEVs in the U.S. market confirmed this perspective. Japan's use of feebate schemes also shows how a particular relative *dis*advantage (HEVs' upfront cost premium) can be strategically overcome. But costs were already sufficiently competitive, based in part on the large-scale manufacturing effort in response to clear *expectations* about market demand, both domestically and in the United States. By comparison, the lack of available *scale economies* in China with its smaller domestic markets, trade restrictions, and lack of component standardization, meant the cost barriers remain prohibitively high from consumers' perspective.

Adopter preferences are not, however, always clearly known nor anticipated, as Japanese manufacturers illustrated by stealing a march on their U.S. rivals in the U.S. domestic market. And perceived relative advantages vary across different consumer segments. The *adoption and use* of HEVs by early adopters in the United States and Japan helped reduce perceived risks with the technology among potential mass market adopters. It also seeded the effects of social influence and imitation that play an ever-stronger role as a technology diffuses. These many *feedbacks* from end users to technology developers favor product diversity and choice over singular visions of narrowly specified end-use technologies. End-user preferences can thus support experimentation and *learning* processes during market deployment. Moreover, product diversity does not have to mean fragmented efforts. The Japanese case evidences that *scale economies* and *knowledge spillover* benefit at the level of components and system integration, even if the final consumer product varied in style, design, and performance.

The case study on HEVs reinforces the importance of *shared expectations* in supporting innovation efforts, but also demonstrates that not all innovation system *actors* have to share the same expectations. In Japan in particular, industry actors, vehicle firms, and particularly the leaders of those firms who were prepared to *take significant market and technology risks*, diverged from the government's emphasis on battery electric or fuel cell vehicles; this divergence proved prescient. This contrasts with U.S. firms' involvement in public-private RD&D programs built not around shared expectations of technological and market potentials, but on resistance to future regulation. In other words, the *advocacy coalition* for HEVs in the United States was weak.

Insights for Energy Technology Innovation Policy

- The specific instruments, mechanisms, and incentives mobilized by public policy to support a technology's development can vary widely. The first mover success of Japan compared to the catch-up roles of HEV developers in the United States and China is partly attributable to the coherent alignment of supporting policies and their adaptive response to changing market conditions.

- Public investments in the RD&D of advanced-vehicle technologies were initially poorly coordinated with market deployment incentives in Japan and the United States. But Japanese policy responded more rapidly to the commercial introduction of HEVs, providing incentives for market deployment that were sensitive to consumer preferences and specifically targeted technology attributes that put the new technology at a relative disadvantage.
- A broad portfolio of technology-push and market-pull policies is needed to support the interrelated stages and processes within the innovation system. But it is essential these policies are aligned. The effectiveness of R&D and deployment incentives is weakened by misaligned market conditions including low fuel prices (e.g., China) or weak fuel-efficiency standards (e.g., the United States). In contrast, Japan's feebate scheme to address the upfront cost premium of HEVs aligned with the high fuel prices that underwrote rising market demand.
- The initial efforts of governments to pick the winning alternative-vehicle technology (electric vehicles, fuel cells) ultimately did not bear fruit. Yet the R&D investments and market deployment efforts were still of benefit to HEVs' eventual success both through learning about consumer preferences and through knowledge spillovers, particularly at the component level.
- Standards are an effective way to establish clear expectations around future market demand, but have to be designed from the outset to respond dynamically to technological improvements. Static standards (like CAFE in the United States) that remain in place over long periods can act as a disincentive for continual performance improvements.

1 Introduction

Hybrid electric vehicles (HEVs) add a parallel direct electric drive train with motor and batteries to the conventional internal combustion engine. Some HEVs are series hybrids and others are parallel hybrids – the main difference being whether the electric motor can be used alone, or whether it can only assist the engine. By combining the two propulsion systems, HEVs reduce the amount of fuel consumed and further save energy by reducing idling and braking losses (Demirdoven and Deutch, 2004). Various technologies can be integrated into an HEV to reduce its demand for energy, thereby reducing the load and enabling a further downsizing of the engine. The extent of improvement in fuel efficiency of HEVs depends greatly on the configuration of the specific hybrid system. These range from "mild" HEVs with modest fuel efficiency improvements to "strong" HEVs producing substantial gains in fuel efficiency. Given potential fuel efficiency gains, governments have been interested in promoting HEVs and other alternatives to conventional internal combustion engines in line with concerns about oil security, air pollution, and global climate change.

This case study considers the development and deployment of hybrid electric vehicles (HEVs) through country reviews of Japan, the United States, and China, focusing particularly on the role of the three respective governments in supporting private sector innovation. Although the drivers of innovation were broadly similar among the three countries, the policy mechanisms and deployment incentives used diverged significantly. In all three countries, the governments originally pushed harder for alternative automotive technologies other than hybrids – pure electric vehicles or hydrogen fuel cell vehicles. In some cases, private firms made R&D choices that do not appear to have been strongly influenced by public policy; in other cases, government policies appear to actually have turned firms away from HEVs.

Once HEVs emerged in the marketplace, however, the government response to them was completely different in the three countries, especially in terms of the extent to which each government was prepared to support HEVs' transition through the early deployment phase of commercial application to facilitate widespread market diffusion.

2 Hybrid Electric Vehicles in Japan, the United States, and China

2.1 Japan

Japan was the first country to commercialize hybrid electric passenger cars. The history of HEV research and development (R&D) in Japan, however, is opaque. Private industry firms, most notably Toyota and Honda, developed HEVs seemingly independently from the Japanese government (and from each other). The industry nonetheless benefited from government support for major HEV components through the government's battery electric vehicle (BEV) R&D program that began in 1972 with the creation of a BEV public-private R&D program.

The program produced a first BEV prototype by 1976. Applications for patents on electric vehicles appeared as early as the mid-1970s (Yarime, Shiroyama, and Kuroki, 2008). Not until 1997 (after Toyota produced its first hybrid) did the government officially broaden this program to explicitly encompass R&D for HEVs. Indeed, the government mainly focused on development and deployment of pure battery electric vehicles until the mid-1990s. This strategy apparently derived from the government's interest to move to nuclear-fueled BEVs, which might have afforded the Japanese greater energy security (Åhman, 2009). Other alternative-fueled vehicles, including compressed natural gas, methanol, and fuel cell vehicles, were also considered and supported as well in the 1997 revision to the program to include HEVs. Meanwhile, a special public-private R&D program focused on improving batteries was established in 1992, and different combinations of battery and vehicle manufacturers joined forces to develop and produce advanced batteries. Although the government mainly focused on BEVs, there were spillovers to HEVs from the government investments in electronic control techniques, permanent magnet motors, and the nickel metal hydride (NiMH) battery programs (Åhman, 2006).

An important feature of Japanese government policy toward HEVs is its consistent and complementary support for the commercialization of alternative-fueled vehicles. The Clean Energy Vehicles Introduction Program that supported deployment was initially restricted to BEVs. But HEVs were quickly added as an eligible technology in 1998 after Toyota and Honda released their first hybrid models, a case of deployment incentive lagging commercial applications.

In the first couple of years of the program more than half the vehicles received major subsidies (Åhman, 2006). The comprehensive deployment support policy for hybrids included preferential acquisition taxes, preferential annual automobile taxes, and subsidies (up to one-half of the incremental cost over a conventional gasoline vehicle). In addition, fuel-efficient and low-emission vehicles received rebates (in a so-called feebate scheme). These feebates were so effective that the government's projected revenue from the fees imposed on inefficient and polluting vehicles were much lower than anticipated because of the consumer popularity of the fuel-efficient, low-emission vehicles (Ito, 2006). As a result of the popularity of the rebate, the Japanese government began to adjust it annually to support a continual shift to more efficient cars. Finally, Japanese fuel taxes undoubtedly contributed to stimulating demand for the more fuel-efficient HEVs.

The best-known and bestselling hybrid vehicle in the market today is the Toyota Prius, but the Ministry of International Trade and Industry (MITI) was reportedly unaware of the extent to which Toyota was developing the Prius during the 1990s (Åhman, 2006), and also therefore presumably unaware of Honda's development of its own HEV, the Insight. It is not clear what motivated Japanese industry to develop HEVs as the government did not encourage or support this particular technological system, but various explanations can be suggested.

First, the 1992 Earth Summit in Rio may have motivated company leaders to develop a new vision for reducing the fuel consumption of popular vehicles. Both Toyota and Honda had strong internal environmental leadership. In 1992, Toyota set up an environmental committee directly chaired by the president, which was "of decisive importance as an institutional framework" (Yarimi et al., 2008: 198). In the same year, it also published its first Global Environmental Charter, followed by an environmental implementation plan in 1993. In September 1993, Toyota initiated a project, Global 21st Century, to develop more fuel-efficient automotive technologies, with the aim of improving fuel efficiency by more than 50 percent.

Second, R&D programs and market incentives in the United States may have spurred Japanese industry on to outcompete the major U.S. manufacturers. In 1993, the U.S. Partnership for Next Generation Vehicles set targets for threefold improvements in fuel efficiency, and in 1990, California's Zero Emission Vehicle mandate signaled future demand in a sizeable market niche for highly efficient vehicle technologies (see later in this chapter for further details). These competitive incentives likely played a role in Japanese automakers' efforts to commercialize electric vehicles.

Third, innovation efforts may have led Japanese industry to discover or revisit the potential benefits of HEV systems. Through its Global 21st Century project, Toyota engineers rediscovered older Toyota hybrid technology, which had been originally developed in 1977 for a sports car. In contrast to the Toyota case, Nissan and Honda were more reactive, waiting for clearer policy signals in support of HEV deployment before focusing on innovation efforts. Using patent applications as a proxy for innovation activity, full-scale development of hybrid vehicles did not occur until after California's Zero Emission Vehicle regulation was strengthened in 1996 (Yarimi et al., 2008). Patent applications on electric vehicles remained very low throughout the 1980s and 1990s, but applications on series hybrid vehicles rose in the early 1990s before falling again after parallel hybrids proved more efficient.

Patent applications for parallel hybrids rose after 1995 and peaked in 2000 (Yarimi et al., 2008).

In summary, a combination of these firms' leaders' acceptance of environmental challenges as legitimate, and a long uninterrupted history of electric vehicle development supported by government, led Japanese firms to realize the potential for HEV technology before most competitors (Åhman, 2009).

2.2 The United States

The U.S. government apparently recognized the potential for HEVs relatively sooner than the Japanese government. Although the U.S. Department of Energy had supported investments in vehicle efficiency technology R&D since the 1970s (Gallagher and Anadón, 2009), the first marquee effort to develop a hybrid electric car was launched in 1993. The Partnership for a New Generation of Vehicles (PNGV) was launched as a public-private R&D partnership by the federal government together with the U.S. Council for Automotive Research, a partnership of Ford, Chrysler, and General Motors. Ultimately, seven federal agencies, U.S. national laboratories, and multiple private companies joined the PNGV effort, which was placed under the leadership of the U.S. Department of Commerce. At the onset of the program, the specific goals were to develop a vehicle that achieved up to three times the fuel economy of mid-sized 1993 cars (a goal of approximately 80 mpg) with no sacrifice in performance, size, cost, emissions, or safety. Industry participation was apparently motivated more by an interest in avoiding new federally mandated fuel efficiency and/or emissions standards by participating in the partnership than by a true interest in innovation (Sperling, 2001). Separately, a cooperative agreement between the Department of Energy and the U.S. Advanced Battery Consortium was formed to develop nickel-metal-hydride (NiMH) battery technology.

The U.S. government relied mainly on this public-private partnership model for technology-push innovation efforts to develop HEVs. No complementary policies to create market demand or niche markets were used during the 1990s to support the deployment of HEVs; the government relied instead on fuel economy performance standards imposed on car manufacturers. But these corporate average fuel economy (CAFE) standards had no effect on demand for HEVs or on encouraging automakers to produce HEVs because the level of the standards was static throughout the 1990s and early 2000s and the standards were insufficiently stringent to be technology forcing. A National Academy of Sciences review of the CAFE standards found no evidence that it had led to the development of HEVs, but it did find that a combination of the standards and higher fuel prices resulted in technological improvements that nearly doubled the fuel economy of new passenger cars (NRC, 2002). The review also concluded that CAFE did not constrain most imported vehicles because they were so much more energy efficient than those produced by U.S. automakers.

U.S. automakers provide different explanations for why they were slow to develop and market HEVs. Some industry representatives have claimed that the firms were derailed by California's Zero Emission Vehicle mandate, and overly focused on pure electric vehicle technology. General Motors spent $1 billion developing its EV1 car, which was leased in California and Arizona starting in 1995. The EV1's range of 100 miles ultimately proved too limiting, and GM discontinued the vehicle in 1999.

After abandoning its electric vehicle (EV) program, General Motors decided to focus on fuel cell vehicles to the apparent exclusion of HEVs. According to Gary Cowager, president of North American operations: "EV1 was frankly what held us back on hybrids. When the hybrid thing started, people said, show me the business case" (Maynard, 2004). DaimlerChrysler also leaned toward fuel cell vehicles, forming a partnership with the leading Canadian fuel cell developer, Ballard, in 1997. (FreedomCar, the successor to the PNGV program that was considered not to have met its cost targets, also largely, but not exclusively, focused on fuel cell vehicles.)

Chrysler had developed a hybrid version of its Dodge Durango that was ready for introduction in 2002, but parent company DaimlerChrysler pulled it back, believing that diesels were a better way to go (Maynard, 2004). Ford Motor Company came the furthest and fastest in developing and deploying HEVs, perhaps driven by the interest of its CEO at the time, Bill Ford, who as a self-avowed environmentalist wanted results sooner than was likely with hydrogen fuel cells. Oddly, General Motors later returned to its electric vehicle roots, committing strongly to its Chevy Volt project, which was touted as the first U.S. plug-in hybrid electric vehicle. The Volt finally came into production in 2010.

Despite the PNGV program and CAFE standards, no hybrid vehicles were produced by U.S. automakers prior to 2005, and none were available for sale in the U.S. market prior to 2001 when Honda and Toyota brought the Insight and Prius, respectively, to the U.S. market. To the surprise of many, American consumers enthusiastically embraced the Japanese-produced hybrid passenger cars as soon as they reached U.S. shores. It is not clear why U.S. automakers were unaware of the evident market demand, but Japanese HEVs had been available for four years before the Ford Motor Company introduced the first U.S.-produced HEV. Both the Insight and the Prius were a big success in the U.S. market, especially the larger second-generation Prius. In 2004, Toyota's redesigned model brought improved styling, passenger room, cargo area, and power, and was well positioned for success when gasoline prices started to rise in 2005 (Gallagher and Muehlegger, 2011). Indeed, U.S. HEV sales, especially the highly fuel-efficient models, surged that year. In 2009, more HEVs were registered in the United States than in any other country, including Japan: 44 percent of global registrations were in the United States, 41 percent in Japan, and less than 3 percent each in Canada, the Netherlands, and the United Kingdom. (HybridCars, 2009).

Tax credits were the only federal deployment policy for hybrid vehicles implemented in the United States in the 2000s aside from the CAFE standards. Originally provided as an income tax deduction to HEV purchasers, in 2006 the government converted this subsidy into a tax credit, which is more valuable for consumers. Numerous states created additional deployment incentives, including sales tax reductions or exemptions, permission to use high-occupancy vehicle lanes with only one driver, free parking spots, reductions in state income taxes, and exemptions from emission testing requirements.

The popularity of HEVs in the U.S. market remains a puzzle. One would expect that the combination of relatively low retail gasoline and diesel prices in the United States, weak deployment incentives, and the higher cost of the HEV would result in weak consumer demand. In contrast, Japanese gasoline prices are typically two to three times higher because of gasoline taxes (see Figure 14.1). Insights into consumer

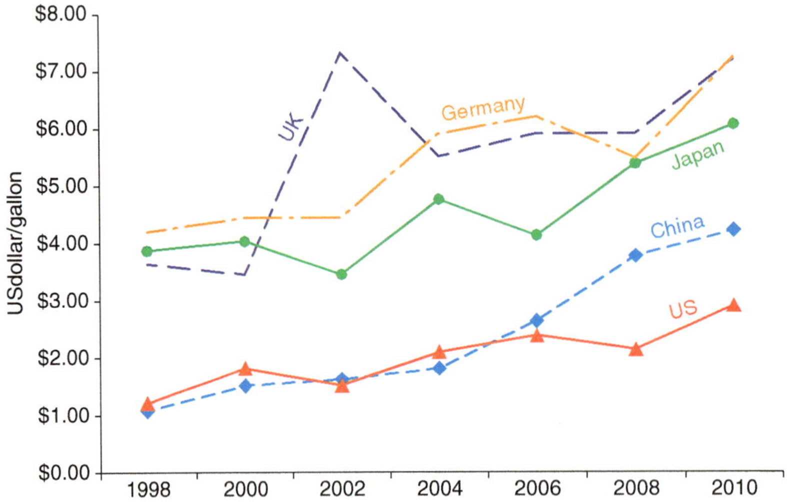

Figure 14.1. Retail gasoline prices (1998–2010).
Source: GTZ (2012).

behavior provide some explanation for the unexpectedly high rate of HEV adoption in the United States. In one study, state-level sales tax reductions, average income, and environmentalism were all found to be significant in explaining state-level market shares of HEVs (Gallagher and Muehlegger, 2011). Sales tax waivers were more significant than income tax credits presumably because of the immediacy of the sales tax reduction at point of sale. The federal income tax credit, which is uncertain in value and more distant in the future, does not seem to have lured consumers. Rising gasoline prices also appeared to affect consumer decisions – consumers conscious of gas prices tended to buy very fuel-efficient "strong" hybrids rather than so-called mild hybrids, which only provide modest fuel economy improvements (Gallagher and Muehlegger, 2011). Other studies have similarly found that retail fuel prices matter to consumers of energy-efficient products (Jaffe, Newell, and Stavins, 1999), that social preferences can be a powerful factor in purchase decisions, and that environmentally motivated consumers were more likely to purchase HEVs irrespective of the differential in price with conventional vehicles (Kahn 2007; Turrentine and Kurani, 2007).

In line with long-standing evidence on the diffusion of innovations (Rogers, 2003), these early adopters willing to take the risk of purchasing an expensive, unproven, potentially unreliable new technology allowed other consumers to observe the quality and performance of HEVs. The outcome of this "information cascade" is observable when later adopters make the same choices as early adopters without having gone through the same investment in "learning by experience" (Geroski, 2000).

2.3 China

Even more strongly than the United States, China is in a catch-up mode of innovation for HEVs. (For further details on the Chinese case, see Gallagher, 2006.) HEVs currently represent only a tiny fraction of Chinese passenger car sales. Only twenty-one hundred HEVs were sold in China in 2008, out of a total of 6.4 million passenger cars produced that year (Bradsher, 2009). Chinese domestic manufacturers are trying to acquire and develop indigenous HEV capabilities, but most do not yet produce HEVs on a commercial basis (though many have developed prototypes).

The standout Chinese firm, BYD, launched its F3DM plug-in hybrid in December 2008. Similar to the case of Toyota, BYD's successful development of electric vehicles was strongly influenced by the founder and CEO of BYD, Wang Chuan-Fu. BYD was established in 1995 as a battery manufacturer, primarily supplying the cell phone market. In 2003, BYD acquired the Shaanxi Qinchan Auto Company and established the BYD Auto Company, and within two years, began producing its first passenger car, the F3. In 2006, BYD produced its first electric car prototype, and two years later launched the F3DM for sale (DM stands for dual mode). The F3DM is advertised to run up to 100 kilometers on a charge before reverting to its gas engine (Kurtenbach, 2009). It can also run as a pure plug-in electric vehicle, or it can be run as an HEV because it has a back-up gasoline engine. The F3DM sells for about $22,000. Wang Chuan-Fu's commitment to environmental sustainability extends beyond fuel efficiency. The company has set a goal of making its batteries 100 percent recyclable, and has developed a nontoxic electrolyte fluid (Gunther, 2009).

A major barrier to the development and deployment of HEVs in China is price. Toyota and First Auto Works (FAW) began assembling the Prius in China in 2005, but the price of the Chinese-assembled Prius was about 300,000 RMB (US$40,000), nearly twice the price of a Prius sold in the United States. One reason for this higher price is the 10 percent to 28 percent tariff paid on imported parts and components. Also, few, if any, Chinese parts and components producers were able to supply the Toyota-FAW facility, so parts were sourced internationally. Domestic manufacturers have been reluctant to produce HEV components because of the small volume and lack of economies of scale. Also, the technological capabilities of the parts and components firms in China may be inadequate to the task of producing to specification. Some have argued for standardization of the key components of HEVs in China to facilitate greater economies of scale and cost reductions for domestic manufacturers.

Many Chinese manufacturers have stated that another big barrier is Japanese firms' ownership of many of the relevant patents for key components, and that lack of Chinese intellectual property has proven to be a constraint. Although a lack of leapfrogging in the development of China's automobile industry was observed until the turn of the century (Gallagher, 2006), BYD's production of the F3DM represents a big leap forward.

China's Ministry of Science and Technology (MOST) provided considerable "technology-push" support through its high-tech "863" applied R&D program. Similar to the cases of the United States and Japan, the MOST program originally focused more heavily on pure BEVs rather than hybrids, and also fuel cell vehicles in line with the U.S. FreedomCar program. Research began on electric vehicles in 1990, and an EV demonstration zone was established in Shantou in 1997, beginning operation in 1998 with a budget of approximately $2.5 million, one-quarter of which was supplied by MOST, and the rest by local government and Chinese firms. Both GM and Toyota donated electric vehicles for testing. Some vehicles were used as taxis, and all were monitored closely (Ming, Jingguang, and Mytelka, 2008; Shantou

EV Project Personal Communication, 2001). This program catalyzed a number of Chinese firms to focus on HEV development and to add some of their own funds as well.

Aside from FAW, auto companies Chery, Dongfeng, and Chang'An all have HEV R&D programs co-funded by MOST. Many other firms are actively developing BEV and fuel cell vehicles as well. As of 2006, Chery was pursuing a strategy of licensing and hiring of foreign design firms. Dongfeng and Chang'An had not succeeded in convincing their foreign partners to collaborate on HEV R&D, but were pressing ahead on their own. The city of Tianjin is pursuing a "cluster" strategy, supporting and co-locating electric vehicle manufacturers with battery developers, together with one of China's top research and testing institutes, the China Automotive Technology and Research Center.

To create market niches and underwrite demand for these technology-push efforts, the Chinese government is allowing six cities to experiment with subsidies to individual consumers who purchase EVs. Beijing and Shenzhen, for example, are both providing purchase incentives for EVs of around RMB 60,000. The central government will provide an additional subsidy of up to RMB 50,000 for the purchase of a PEV and RMB 60 000 for the purchase of a BEV. Hybrid vehicles are being subsidized at lower rates (Wang, 2013).

Yet in terms of consumer demand, the relatively low gasoline and diesel prices in China create a strong disincentive for purchase of HEVs. The Chinese government still controls the price of retail fuels, and until 2006 often held domestic prices below world prices (see Figure 14.1). Chinese consumers lack awareness of HEV technology, and for those who are aware, the lack of product diversity is a drawback. The Chinese government imposed its first passenger car fuel efficiency standards in 2005, and they were strengthened in 2008. These standards are imposed by weight class. The government also imposed purchase fees on a sliding scale based on engine size. Both of these policies theoretically favor HEVs, but neither is strong enough to overcome the price hurdle. As of March 2013, approximately twenty-eight thousand electric vehicles (excluding electric bicycles) were registered in China, of which about 80 percent were public buses.

3 Comparative Analysis of Factors Influencing HEV Development

Political and economic factors were initially the main drivers for government technology policy and investments in alternative vehicle technologies in all three countries reviewed. In the United States and Japan, concern about energy security in the wake of the oil crises of the 1970s spurred investments into battery electric vehicles and hydrogen fuel cell vehicles. Japan committed itself to nuclear energy during the 1980s and 1990s, offering a seemingly secure pathway for electrifying the vehicle fleet. In the United States, relatively abundant resources of coal could supply the electricity for electric vehicles.

In the early 1990s, environmental factors emerged as a driver for advanced vehicle technologies when California began to push for zero-emission vehicles. For China, energy security has been the dominant driver of its policy for advanced vehicle technologies, although there is reason to believe that its new focus on a low-carbon economy might also provide additional impetus for the development and deployment of alternative vehicles and fuels.

But in all cases, governments were either not sufficiently knowledgeable about the costs of BEVs (and the associated technical constraints), or they thought these costs could be overcome with effort. Although none of the governments supported HEVs initially, public investments into BEVs were the basis of important spillovers to the eventual development of HEVs because BEVs and HEVs share many components and systems.

The role of private firms differs substantially across the three countries. U.S. firms appear to have lacked the foresight or capacity of Japanese firms to develop HEVs even in the absence of government R&D programs. This is ironic because the conventional wisdom is that Japanese firms work closely and harmoniously with their government and received much subsidization for their hybrid vehicle programs. An alternative explanation is that U.S. firms were overly responsive to policy. By letting the California zero-emission vehicle mandate steer their R&D programs away from hybrids, the U.S. firms lost ground against the Japanese firms. Even when the public-private PNGV program was established, U.S. firms failed to accelerate investments and research into HEVs. Given their disinterest, perhaps the best explanation is that the U.S. firms were too enthralled by the profits to be made with large inefficient vehicles in the short term to see that there might be a time when consumers would demand smaller, more fuel-efficient vehicles.

Indeed, internal corporate leadership appears to have been a major factor explaining the relative success among the major national automotive manufacturers. The personal commitment of Toyota's president during the early 1990s clearly set Toyota's course as did Bill Ford's interest in environmentalism and fuel efficiency, seemingly at odds with the company culture at times. In China's BYD Auto, the founder and CEO Wang Chuan-Fu capitalized on the company's expertise in batteries to fulfill a vision of selling an environmentally sustainable car.

In all three countries, policy for government investments in the R&D and demonstration of advanced vehicle technologies was initially poorly coordinated with policy for the early deployment of these technologies. Japan and the United States reactively established policies to support the early deployment of HEVs once they were introduced to the market. From that point, Japan implemented much more effective policies to support commercialization than did the United States.

In the United States, U.S. firms routinely cited lack of consumer demand as the main reason for not investing in more fuel-efficient technologies, and indeed consumer demand would have been stronger if policies like fuels taxes had existed to support the purchase of more fuel-efficient cars. Hybrid car consumers are clearly responsive to increased gas prices and other fiscal incentives, including sales tax reductions or exemptions and feebate schemes, and they also appear to buy hybrids out of concern for the environment or energy security. Higher fuel prices in Japan associated with high fuel taxes were correlated with strong consumer interest in HEVs from the moment of their introduction. The Japanese government also displayed flexibility and adaptive policy making, correcting its purchase incentive system quickly once hybrids were available to greater spur the early deployment of HEVs.

The Chinese government appears to have studied the U.S. and Japanese cases carefully. While it was perhaps unduly influenced by the technological choices in the United States in terms of its RD&D policy, it supported HEVs all along. Similar to the situation in the United States, the Chinese government was reluctant to increase fuel prices for many years, but is now displaying a modest willingness to let them rise. The Chinese government has also been more aggressive than the U.S. government at implementing fiscal policies akin to the Japanese ones that support the purchase of highly fuel-efficient vehicles.

FURTHER READING

Gallagher, K. S., 2006. *China Shifts Gears: Automakers, Oil, Pollution, and Development.* Cambridge, MA: MIT Press.

Ozaki, R. & Sevastyanova, K., 2011. Going Hybrid: An Analysis of Consumer Preference Motivations. *Energy Policy*, 38(5): 2217–27.

Pohl, H. & Yarime, M., 2012. Integrating Innovation System and Management Concepts: The Development of Electric and Hybrid Electric Vehicles in Japan. *Technological Forecasting and Social Change*, 79(8): 1431–46.

REFERENCES

Åhman, M., 2006. Government Policy and the Development of Electric Vehicles in Japan. *Energy Policy*, 34(4) (March): 433–43.

2009. Personal Communication.

Bradsher, K., 2009. China Vies to Be World's Leader in Electric Cars. *New York Times*, April 1.

Demirdoven, N., & Deutch, J., 2004. Hybrid Cars Now, Fuel Cells Cars Later. *Science*, 305(13) (August): 974–76.

Gallagher, K. S., 2006a. Limits to Leapfrogging in Energy Technologies: Evidence from the Chinese Automobile Industry. *Energy Policy*, 34(4): 383–94.

2006b. *Roundtable on Barriers and Incentives for Hybrid Vehicles in China: A Roundtable Report*, based on roundtable held in Beijing, May 19, Cambridge, MA: Belfer Center.

Gallagher, K. S. & Díaz Anadón, L., 2009. DOE Budget Authority for Energy Research, Development, and Demonstration Database, *Energy Technology Innovation Policy*, John F. Kennedy School of Government, Harvard University, June 2, available at energytechnologypolicy.org.

Gallagher, K. S. & Muehlegger, E., 2011. Giving Green to Get Green? Incentives and Consumer Adoption of Hybrid Vehicle Technology. *Journal of Environmental Economics and Management*, 61(1): 1–15.

Geroski, P. A., 2000. Models of Technology Diffusion. *Research Policy*, 29: 603–25.

GTZ 2012., International Fuel Prices. 7th Edition, GTZ.

Gunther, M., 2009. Warren Buffet Takes Charge. CNNMoney.com, April 13.

HybridCars.com 2009, "Analysis of 2009 Sales Numbers" available at http://www.hybridcars.com/december-2009-dashboard/ (accessed August 7, 2013).

Ito, S. 2006., Personal Communication, Automotive Division, Ministry of Economy, Trade, and Industry, Japanese Government.

Jaffe, A., Newell, R., & Stavins, R., 1999. The Induced Innovation Hypothesis and Energy-Saving Technological Change. *Quarterly Journal of Economics*, 117: 941–75.

Jie, M. and Y. Hagiwara, 2013. "In Ghosn We Trust Tested as Nissan Electric Push Falters," Bloomberg, March 21.

Kahn, M. E., 2007. Do Greens Drive Hummers or Hybrids? Environmental Ideology as a Determinant of Consumer Choice. *Journal of Environmental Economics and Management*, 54(2): 129–45.

Kurtenbach, E., 2009. China Automakers Focus on Hybrid and Electric Vehicles. *USA Today*, April 20.

Maynard, M., 2003. *The End of Detroit: How the Big Three Lost Their Grip on the American Car Market.* New York: Currency/Doubleday, 284–90.

Ming, P., Jingguang, L., & Mytelka, L., 2008. Hydrogen and Fuel-Cell Activities in China. In L. Mytelka & G. Boyle (eds.) *Making Choices about Hydrogen*. Tokyo, New York, Paris: United Nations University Press.
NRC, 2002. *Effectiveness and Impact of Corporate Average Fuel Economy Standards. National Research Council Committee on the Effectiveness and Impact of Corporate Average Fuel Economy Standards*. Washington, DC: Transportation Research Board.
Rogers, E. M., 2003. *Diffusion of Innovations*. New York: Free Press.
Shantou EV Project Director Wen Zong Kong, Personal Communication, 2001. *Interview with Deputy Director and Senior Engineer, Vice Director, and General Manager of Shantou EV Project, Shantou*, Guandong, September 6–7.
Sperling, D., 2001. Public-Private Technology R&D Partnerships: Lessons from US Partnership for a New Generation of Vehicles. *Transport Policy*, 8: 247–56.
Turrentine, T. and Kurani, K., 2007. Car Buyers and Fuel Economy. *Energy Policy*, 35: 1213–23.
Wang, L., 2013. *Personal communication with Dr. Wang Lifang*, Chinese Academy of Sciences Institute of Electrical Engineering.
Yarime, M., Shiroyama, H., and Kuroki, Y., 2008. The Strategies of the Japanese Auto Industry in Developing Hybrid and Fuel-Cell Vehicles. In L. Mytelka & G. Boyle (eds.), *Making Choices about Hydrogen*. Tokyo, New York, Paris: United Nations University Press, 187–212.

ACKNOWLEDGMENTS

I gratefully acknowledge the comments and suggestions of Charlie Wilson and Arnulf Grubler, as well as the other lead authors of Chapter 24 on innovation systems in the *Global Energy Assessment*.

15 Solar Photovoltaics: Multiple Drivers of Technological Improvement

Gregory F. Nemet

Author's Summary

A variety of factors, including government activities, have enabled the two orders of magnitude reduction in the cost of solar photovoltaics (PV) over the past five decades. Despite this achievement, the technology remains too expensive compared to existing electricity sources, such that widespread deployment depends on substantial future improvements. No single determinant predominantly explains the improvement to date. Research and development (R&D), economies of scale, learning by doing, and knowledge spillovers from other technologies have all played a role in reducing system costs.

Improvements in electrical conversion efficiency have accounted for about one-third of the decline in cost over time. R&D, especially public sector R&D, has been central to this change. Deployment of PV has also benefited from a sequence of niche markets in which users of the technology were less price sensitive and had strong preferences for characteristics such as reliability and performance, which allowed product differentiation.

Governments have played a large role in stimulating increasing demand for PV, which has further reduced costs by enabling opportunities for economies of scale in manufacturing. Japan's program was especially innovative in that it not only adopted a long time horizon, but also set a declining subsidy that fell to zero after the ten years of the program. This provided not only a clear signal about market demand, but also clear expectations of future levels of subsidy. The Renewable Energy Law in Germany in the 2000s successfully replicated many of the features of Japan's program.

The current emergence of new technological generations, including thin film and organic PV, will require some reallocation in emphasis from market support to technical support through R&D, recognizing the sequence of innovation efforts from R&D to market formation to local learning by doing in the history of the dominant crystallized silicon PV.

Editors' Guide

Relationship of Case Study to Energy Technology Innovation System Framework

A common thread throughout this case study of solar PV is the hundredfold reduction in costs since the first cells were produced in the 1950s. Cost is a major feature, and often *the* major feature, of *adopter preferences* for technologies. Similarly, falling costs in absolute terms, and relative to the costs of alternative technologies providing the same or similar energy service, are a central element of the *relative advantage* of a technology. Yet adopter preferences are not homogeneous. The case study also emphasizes the successive *market niches* from powering satellites to powering off-grid homes, each with its distinctive set of preferences for particular performance (or cost) characteristics. These market niches provided the context for *learning* processes in commercial settings that help further improve relative advantage for a wider set of potential adopters who are more cost sensitive.

Solar PV can not yet compete on a level playing field with long-established incumbent technologies (particularly while these impose but do not bear environmental and social costs), but the progress of PV toward this point of "grid parity" is clear. *Economies of scale* captured through the mobilization of *resources* to grow the manufacturing base explain almost half of observed cost reductions. Much of the success of multiyear demand-side programs (in Japan and Germany) is attributable to the clear and long-term confidence in future demand that enabled large investments in manufacturing facilities to bring down costs through *economies of scale*. Capturing scale economies is therefore contingent on robust *shared expectations* for future markets, a classic "market-pull" driver associated with the *adoption and use* of a technology.

The case study also highlights the importance of differentiating cost improvements across the various components of an installed technology. Although balance of system components seems to have improved apace with the PV module costs, the components' dispersion is higher and their sensitivity to local conditions greater. Opportunities for *knowledge spillovers* are also therefore lower, particularly if compared to the documented transfers of materials and techniques into the PV module manufacturing industries.

This multi-decadal progress in the cost and performance of solar PV is characterized by persistent interactions between the *knowledge generation* (R&D), *learning*, and *knowledge spillover* processes that lead most directly to cost improvements, and the *actors and institutions* and *resources* upon which they depend. *Shared expectations* are a key institution identified as critical for providing the necessary confidence in future market demand to attract inward investment for growing the manufacturing base and supporting widespread diffusion. Both government and market *actors* played a central role in developing solar PV technologies but with a shifting emphasis from government-sponsored R&D breakthroughs to private industry mediation of *knowledge spillovers* and localized *learning* in PV system installation.

Insights for Energy Technology Innovation Policy

- There are multiple, interacting drivers of cost reduction, including R&D, demand subsidies, learning by doing in localized installation industries, and knowledge spillovers between and within industries. There is no one-size-fits-all innovation policy, nor a singular entry point through either technology-push or market-pull approaches.
- Timing also matters. Discrete innovations share a common life cycle through the innovation stages from R&D to diffusion and senescence. Innovation systems also develop and change alongside maturing technologies. This is illustrated in the case study by the main drivers of cost improvements shifting from R&D to market demand and shared expectations and then to learning. As this sequence continues for the original crystalline silicon technology, it potentially begins again for newer thin film technologies. Consequently, the timing and target of innovation policy must be sensitive to differences in the maturity of potentially overlapping technological generations. The success of new technological generations may require renewed R&D support even while markets for existing technologies are expanding.
- R&D support needs a long-term commitment, whether through budgets or grants spanning multiple years, or supporting policies as in the cases of Japan's Sunshine Program or the United States' Project Independence, both of which demonstrated commitment by elevating PV developments to a serious national priority.
- The investments needed to grow an industry's manufacturing base to capture available scale economies and so bring down costs depend on investors and other market actors sharing clarity and confidence in the long-term future of market demand. If this market demand is created or shaped by public policy, it is critical that this policy is stable, credible, and visible to allow shared expectations within an industry to coalesce. Inflated levels of subsidies are not just unsustainable over successive economic and election cycles, but they can also risk inflating costs by pushing up end users' willingness to pay for a technology (to obtain the subsidy).
- Niche markets have been a crucial aspect of the innovation history of solar PV. Excepting the current emphasis on grid PV, these niche markets share the characteristic of being variously enhanced or supported *but not bankrolled* by public policy. This independence from public subsidy to internalize environmental or social benefits has proven important, as solar PV has had to rely on privately appropriable performance advantages to drive diffusion. These are more enduring and less exposed to stop-start interest and initiatives than are cost advantages wholly dependent on public policy. Nevertheless, through procurement, standards, market regulations, or otherwise, policy can play a role in supporting these niches indirectly.

1 Introduction

A variety of factors, including government activities, have enabled the two orders of magnitude reduction in the cost of PV over the past five decades. Research and

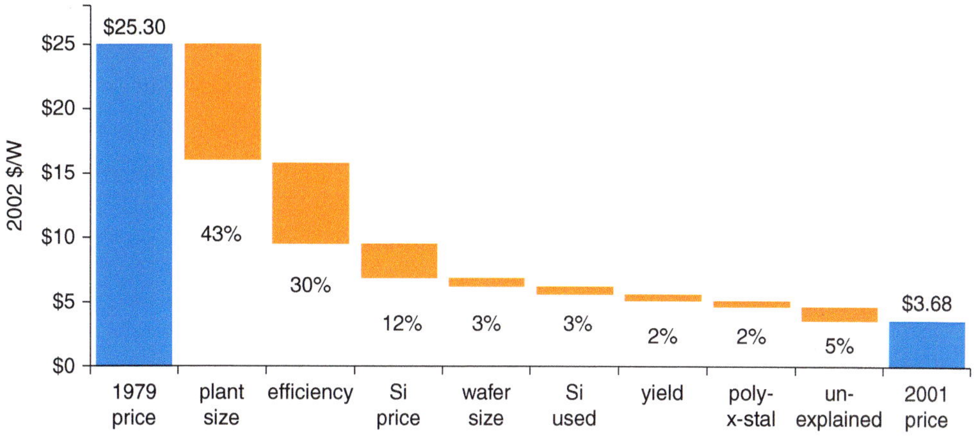

Figure 15.1. Portion of cost reduction in PV modules accounted for by different factors.
Notes: Si = Silicon. Poly-x-stal = poly-crystalline.

development (R&D), economies of scale, learning by doing, and knowledge spillovers from other technologies have all played a part in driving technological improvement and reducing PV system costs. Not all of these factors, and the interactions between them, were important for the entire sequence of the technology's development. Certain factors dominated for periods defined by shifts of emphasis in the innovation efforts of both the private sector and governments. From the 1970s to 1985, emphasis was on R&D to improve efficiencies and manufacturing techniques of crystallized silicon PV. In the 1990s and early 2000s, the emphasis shifted to long-term demand programs that enabled economies of scale in manufacturing. And in the 2000s, innovation efforts focused on stimulating local learning by doing to reduce installation costs in addition to continuous improvements through manufacturing scale. This sequence from R&D to market formation to local learning by doing for the historically dominant crystallized silicon PV may now be repeated for emerging alternative solar technologies including thin film and organic PV.

2 R&D and Efficiency Improvements

The average electrical conversion efficiency of PV systems has more than doubled since the first commercial systems in the 1970s. Efficiency has been important to cost reductions, accounting for about a third of the decline in cost over time (see Figure 15.1). R&D, especially public sector R&D, has been central to this change. Data on the highest laboratory cell efficiencies over time show that of the sixteen advances in efficiency since 1980 (Surek, 2003), only six were accomplished by firms that manufacture commercial cells. Most of the improvements were accomplished by universities and government laboratories, none of which would have learned from experience with large-scale production; government and university R&D programs produced ten of the sixteen breakthroughs in cell efficiency. Almost every one of the twenty most important improvements in PV occurred during a ten-year period between the mid-1970s and the mid-1980s (Green, 2005), most of them in the United States, where more than a billion dollars were invested in PV R&D during that period (Taylor et al., 2007).

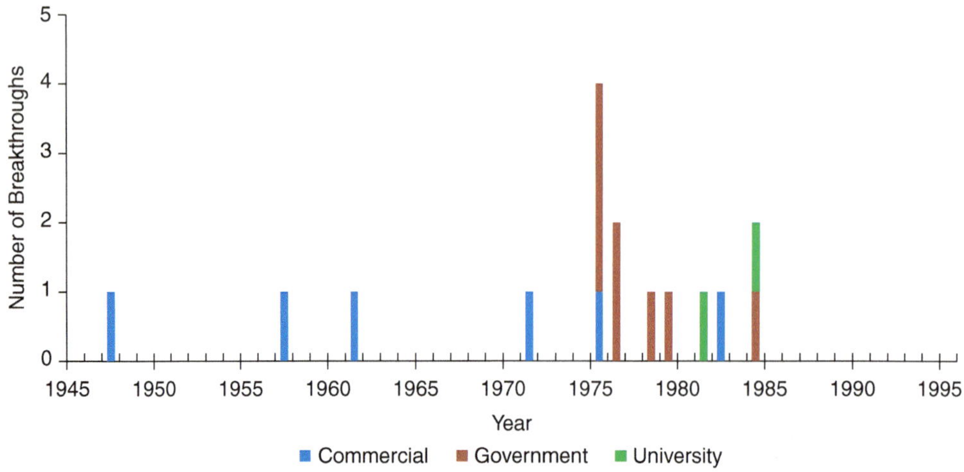

Figure 15.2. Frequency of breakthroughs in PV technology by origin.
Sources: Husmann (2011); Nemet and Husmann (2012).

As a second perspective on R&D outcomes, Husmann used a combination of expert opinion and patent citations to identify seventeen breakthroughs in PV technology from the 1940s through the early 1990s (Husmann, 2011; Nemet and Husmann, 2012). The timing and type of organization creating the breakthrough are shown in Figure 15.2. Examples of breakthroughs in PV technology include: Boron dopants (1947), antireflective coatings (1961), hydrogen passivation (1975), ethylene vinyl acetate (EVA) as a laminating material (1975), and reactive ion etching (1976). The earliest breakthroughs occurred in commercial settings before a substantial PV R&D program began in the mid-1970s, and included inventions by Bell Laboratories, Hoffman Electronics, and Westinghouse Electric. Most breakthroughs, however, occurred during the period of highest R&D from the mid-1970s to the mid-1980s.

3 Sequential Niche Markets

Deployment of PV has benefited from a sequence of niche markets. Customers in these markets were less price sensitive and had strong preferences for characteristics such as reliability, grid independence, and performance (including flexibility and weight) that allowed product differentiation. Governments have played a large role in creating or enhancing these niche markets. In the 1960s and 1970s the U.S. space program and the Department of Defense accounted for more than half the global market for PV. The high cost of electricity in space allowed PV to be competitive even at an early stage and even when electricity was above \$200/kWh (at least three orders of magnitude above current U.S. wholesale rates). Subsequent niche markets included telecom repeater stations, off-grid homes, and especially consumer electronics such as toys, calculators, and watches. Each was important independently of government decision making and innovation policy. This allowed the PV industry to expand from the mid-1980s until the late 1990s, during which time energy prices fell and alternative energy sources lost their importance as a social priority. From the 1990s onward, households with strong preferences for environmental protection created larger markets, especially as rooftop PV was a conspicuous, socially visible

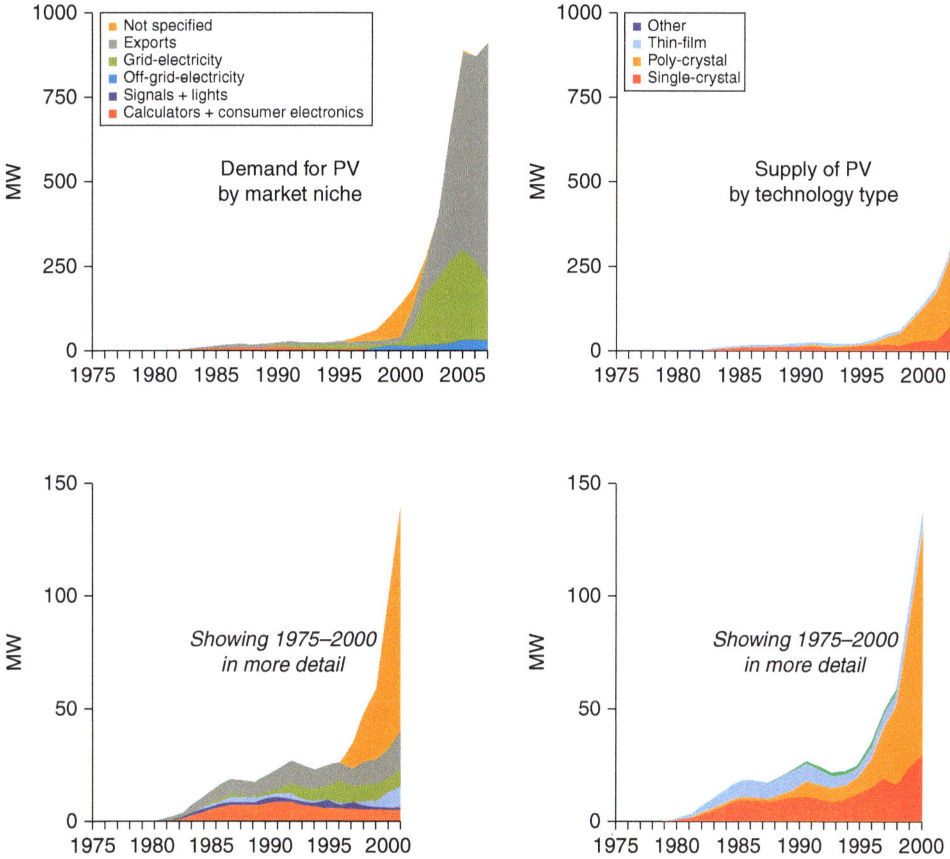

Figure 15.3. Categories of demand niches and technology types for PV in Japan.
Notes: Left panels show size of different market niches or demand for PV; Right panels show PV technologies supplying this demand (units shown in capacity terms). Top panels show full time series and y-axis; bottom panels show truncated time series and y-axis to expand on pre-2000 data.
Sources: Watanabe and colleagues (2002); JPEA (2010).

form of proenvironmental action. Figure 15.3 shows the progression of sequential PV niche markets in Japan. Especially important at early stages were calculators, consumer electronics, signals, and off-grid applications.

4 Market Demand and Expectations

Increasing demand for PV has reduced costs by enabling opportunities for economies of scale in manufacturing. Nemet (2006) assembled empirical data to populate a simple engineering-based model identifying the most important factors affecting the cost of PV modules over the past three decades. That study found that three factors account for almost all of the observed cost reductions: (1) a two orders of magnitude increase in the size of manufacturing facilities that provided opportunities for economies of scale; (2) a doubling in the electrical conversion efficiency of commercial modules; and (3) a fall in the price of the primary input material, purified silicon. Because investments in larger facilities take time to pay off, economies of scale depend on expectations of future demand. As a result, public programs that reduce uncertainty

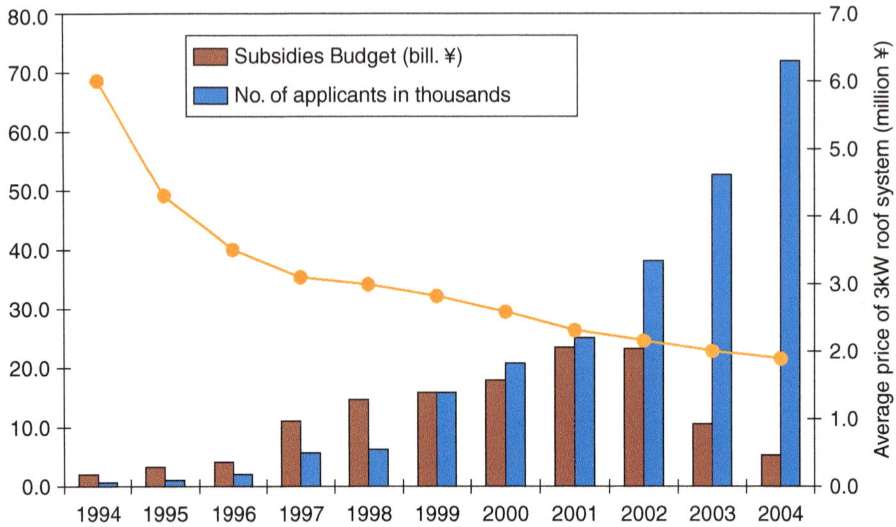

Figure 15.4. Japan subsidies and applications for rooftop PV systems.
Notes: Bars show subsidies and applications on left y-axis; Line shows average prices on right y-axis.
Source: Jäger-Waldau (2004).

by setting clear long-term expectations, such as Japan's Sunshine Program in the 1990s (Shum and Watanabe, 2007), are more effective at enabling scale economies than generous subsidies that can suddenly disappear, such as California's incentives for wind and solar in the early 1980s. Japan's program was especially innovative in that it took not only a long time horizon but also set a declining subsidy such that it fell to zero after the ten years of the program (Figure 15.4). This provided not only expectations of demand but also clear expectations of future levels of subsidy.

Government programs with longer time horizons, such as Germany's feed-in tariffs in the 2000s and the California Solar Initiative launched in 2006, create similar opportunities (Nemet, 2007; Peevey and Malcolm, 2006). Germany's program is especially notable in that production there has become sufficient to create external economies of scale with the emergence of machine tool manufacturers that now produce equipment specifically for the PV industry (Neuhoff et al., 2007). Similarly, production of lower purity, and thus cheaper solar-grade silicon is now profitable because plants can be built at large scales. Economies of scale in unit size are also now observable: large installations show much lower costs per watt than do small residential systems (Wiser, Barbose, and Peterman, 2009). Increasing installation size is likely to become more important as economies of scale reduce module manufacturing costs, leaving installation costs as an increasingly large share of system costs (with total costs comprising module costs and "balance of system" costs including hardware, mounting, and installation).

5 Learning by Doing and Knowledge Spillovers

Learning from experience in production has played a rather small role in reducing module costs (Nemet, 2006). However, learning by doing may play a much more important role in reducing installation costs (Benthem, Gillingham, and Sweeney, 2008). An important aspect of this learning is that it is a local phenomenon, whereas module production is truly global (Shum and Watanabe, 2008). As installation costs become a larger portion of total system costs, the extent of this learning by doing will become increasingly important. Whether or not the benefits of this learning are appropriable by private installers will determine whether governments need to play a role in promoting learning investments. For example, there may be benefits to free riding on the experiences of others, which may slow down the pace of learning.

Like many technologies, PV has benefited from the adoption of innovations that originated in other industries. These include the use of excess purified silicon from the chip-making industry; the use of wire saws from radial tires to slice multiple silicon wafers; electronics connectors to ease installation; screen printing techniques from lithography; as well as an array of manufacturing techniques taken from microprocessors in the case of crystallized silicon PV, and from LCD displays in the case of thin film PV (Keshner and Arya, 2004).

The global aspect of module manufacturing suggests that interfirm and international technology spillovers are likely to be more important in module production than in installation. International spillovers can be found in trade statistics as knowledge embodied in devices. The distinction between the value of capital and of knowledge in trade data can be difficult to observe directly. Recent efforts have augmented trade data with patent citation data to better identify flows of knowledge from entities in one country to those in another (Dechezleprêtre et al., 2011; Popp, Hascic, and Medhi, 2011).

6 Interactions between R&D and Demand

Certain periods in the history of PV have seen particular drivers of innovation dominate. R&D led to improved efficiencies and manufacturing techniques from the 1970s to 1985. Market formation and demand-side programs underwrote manufacturing economies of scale in the 1990s. And, as described further later in this chapter, local learning by doing helped reduce installation costs in the 2000s. But although particular factors dominate in particular periods, interactions among the various factors have also been important. One such "virtuous cycle" was particularly relevant in Japan in the 1990s, when strong demand-side policies and support for R&D existed simultaneously (Watanabe, Wakabayashi, and Miyazawa, 2000). In this case, it was not so much efficiency breakthroughs and alternative cell designs that drove improvements, but rather support from the Japanese government via the Ministry of Economy, Trade and Industry (METI) that enabled coordination of expectations and sharing of best practices, which in turn supported manufacturing improvements. The influence of the U.S. government may have played a similar role in enabling the 1970s breakthroughs, not just through providing resources for R&D, but by creating a sense of commitment that convinced many to work on the technical and market challenges associated with commercializing this nascent technology (Laird, 2001).

7 Cost Trajectories

Changes in demand and in market structure have affected the cost of PV as well, even if they don't directly affect the technical characteristics of the technology. Changes in industry concentration in the 1980s and 1990s affected market power and led to a changing relationship between prices and costs over time. The very high levels of

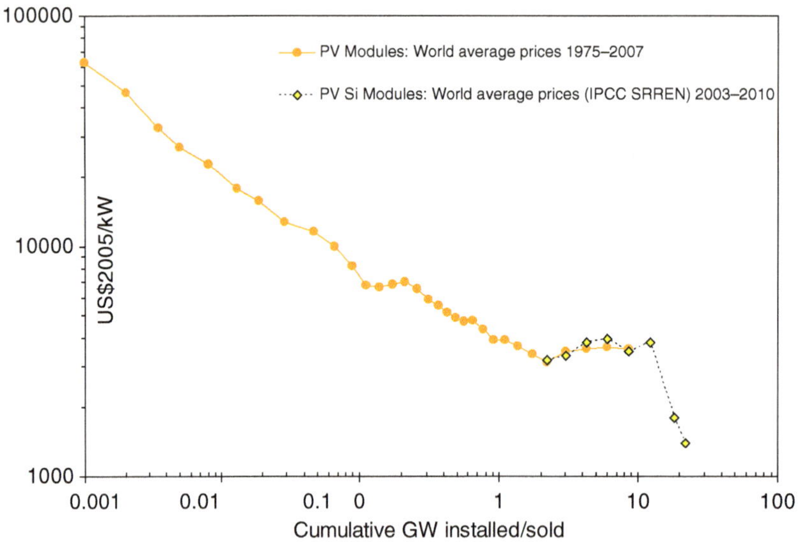

Figure 15.5. Prices for PV modules (1975–2010) and PV systems (1998–2009).
Sources: Edenhofer and colleagues (2011); Nemet and Baker (2009); Wiser and colleagues (2009).

demand in the 2005–2007 period were in part to blame for the high prices during that period that led to a reversal in the multi-decade downward cost trajectory. This can be clearly seen in Figure 15.5, which shows the learning curve of PV module and system prices as a function of cumulative deployment. A rough estimate suggests that half of the cost increase as cumulative capacity grew from 2 GW to around 10 GW was due to higher material costs and the rest likely due to higher willingness to pay, as aggressive subsidy programs brought new consumers to the market for PV. The most recent data from the IPCC's Special Report on Renewable Energy (SRREN), also shown in Figure 15.5, indicates the downward cost trend for PV modules has resumed.

Most cost assessments of PV have focused on the evolution of the core technology, the modules that convert sunlight to electricity. However, modules no longer dominate systems costs. As the prices of modules have fallen, the rest of the components that comprise a PV system have accounted for an increasing share of the overall costs. A study of balance of system (BOS) prices in the 1990s found that the rate in technology improvement in BOS components has been quite similar to that of modules, which some may find surprising given the heterogeneous set of components and activities that fall under the rubric of BOS (Schaeffer et al., 2004). The dispersion in the data for BOS components is higher than that for modules. The cost of inverters, the largest hardware component of BOS, improved more slowly than modules. However non-inverter costs, including installation, wiring, and mounting systems, improved even more quickly.

More recently, work at Lawrence Berkeley National Laboratory has documented the cost of U.S. installed systems from 1998 to 2010 (Barbose et al., 2011; Wiser et al., 2009). System costs have fallen from more than $10,000/kW in 1998 to $6,200/kW in 2010, with an unusually large decline during 2010. Further detail is shown in Figure 15.6, which includes the number of systems and total capacity installed each year.

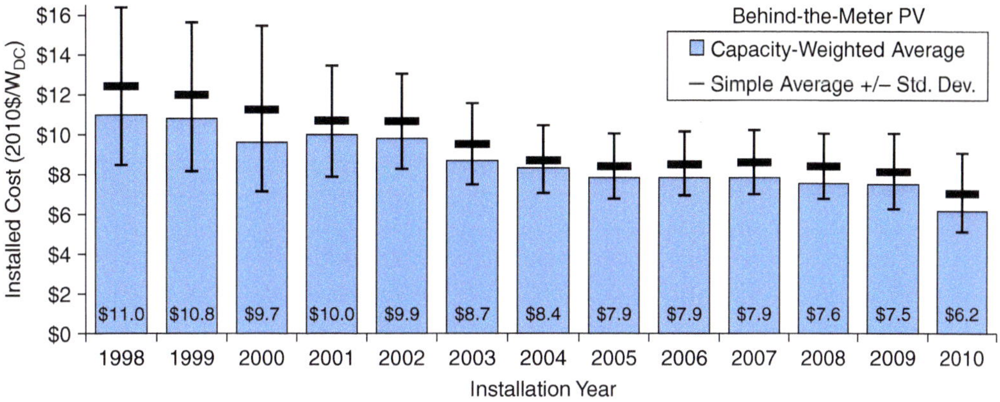

Figure 15.6. Installed cost of PV systems (1998–2007).
Source: Barbose and colleagues (2011).

8 Conclusions and Prospects for the Next Generation of PV Technologies

The history of PV technology is in many ways a success story when one considers the factor of 100 reduction in costs since the first cells in the 1950s. This story is important not only for its lessons for other technologies, but also for the implications for future developments of PV.

PV has the potential to be very important for climate change mitigation and energy security yet still costs more than conventional power sources. Several observations are prominent from the large set of studies documenting these developments.

- The success of PV is attributable not mainly to one area of emphasis but to a wide array of supporting policy instruments, including R&D, demand subsidies, industrial development, and coordination.
- R&D support needs long-term commitment. A commitment can convey credibility through budget consistency, grants that span multiple years, or supporting policies with their own longevity, such as Japan's Sunshine Program. An alternative means for establishing the credibility of long-term commitments was Nixon's Project Independence in the United States, which made clear that leadership considered alternative energy a serious national priority.
- A sequence of niche markets with sequentially declining willingness to pay has been essential for the development of the PV industry. Niche markets have been most effective when they were not government supported. For example, remote telecommunications applications, off-grid residences, and consumer electronics helped the industry grow and the technology improve even though government programs did not directly support them. In this sense, PV could be considered fortunate, as natural niche markets in the absence of government support do not exist for all technologies.
- Long-term demand subsidies with clear timelines have been crucial to stimulating deployment and reducing costs because the benefits of demand subsidies come from expectations about markets in the long term. Clear expectations encourage private sector investments with long-term payoffs, such as large manufacturing facilities and technology development.

Based on these insights on successful innovation policies, what are the prospects for the new technological generations of PV? Recent work suggests that their success may require renewed R&D support even while markets for the existing technology are expanding (Nemet and Baker, 2009). As with other technologies, the need for R&D investment does not end once commercial markets take off. PV based on crystalline silicon has remained the overwhelmingly dominant technology for three decades despite the advantages of thin film technologies that use less raw material and are more amenable to mass manufacturing techniques. Given that installation methods, inverters, and maintenance are quite similar across types of PV, there is likely more opportunity for intergenerational spillover in system installation than in module production. So an important innovation-related policy question is the extent to which learning and improvement in crystalline silicon PV modules will be transferable to thin film modules.

Innovation policy oriented toward the longer term must address subsequent generations of PV technology, such as those based purely on organic materials. Nemet and Baker (2009) found that production-related effects on technological advance – learning by doing and economies of scale – are not as critical to the long-term potential for cost reduction in organic PV as is the investment in and success of R&D. The emergence of new technological generations will require some reallocation in emphasis from market support to technical support through R&D, recognizing the sequence of innovation efforts from R&D to market formation to local learning by doing in the history of the dominant crystallized silicon PV.

FURTHER READING

The following references give useful additional detail to the innovation history of solar PV: Barbose and colleagues (2011); Laird (2001); Nemet and Husmann (2012); Nemet (2006).

REFERENCES

Barbose, G., Darghouth, N., Wiser, R., & Seel, J., 2011. *Tracking the Sun IV: An Historical Summary of the Installed Cost of Photovoltaics in the United States from 1998 to 2010*. Berkeley, CA: Lawrence Berkeley National Laboratory.

Benthem, A. V., Gillingham, K., & Sweeney, J., 2008. Learning-by-Doing and the Optimal Solar Policy in California. *The Energy Journal*, 29(3): 131.

Dechezleprêtre, A., Glachant, M., Haščic, I., Johnstone, N., & Ménière, Y., 2011. *Invention and Transfer of Climate Change Mitigation Technologies: A Global Analysis*. Paris: CERNA, Mines ParisTech.

Edenhofer, O., Pichs Madruga, R., Sokona, Y., Seyboth, K., Matschoss, P., Kadner, S., Zwickel, T., Eickemeier, P., Hansen, G., Schlömer, S., & Von Stechow, C., 2011. *IPCC Special Report on Renewable Energy Sources and Climate Change Mitigation*. Cambridge, UK and New York, NY: Cambridge University Press.

Green, M. A., 2005. Silicon Photovoltaic Modules: A Brief History of the First 50 Years. *Progress in Photovoltaics: Research and Applications*, 13(5): 447–55.

Husmann, D., 2011. *U.S. Photovoltaic Breakthroughs from 1947 To 1993: Their Identification, Origin, and Commercialization*. M.S. Masters thesis, University of Wisconsin, Madison.

Jäger-Waldau, A., 2004. *Research, Solar Cell Production and Market Implementation of Photovoltaics*. Ispra, Italy: European Commission, DG JRC, Institute for Environment and Sustainability, Renewable Energies Unit.

JPEA, 2010. *Shipment of PV Modules in Japan*. Japan Photovoltaic Energy Association (JPEA). http://www.jpea.gr.jp/pdf/qlg2010_Eng.pdf.
Keshner, M. S. & Arya, R., 2004. *Study of the Potential Cost Reductions Resulting from Super-Large-Scale Manufacturing of PV Modules*. Golden, CO: National Renewable Energy Laboratory.
Laird, F. N., 2001. *Solar Energy, Technology Policy, and Institutional Values*. New York: Cambridge University Press.
Nemet, G. F., 2006. Beyond the Learning Curve: Factors Influencing Cost Reductions in Photovoltaics. *Energy Policy*, 34(17): 3218–32.
2007. *Policy and Innovation in Low-Carbon Energy Technologies*. PhD Dissertation, University of California – Berkeley.
Nemet, G. F. & Baker, E., 2009. Demand Subsidies versus R\&D: Comparing the Uncertain Impacts of Policy on a Pre-Commercial Low-Carbon Energy Technology. *The Energy Journal*, 30(4): 49–80.
Nemet, G. F. & Husmann, D., 2012. Historical and Future Cost Dynamics of Photovoltaic Technology. In A. Sayigh (ed.) *Comprehensive Renewable Energy*. Oxford: Elsevier.
Neuhoff, K., Nemet, G., Sato, M., & Schumacher, K., 2007. *The Role of the Supply Chain in Innovation: The Example of Photovoltaic Cells*. Cambridge, UK: University of Cambridge – Electricity Policy Research Group.
Peevey, M. R. & Malcolm, K., 2006. *Interim Order Adopting Policies and Funding for the California Solar Initiative*. California Public Utilities Commission.
Popp, D., Hascic, I., & Medhi, N., 2011. Technology and the Diffusion of Renewable Energy. *Energy Economics*, 33(4): 648–62.
Schaeffer, G. J., Seebregts, A. J., Beurskens, L. W. M., Moor, H. H. C. D., Alsema, E. A., Sark, W., Durstewicz, M., Perrin, M., Boulanger, P., Laukamp, H., & Zuccaro, C., 2004. *Learning from the Sun; Analysis of the use of experience curves for energy policy purposes: The case of photovoltaic power. Final report of the PHOTEX project*, ECN Renewable Energy in the Built Environment.
Shum, K. L. & Watanabe, C., 2007. Photovoltaic Deployment Strategy in Japan and the USA – An Institutional Appraisal. *Energy Policy*, 35(2): 1186–95.
2008. Toward a Local Learning (Innovation) Model of Solar Photovoltaic Development. *Energy Policy*, 36(2): 508–21.
Surek, T., 2003. Progress in U.S. Photovoltaics: Looking Back 30 years and Looking Ahead 20. *3rd World Conference on Photovoltaic Energy Conversion*. Osaka, Japan.
Taylor, M., Nemet, G., Colvin, M., Begley, L., Wadia, C., & Dillavou, T., 2007. *Government Actions and Innovation in Clean Energy Technologies: The Cases of Photovoltaic Cells, Solar Thermal Electric Power, and Solar Water Heating, CEC-500–2007–012*. Sacramento, California Energy Commission.
Watanabe, C., Griffy-Brown, C., Zhu, B., & Nagamatsu, A., 2002. Inter-Firm Technology Spillover and the "Virtuous Cycle" of Photovoltaic Development in Japan. In A. Grubler, N. Nakicenovic, & W. Nordhaus (eds.) *Technological Change and the Environment*. Washington, DC: Resources for the Future.
Watanabe, C., Wakabayashi, K. & Miyazawa, T., 2000. Industrial Dynamism and the Creation of a "Virtuous Cycle" between R&D, Market Growth and Price Reduction – The Case of Photovoltaic Power Generation (PV) Development in Japan. *Technovation*, 20(6): 299–312.
Wiser, R., Barbose, G., & Peterman, C., 2009. *Tracking the Sun: The Installed Cost of Photovoltaics from 1998–2007*. Berkeley, CA: Lawrence Berkeley National Laboratory.

PART V

ACTORS AND INSTITUTIONS IN THE ENERGY TECHNOLOGY SYSTEM

16 A Comparative Assessment of Wind Turbine Innovation and Diffusion Policies

Lena Neij and Per Dannemand Andersen

Authors' Summary

Wind turbines have become a mainstream technology, a first choice for many when investing in new electricity generation facilities. This comparative case study addresses how governmental policy has been formulated to support wind turbine innovation and diffusion processes. Three innovation stages and corresponding innovation strategies are identified. The first stage is the early movers of the 1970s and early 1980s, including pioneer countries such as Denmark, the United States, Germany, and the Netherlands. The second stage is the booming markets of the 1990s, guided by the successful Danish innovation path of the 1980s. The third stage is the emerging markets of the 1990s and 2000s, including countries such as India and China. Within these different stages, common elements in government policy strategies can be identified as essential for a sustainable and successful innovation process. These can be summarised as:

- support diversity in technology and market formation
- research, development, and demonstration (RD&D) is necessary but not sufficient
- quality assurance is essential for new technologies
- support interaction and networking
- ensure support is stable, continuous, and flexible

Taken together, these elements of a successful innovation approach show that government policy needs to support the development of the entire innovation system: not just the development of turbines and associated infrastructure, but also the involvement of actors, networks, and market institutions. The case of wind energy development in Denmark presents a good example of how and why such a systemic approach can succeed.

Editors' Guide

Relationship of Case Study to Energy Technology Innovation System Framework

Support for diversity in both technology and market formation activities proved essential. Diversity is the breeding ground for *entrepreneurial experimentation*. Diversity also enhances the potential for *knowledge spillovers* from one application (or market) to another, as is now evident in the emerging offshore wind sector for which tested and established onshore technologies have been adapted. The wind power innovation history also exemplifies *knowledge spillovers* from early to later adopting markets. This is clearly seen in the technology development joint venture between Spanish and pioneer Danish manufacturers that led to the rapid growth of the wind market in Spain through the 2000s. *Knowledge spillovers* embedded in technology transfer arrangements are now similarly evident in China and India, although both countries have *mobilised significant resources* in domestic market formation activities including the development of local institutions and manufacturing capabilities.

Quality assurance is an example of an important institution. Quality assurance is essential to underpin market confidence and the *shared expectations* that form around the magnitude and certainty of a new technology's prospects. In the U.S. case, the lack of operational testing ultimately led to technical problems that undermined investor confidence. Quality assurance can rest on a formal standards and certification process, but the involvement of a number of *actors and institutions* is necessary to make this work. In the Danish case, the testing station at Risø not only served as the certification institution, but also facilitated *exchange and interaction* among diverse actors, from turbine manufacturers and developers to independent technical experts and landowners. This in turn helped reinforce the confidence and *shared expectations* in the market. Although the Netherlands had a similar institutional setup, it was based within a competitive market setup that limited these incentives for mutually productive *exchange and interaction*.

The *learning* benefits associated with early experiences of wind turbine deployment meant performance improvements across a range of technology attributes including cost but also reliability, efficiency, noise reduction, and grid stability. This learning process extended over many years also enabled *economies of scale* to be successfully captured in the 1990s, unlike the earlier, premature efforts to rapidly up-scale turbines on an insufficient knowledge base.

Insights for Energy Technology Innovation Policy

- There are four central policy lessons from the innovation history of wind power: (i) combine both technology-push (RD&D) with market-pull incentives (market formation and institution building); (ii) sustain diversity of both technologies and market niches through the formative years of the industry; (iii) maintain continuous and stable policy incentives; (iv) support the entire innovation system, including the diverse actors and institutions involved.

- There are inherent difficulties foreseeing the performance and improvement trajectories of any technology and also the pivotal roles particular actors or institutions play. Public policy therefore plays an essential role in sustaining diversity. Supporting or protecting distinct market niches is one route to nurturing both technological diversity and actor diversity. Public institutions can also play a key role in coordinating and encouraging exchange and interaction between market actors, with the Danish test station a case in point.
- The history of wind turbines development is long, beginning in the 1880s and experiencing many bumps and failures since then. Innovation successes have been built on stable and continuous support, which has sustained the knowledge generated and learnt along the way.

1 Introduction: Early History of Wind Energy

Wind turbines have become a mainstream energy technology, a first choice for many when investing in new electricity generation facilities. Wind energy represents the potential for a greener development pathway. The global commercial reach of leading wind turbine manufacturers is a marker of successful innovation processes. What, then, characterises the success story of wind turbine innovation? And what role has public policy played?

The first modern electricity-producing wind turbines were developed in the 1880s, but not until the 1970s were wind turbines seen as an important innovation path for future development. The early years of wind turbine development were characterised by experimentation and development of different types of wind turbine concepts (see Figure 16.1). Wind turbines came to be partially successful in rural electrification, specifically in the time up to and during World War 1. In the 1950s, the first grid-connected wind turbine, with a three-bladed rotor, was built. Research and development (R&D) efforts formed the basis of Juul's pioneering design of the modern electricity-producing wind turbine, the 200kW Gedser turbine, installed in 1957 and operational until 1967 (see Figure 16.1c). This in turn formed the basis of the so-called Danish wind turbine concept: three-bladed designs and the rotor position actively maintained upwind using an electrical motor. In 1961, a United Nations conference on new sources of energy was held in Rome at which different wind energy concepts and turbine designs were discussed (United Nations, 1964). A 1955 textbook, *The Generation of Electricity by Wind Power* by E. W. Golding, was republished in new editions and widely used during the 1970s and 1980s. Overall, the extended experimentation throughout the early and mid-twentieth century accumulated a platform of knowledge and concepts that the modern wind industry could build on.

2 Early Innovation Policies in the 1970s and Early 1980s

In the 1970s, the pioneer countries in supporting wind energy through innovation policies included Denmark, the United States, Germany, the Netherlands, Great

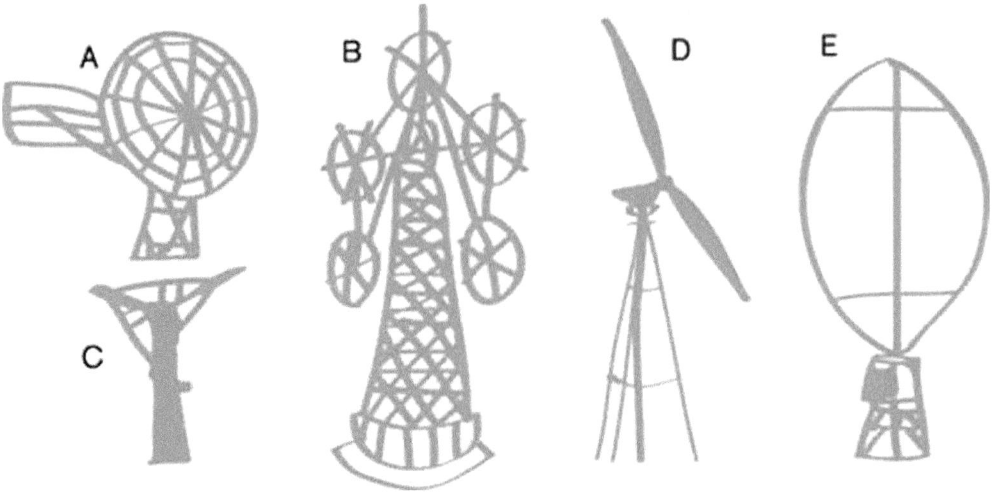

Figure 16.1. Historical wind turbine concepts.
Notes: (A) The world's first automatically operated wind turbine (12kw Turbine), Cleveland 1888; (B) Hermann Honnef Wind Turbine, Germany 1932; (C) Johannes Juul's 200Kw Gedser Wind Turbine, Denmark 1957; (D) Ulrich Hütter Wind Turbine, Germany 1959; (E) A Darrieus Wind Turbine (named after French inventor Georges Darrieus) patented in 1931.

Britain, and Sweden. However, two different approaches to technology development were taken (e.g., Gipe, 1995; Sawin, 2001). One line of public R&D programmes focussed on large, megawatt-scale turbines (see Figure 16.2). The R&D programmes were science based, developing advanced turbines with high efficiencies to compensate for higher construction costs. Knowledge development was, however, complex. R&D resources ploughed into these programmes in countries like Sweden, Germany, and the United Kingdom resulted in only a few 2–4MW turbines with limited operating hours (Gipe, 1995; Heymann, 1998). The two-bladed 3MW Growian turbine with 50m blades was an exemplar of unsuccessful rapid up-scaling. It managed 420 hours of operation during the period 1983–87, after which it was demolished because of numerous severe design faults.

In addition, market-pull drivers for these megawatt-scale turbines was limited. The intended owners and users were established utilities, but their interest in large and unproven wind turbines was limited, as were the market formation activities supported by governments (see, e.g., Åstrand and Neij, 2006; Kamp, Smits, and Andriesse, 2003). The relative failure of this line of development focussing on large-scale wind turbines did not, however, have larger financial consequences as investments were relatively small.

The other line of R&D programmes focussed on smaller turbines for specific niche markets. In particular, the Danish, Dutch, and U.S. wind power policies supported the development of small, simple, and reliable machines. In Denmark, the government also invested in market formation activities. As early as 1979, the Danish R&D programmes were combined with investment and production subsidies (similar to the feed-in tariffs for renewable electricity used widely today). The open availability of subsidies had unforeseen effects as smaller actors, such as farmers and entrepreneurs, entered the market. Their involvement in the innovation process proved essential.

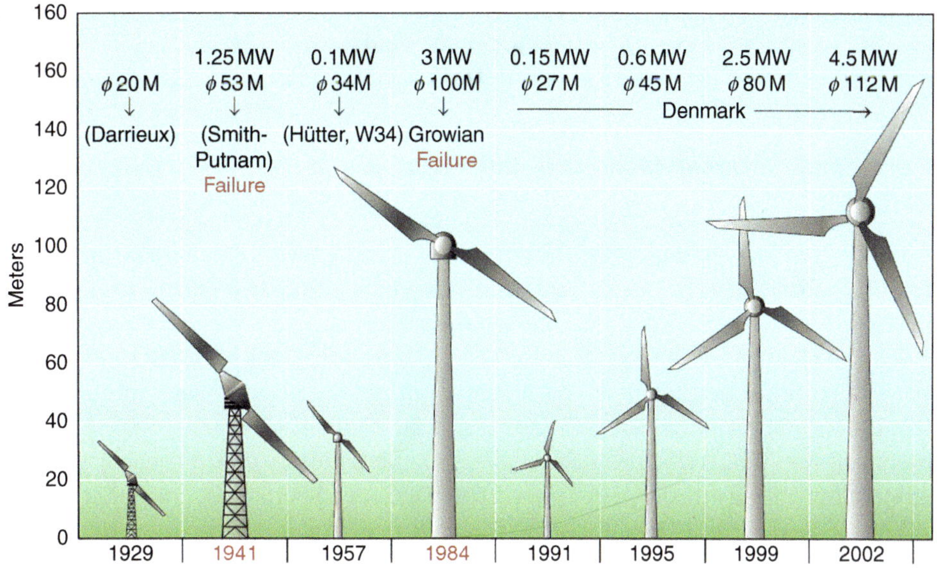

Figure 16.2. Size of largest wind turbines including failures.
Source: original artwork, copyright International Institute for Applied Systems Analysis (IIASA).

The combined technology-push and market-pull of early Danish innovation was symptomatic of a broader systemic approach to wind power. The development of a domestic market was promoted through wind resource mapping, grid connection regulations, guidelines and regulation on municipal planning (governing the permitting process), and information activities. Considerable emphasis was also placed on technological reliability. In 1978, the Test Station for Wind Turbines was established at the Risø National Laboratory, and from 1979 the Danish Energy Agency authorised Risø to develop a certification process for wind turbines that it could then issue and administer. The Risø National Laboratory's role in testing turbines, together with its more traditional forms of R&D, led to a fruitful three-way interaction between actors in industry, policy, and research, with Risø providing manufacturers with essential technical and scientific support.

Like Denmark, the Netherlands introduced subsidies and set up a test station in 1981. However, the Dutch approach focussed on supporting a competitive market, and manufacturers did not receive any support on how to improve the turbines (Kamp et al., 2003). This competitive approach in the Netherlands also meant more limited interaction between actors, and so less of the essential learning and feedback processes seen in Denmark.

In the United States, a number of subsidy schemes were introduced. As a result, the market for wind turbines took off rapidly, the number of wind turbine manufacturers increased, and the importing of turbines escalated, especially from Denmark (Loiter and Norberg-Bohm, 1999). However, the market boom required U.S. manufacturers to develop wind turbines quickly and with inadequate operational testing at a time when U.S.-manufactured turbine designs were still diverse, advanced, and largely unproven (Heymann, 1998). Many of the U.S.-manufactured wind turbines experienced severe operating problems. In 1985, only 38 percent worked properly compared to a figure of 98 percent for well-operating Danish-produced turbines (Heymann, 1998). In 1986, the Californian market crashed because of the poor performance of many installed turbines and a reduction in government subsidies. Up until then, California had installed 1.2GW wind power, representing 90 percent of the global installed capacity at that time (U.S. DOE, 2008).

3 Booming European Markets in the 1990s

The experiences of wind energy development in the 1970s and early 1980s guided subsequent activity in the 1990s. The successful Danish innovation path of the 1980s was the guiding star for many countries. Germany refocused on the production of less advanced small and medium-sized turbines in line with the Danish wind energy concept. Spain's approach for overcoming technical barriers to entering the wind turbine market involved a joint venture, Gamesa Eolica, between the Spanish turbine manufacturer, Gamesa, and the established Danish wind turbine manufacturer, Vestas. Established in 1994, Gamesa Eolica paid licensing fees to manufacture Vestas turbines in Spain. Like Germany, Spain came to manufacture turbines based on the Danish concept, benefitting from technology transfer and spillover of know-how. Like Denmark, Germany and Spain introduced investment and production incentives to support market formation. In 1990, a feed-in law was introduced in Germany mandating utilities to purchase independently generated electricity under terms favourable for the wind turbine operators (Lauber and Mez, 2006). In 1994, Spain introduced a feed-in law with a premium price paid by the electricity utilities.

Resources mobilised for wind power in the 1990s in Denmark, Germany, and Spain supported knowledge development and diffusion as well as market formation. With a few exceptions, technology development focussed on scaling up the smaller turbines that had dominated the market in the 1980s to improve wind capture and reduce the cost of generated electricity. The process of up-scaling more or less standardised turbines had begun in Denmark in the 1980s as turbine sizes increased from 22kW to 500kW. In the late 1990s, commercialised wind turbines had already reached 3MW (see Figure 16.2). Knowledge development efforts also led to major performance improvements: turbine availability reached 98 percent (i.e., downtimes of less than 2 percent because of technical reasons); efficiency increased dramatically; turbine noise was reduced; and impacts on grid stability were managed, allowing penetration rates of more than 10 percent of total electricity production. In all, developments showed that wind energy could be an important contributor in future energy supply systems.

The development of wind turbines and wind-generated electricity in the 1980s and 1990s also resulted in cost reductions attributable to various innovation processes (Neij et al., 2003). In Denmark, the cost of installed wind turbines (i.e., cost per kW) reduced by 5 percent from 1980 to 1985 and by 50 percent over the longer period from 1980 to 2000 (Neij et al., 2003). Because of successful up-scaling and improved efficiency, the cost reduction in generated electricity was even larger, falling by 70 percent during the period from 1980 to 2000 in the case of Danish-produced wind turbines (Neij et al., 2003). The development of larger turbines did, however, face economic and logistical constraints. Until the mid 1990s, up-scaling was limited by the cranes available for the erection of the turbines, and the process of up-scaling with its greater use of materials challenged further cost reductions.

In recent years, prices of wind turbines have increased. One explanation is that rapidly increasing demand for wind turbines on the world market has made it possible for turbine manufacturers to increase the price tag of their products. Another explanation is that the world prices of industrial commodities such as steel and copper have increased, driving up turbine costs. Moreover, cost is no longer declining as a result of turbine up-scaling; the maximum size of onshore wind turbines has stabilised since the early 2000s.

4 Emerging Markets in the 1990s and 2000s

India and China also developed highly supportive policies and markets for wind energy in the 1990s. Both countries mandated grid compatibility with wind development, and developed national legislation and financial incentives to support wind energy (GEF, 2008; Lewis, 2007; Liu, Gan, and Zhang, 2002). Technology and knowledge development was supported through licensing arrangements with foreign wind turbine manufacturers and through domestic R&D programmes. The development of international networks further supported knowledge development. In India the wind turbine manufacturer, Suzlon, located research centres in the Netherlands and Germany and established its international headquarters in Denmark. In China the wind turbine producer, Goldwind, developed know-how by sending employees abroad to obtain advanced training. In parallel, India and China developed duties and related taxes to encourage domestically manufactured wind turbines and components (Lewis, 2007). China even mandated domestic products as part of its market formation incentives.

Consequently, wind turbines diffused rapidly from the mid 1990s. But industry development faced a number of difficulties due to poor wind resource data, weak grid systems, poor installation practices, and unsatisfying performance of locally produced wind turbines (Lewis, 2007; Rajsekhar, Van Hulle, and Jasen, 1999). As a result, the cost of generated electricity was higher than expected. To overcome performance problems, India and China developed certification programmes for domestically produced turbines (Lewis, 2007).

The innovation path in India and China has been characterised by a massive support for domestic market formation, using spillovers of know-how from foreign manufacturers and support for local knowledge development related to improved wind resource data, grid development, and improved technology performance. The process of licensing the production of foreign developed wind turbines resulted in the introduction of somewhat old and small models because of the requirement for domestic manufacturing units. But this emphasis on domestic manufactures allowed the use of local labour assets and the production of wind turbines at a low cost. Domestic production also provided cost reductions related to logistics and transaction costs. In all, the massive and aggressive policy structures of India and China have allowed for an impressive increase in domestically produced wind turbines since the early 2000s.

5 Elements of Effective Innovation Policies

From this comparative analysis of the innovation paths of wind energy in different countries, some key elements in government policy strategies can be distilled as essential for a sustainable and successful innovation process. These elements are set out in detail later in this chapter and summarised here as follows:

- support diversity in technology and market formation
- research, development, and demonstration (RD&D) are necessary but not sufficient
- quality assurance is essential for new technologies
- support interaction and networking
- ensure support is stable, continuous, and flexible

Taken together, these elements of a successful innovation approach show that government policy needs to support the development of the entire innovation system: not just the development of turbines and associated infrastructure, but also the involvement of actors, networks, and market institutions. The case of wind energy development in Denmark presents a good example of how and why such a systemic approach can succeed (Borup et al., 2009).

5.1 Diversity in Technology and Market Formation

Entrepreneurial experimentation in the early to mid twentieth century provided important experience of wind turbine design and a necessary foundation for strong policy support in the early 1970s. However, in the 1970s many countries focussed financially or even exclusively on large and advanced turbines as the single option to meet energy demand in the immediate future. Yet the support of these large, complex wind turbines came to have no impact on the development of commercial turbines in the 1980s as the risks and costs were too high and few actors were willing to commit investment. Wind turbine deployment was driven not by the traditional utilities and energy companies, but by small and medium-sized enterprises and individuals. These actors, not envisioned as core to the development of wind energy from the start, chose to invest in smaller, more reliable, and less costly turbines. This illustration of the difficulties in foreseeing a technology's early evolution emphasises the importance of supporting a diversified innovation portfolio in terms of both technologies and actors. By making it possible for all types of actors to gain support (e.g., investment subsidies), countries like Denmark and the United States successfully supported early market formation. Other governments that foresaw utilities as the main users fruitlessly targeted funding at incumbent actors who were insufficiently interested in a new technology.

5.2 Targeted Research, Development, and Demonstration (RD&D)

In many countries, wind energy innovation was initially supported through research, development, and demonstration (RD&D) as the first stage in a linear innovation process. But RD&D funding alone did not give rise to any commercial turbines. In contrast, Denmark and the United States started to support market formation in the late 1970s and early 1980s, allowing small and medium-sized enterprises and individuals to access the subsidy programmes. These broad user-oriented initiatives contributed to important learning by using, providing essential feedback to the development of wind turbines.

Initial RD&D expenditures in Denmark were approximately EUR 47 million up to 1990 whereas equivalent expenditures in Germany during the same time period were more than five times higher at EUR 227 million (Neij et al., 2003) and early U.S. expenditures more than twenty times higher (Heymann, 1998). Yet Denmark more effectively supported the innovation path of wind turbines. How RD&D resources were targeted, as well as their magnitude, was critically important.

5.3 Quality Assurance

The certification process developed in Denmark as early as the late 1970s was an essential part of the successful development of Danish wind turbines. The Netherlands also developed a test station to guarantee wind turbine quality, but this was less successful. Countries without testing and certification processes for emerging wind turbine models experienced severe problems in market trust, as in the case of early developments in the United States. In recent years, India and China have developed technology certification programmes to ensure high-quality domestic wind turbines.

5.4 Interaction and Networking

Interaction and networking between innovation actors is important. A successful model and example of early experience in policy support for networking was the development of the Danish test station that supported interaction and knowledge transfer between turbine owners (users), manufactures, and researchers. This contrasts with experience in the Netherlands where the test station supported competition rather than cooperation among domestic producers.

5.5 Stable, Continuous, but Flexible Support

The history of wind turbine development is long, with many technology and policy failures along the way. Denmark provided a continuous and stable system of support from the mid 1970s, as did Spain from the early 1990s. This stability in innovation and diffusion policies underwrote knowledge development and market formation activities that in turn supported successful innovation outcomes.

From the relative failure of innovation efforts in the 1980s, Germany rebuilt its innovation and diffusion policies in the 1990s on the basis of the successful Danish approach, applying production incentives in the form of a feed-in law to support market formation. By being flexible and adapting its support structure, Germany became a major player in global wind markets. Over the years, many other countries have come to develop their support structures in a similar way.

FURTHER READING

An excellent, comprehensive account of early developments in the wind industry with country comparisons can be found in Gipe (1995).

REFERENCES

Åstrand, K. & Neij, L., 2006. An Assessment of Governmental Wind Power Programmes in Sweden – Using a System Approach. *Energy Policy*, 34: 277–96.

Borup, M., Andersen, P. D. A., Gregersen, B., & Tanner, A. N., 2009. *Ny Energi og Innovation i Danmark*. Copenhagen: DJØF's Forlag.

GEF, 2008. *Promotion of Wind Energy: Lessons Learned from International Experience and UNDP-GEF Projects*. GEF-UNDP.

Gipe, P., 1995. *Wind Energy Comes of Age*. John Wiley & Sons Inc.

Heymann, M., 1998. Signs of Hubris – The Shaping of Wind Technology Styles in Germany, Denmark and the United States 1940–1990. *Technology and Culture*, 39(4): 641–70.

Kamp, L., Smits, R., & Andriesse, C., 2003. Notions on Learning Applied to Wind Turbine Development in the Netherlands and Denmark. *Energy Policy*, 32(14): 1625–37.

Lauber, V. & Mez, L., 2006. Renewable Electricity Policy in Germany, 1974–2005. *Bulletin of Science, Technology & Society*, 26(2): 105–20.

Lewis, J. I., 2007. Technology Acquisition and Innovation in the Developing World: Wind Turbine Development in China and India. *St Comp Dev*, 42: 208–32.

Liu, W.-Q., Gan, L., & Zhang, X.-L., 2002. Cost-Competitive Incentives for Wind Energy Development in China: Institutional Dynamics and Policy Changes. *Energy Policy*, 30: 753–65.

Loiter, J. M. & Norberg-Bohm, V., 1999. Technology Policy and Renewable Energy: Public Roles in the Development of New Energy Technologies. *Energy Policy*, 27: 85–97.

Neij, L., Dannemand Andersen, P., Durstewitz, M., Helby, P., Hoppe-Kilpper, M., & Morthorst, P. E., 2003. *Experience Curves: A Tool for Energy Policy Assessment*. Lund, Sweden: Environmental and Energy Systems Studies, Lund University.

Rajsekhar, B., Van Hulle, F., & Jasen, J. C., 1999. Indian Wind Energy Programme: Performance and Future Directions. *Energy Policy*, 27: 669–78.

Sawin, J., 2001. *The Role of Government in the Development and Diffusion of Renewable Energy Technologies: Wind Power in the United States, California, Denmark, and Germany, 1970–2000*. PhD Thesis, Fletcher School of Law and Diplomacy, Medford, Massachusetts.

United Nations, 1964. Volume 7: Wind Power. *Proceedings of the United Nations Conference on New Sources of Energy*. August 21–31, Rome: United Nations.

U.S. DOE, 2008. *20% Wind Energy by2030 – Increasing Wind Energy's Contribution to U.S. Electricity Supply*. Washington, DC: U.S. Department of Energy (U.S. DOE).

17 The Role of Standards: The Japanese Top Runner Programme for End-Use Efficiency

Osamu Kimura

Author's Summary

In 1998, Japan launched the Top Runner Programme to improve the energy efficiency of a range of end-use products. In this programme, the most energy-efficient product on the market during the standard-setting process defines the Top Runner Standard for all corresponding manufacturers and products. The Top Runner Programme started with nine products: room air conditioners, fluorescent lighting, television sets, copying machines, computers, magnetic disk units, video cassette recorders, refrigerators, and passenger and freight vehicles. The scope was reviewed every two to three years and by 2012 had gradually expanded to include twenty-three products. It is now considered one of the major pillars of Japanese climate policy. This case study examines twelve years of experience with the programme (from 1998 to 2010). It provides an overview of the structure of the Top Runner Programme, illustrates its impacts on efficiency and energy savings, discusses issues associated with the programme, and concludes with implications for efficiency programmes in other countries.

Editors' Guide

Relationship of Case Study to Energy Technology Innovation System Framework

The Top Runner Standards provided a clear direction for product development aiming at higher energy efficiency and the elimination of low-efficiency products from the market. These *shared expectations*, held by manufacturers, policy makers, and end users, helped underwrite innovation investments on the part of technology developers by providing clear and transparent confidence in future market demand.

This was certainly facilitated by the configuration of relevant *actors and institutions* in Japan. First, the Japanese market is dominated by a relatively few number of domestic manufacturers with high technological competencies. Second, the *public policy* process is multilateral, centred on *informal networks* to enable *exchange and interaction* among all relevant market actors in both public and private sectors (industry, academics, consumer groups, national and local government, and even the media). Third, those bound by the regulations faced clear incentives to comply, both to maintain competitiveness against foreign competitors and to avoid sanctions (attuned to domestic cultural aversion to being publicly "named and shamed"). Fourth, this regulatory environment is stable and credible. Collectively, these overcame any potential *resistance* to the objectives and direction of the efficiency programme, and strengthened the *shared expectations* for its successful outcomes.

Other preconditions for the observed innovation successes relate to the characteristics of the technologies and markets involved. The potential for technological improvement (through *knowledge generation* as well as *learning*) was clearly identified and incorporated into the standards. Manufacturers could continue to offer a differentiated range of products to end users as long as their average efficiency complied with the standards. Technological innovation was thus consistent with heterogeneous *adopter preferences* for different products' *relative advantage*.

Yet as the case study notes, the *adoption and use* of the efficient end-use technologies were a relatively neglected element of the innovation system. It remains unclear whether the more efficient products are financially attractive once upfront cost increases are accounted for (to the extent these are attributable to the standards). And, it is also not clear how the efficiency gains interacted with other performance attributes of the products from the end users' perspective. In the case of room air conditioners, for example, efficiency improvements pursuant to the Top Runner Programme substituted for higher heating capacities as the focus of product innovation.

Insights for Energy Technology Innovation Policy

- The dynamic, continuously adjusted standards of the Top Runner Programme were (and continue to be) successful in accelerating the energy efficiency improvement of many end-use products in Japan, ranging from room air conditioners to passenger vehicles. The dynamic tightening of the standards was transparently structured into the policy design (through regular two to three yearly reviews) and involved a multi-stakeholder standard-setting process that ensured industry buy-in.
- Standards were based on currently available products taking future potential improvements into account. This balance between realism (and acceptability) and ambition (and energy savings) worked well in most cases, although for short-lived products with rapid innovation life cycles and product turnover (e.g., fluorescent lighting, LCD displays), expectations for future improvements proved underestimates and standards were quickly met. This points to the difficulties of foreseeing innovation outcomes, particularly in rapidly innovating fields.
- Standards were set to ensure long-term visibility but took into account the development lead times and turnover rates of different products (e.g., vehicle efficiency standards were set eleven years in advance and revised eight years in advance, whereas computer / hard drive standards were set six years in advance and revised four years in advance). Again, this emphasises the importance of matching policy to distinctive technological conditions.

- Innovation policy can be dynamically adjusted not just in stringency but also in scope. The initial series of nine end-use products covered by the Top Runner Programme expanded to twenty-three products during the twelve-year period reviewed in this case study (1998–2010). The expansion of policy "tested" on narrowly targeted initial markets is facilitated by learning spillovers, particularly if the same market actors are involved.

1 Introduction

Japan started the Top Runner Programme in 1998 to improve the energy efficiency of end-use products. As part of the Energy Conservation Law, the programme set mandatory efficiency standards for a variety of appliances, equipment, and automobiles based on the most efficient products ("Top Runner" products) on the market. It aims to develop "the world's best energy-efficient products." Started with nine products in 1998, it expanded to twenty-three products in 2012. It is now considered one of the major pillars of Japanese climate policy. The purpose of this case study is to examine twelve years of experience with the Top Runner approach (1998–2010). Although it is often cited as an innovative method to stimulate efficiency improvements of targeted products, not only by the Japanese government but also by some European countries (e.g., Jänicke, 2008), there has been very limited evaluation, except for a couple of analyses by European researchers (Nordqvist, 2006; Tojo, 2005). This case study evaluates the effectiveness of the Top Runner approach, focussing on selected products. The empirical material includes government documents and a number of in-depth interviews with manufacturers in Japan. The case study provides an overview of the structure of the Top Runner Programme, illustrates its impacts on efficiency and energy savings, discusses issues associated with the programme, and concludes with implications for efficiency programmes in other countries.

2 Overview of the Top Runner Programme

2.1 Products and Standards

Mandatory energy-efficiency standards for appliances and automobiles have been in effect in Japan since 1980. At first, they failed to induce sufficient energy efficiency improvement as they were rarely revised and were largely based on negotiation with industry without any explicit standard-setting method. When the Kyoto Protocol was adopted in 1997, Japan was required to further accelerate energy conservation efforts to achieve its 6 percent reduction in greenhouse gas emissions by 2008–12 compared to 1990 levels. The Top Runner Programme was expected to be an effective means of contributing to this goal by setting ambitious efficiency targets and reducing energy consumption in the residential sector. In 1998, Top Runner was adopted in the revision of the Energy Conservation Law as a new method to set targets for selected products. The Top Runner Programme includes products that fall within three categories:

1) products with large domestic shipments;
2) products that consume a substantial quantity of energy in the use phase;
3) products with substantial scope for improving energy efficiency.

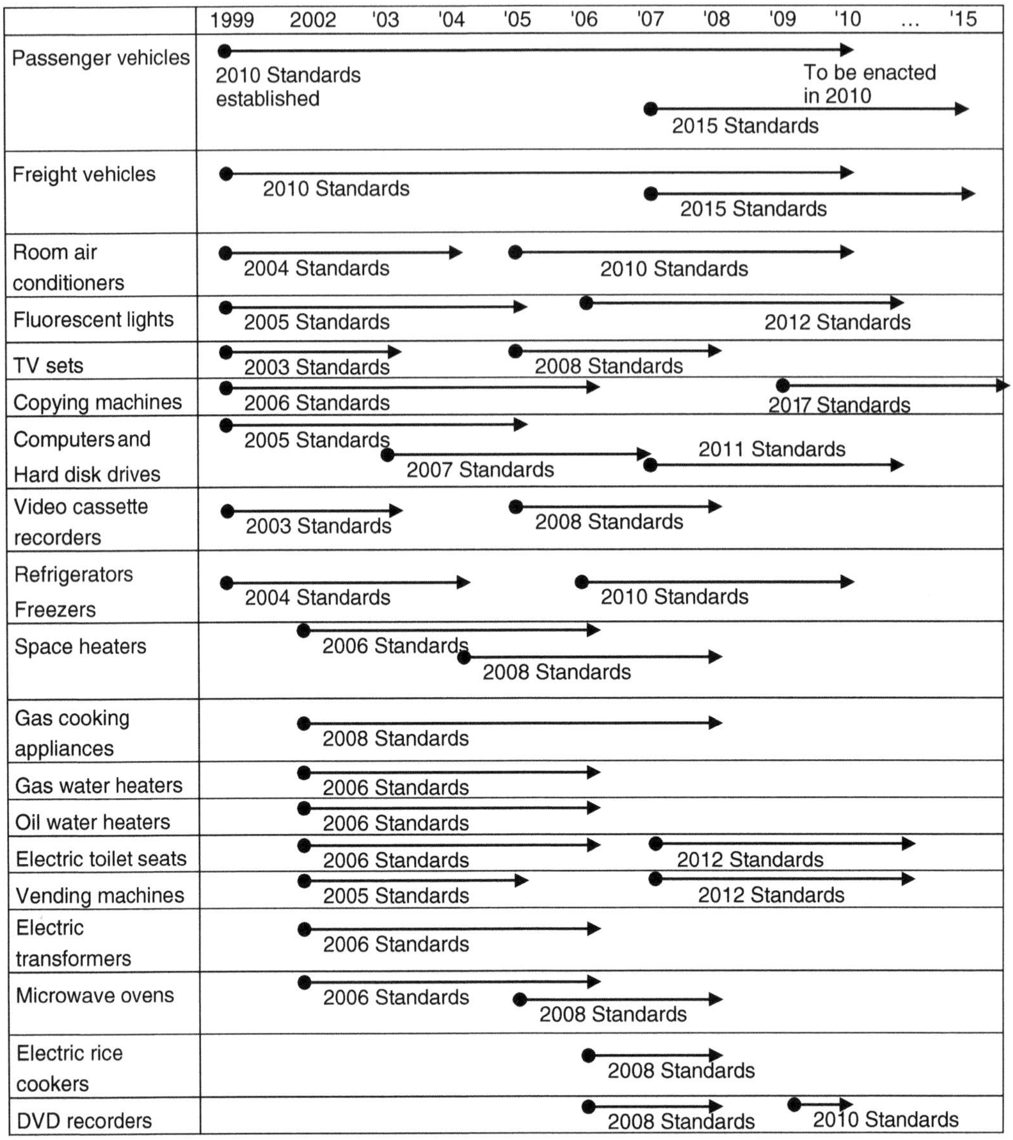

Table 17.1. Scope and target year of the Top Runner Standards.
Notes: the starting/ending points of an arrow show the year of decision/enactment of the standards.
Sources: METI (2007), ECCJ (2009a), ECCJ (2010), METI (2011).

The programme started in 1998 with nine products: room air conditioners, fluorescent lamps, televisions, copying machines, computers, magnetic disk units, video cassette recorders, refrigerators, and passenger and freight vehicles. The scope was reviewed every two to three years and by 2009 had gradually expanded to twenty-one products (see Table 17.1). The electricity consumed by the Top Runner targeted products sums up to more than 70 percent of residential electricity consumption (METI, 2004).

Because the most energy-efficient product on the market during the standard-setting process set the Top Runner Standards, the standards are based on market

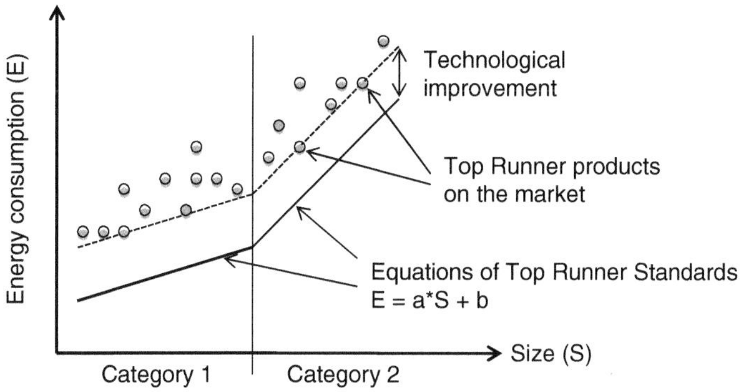

Figure 17.1. Schematic overview of standard setting in the Top Runner Programme.
Note: The Top Runner products are identified and become the basis of the standards, but the standards are decided with consideration of future potential technological improvements.

data. But the programme also takes into consideration technological analysis; that is, Top Runner considers technological potential for efficiency improvements in the future (illustrated in Figure 17.1). For example, the Top Runner Standards for room air conditioners smaller than 4kW for 2010 were defined as a 3 percent to 4 percent improvement over the Top Runner products in 2005 as such technological improvement was assessed to be feasible by stakeholders' discussions in the Air Conditioner Evaluation Standard Subcommittee (Air Conditioner Evaluation Standard Subcommittee, 2006).

Another important feature of Top Runner is that standards are differentiated based on various parameters. While it seems quite reasonable to differentiate the standards based on size and weight, technology type is also considered in the categorisation. For example, liquid crystalline displays have different standards from cathode ray tube displays because different technologies used have large implication for their energy efficiency. Hybrid vehicles are also excluded from the standard setting because conventional internal combustion engine vehicles cannot meet their high efficiency without adopting the same technology.

In order to comply with the standards, producers must make sure that the weighted average efficiency of the products they sell in a target year achieves the standards. Therefore not all of a manufacturer's products have to meet the target, but the average of all products has to. This flexibility enables producers to provide a wide range of models to meet market demand while guiding the overall market to higher energy efficiency.

In sum, the major characteristics of the Top Runner approach can be described as:

- the Top Runners set the standard, with consideration of future potential improvements;
- differentiated standards are set based on a range of parameters;
- compliance with the standard is evaluated by corporate average product sales.

2.2 Institutional Setting

In Japan, energy efficiency standards are discussed and determined by the Ministry of Economy, Trade and Industry (METI) and its advisory committees, which comprise

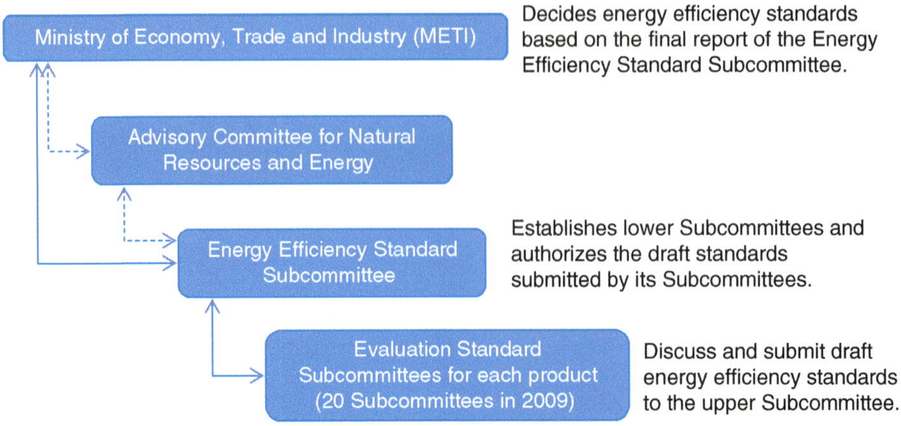

Figure 17.2. Institutional framework for setting the Top Runner Standards.
Source: ECCJ (2008).

representatives from academia, industry, consumer groups, local governments, and mass media. The Advisory Committee for Natural Resources and Energy is in charge of overall energy policy including energy efficiency policy. The Energy Efficiency Standards Subcommittee, one of the subcommittees under the advisory committee, is responsible for setting energy efficiency standards. This subcommittee, organised by METI, establishes an evaluation standard subcommittee for each of the targeted products, and evaluation standard subcommittees make draft standards. Figure 17.2 summarises this institutional hierarchy. Because detailed market and engineering information on the targeted products is required, there is strong involvement of industry associations in the standard-setting process. The Energy Efficiency Standard Subcommittee authorises the draft standards submitted by evaluation standard subcommittees, which are finally decided by METI. It is usually takes a year or two to set standards for one product (ECCJ, 2008).

METI also considers revision of standards when the target year arrives (see Table 17.1). When the target year arrives, the producers submit a report on their sales and the energy efficiency of their products, and METI evaluates their compliance. In case of noncompliance, the Top Runner Programme uses a "name and shame" approach. The ministry first makes a recommendation to the noncompliant producer to improve its performance. If this is not sufficient, METI makes the recommendation public. Finally, METI can order the producer to meet the recommendations (ECCJ, 2008). So far, this approach has worked very well. Although there are no records of compliance rates, so far no producers have been publicly named as noncompliant. This is due to both the structure of the Japanese appliance market, which is dominated by a limited number of domestic producers, and Japanese culture in which government pressure and public shaming acts as a strong compliance mechanism.

3 Impacts of the Top Runner Programme

The Top Runner approach requires substantial energy efficiency improvements for each targeted product. The rates of energy efficiency improvement required by Top Runner standards range from 16 percent to 80 percent, which have been achieved in all products

Table 17.2. Energy efficiency improvement of major products with top runner standards.

Product	Estimated improvement with top runner standards *	Result
Room air conditioners	66.1% increase in COP (FY 1997 vs. 2004 freezing year)	67.8%
Refrigerators	30.5% decrease in kWh/year (FY 1998 vs. FY 2004)	55.2%
TV sets using cathode ray tube displays	16.4% decrease in kWh/year (FY 1997 vs. FY 2003)	25.7%
Computers	69.2% decrease in kWh/year (FY 2001 vs. FY 2007)	80.8%
Fluorescent lights	16.6% increase in lm/W (FY 1997 vs. FY 2005)	35.7%
Vending machines	33.9% decrease in kWh/year (FY 2000 vs. FY 2005)	37.3%
Gasoline passenger vehicles	22.8% increase in km/L (FY 1995 vs. FY 2010)	22.8% (FY 1995 vs. FY 2005)

* Estimated improvement of weighted average energy efficiency of all categories within each product group.
Notes: FY = fiscal year; COP = coefficient of performance (ratio between power input and power output).
Source: ECCJ (2010).

so far, often with large excess (Table 17.2). These results highlight the programme's success in meeting targets and spurring significant efficiency improvements.

Although the contribution of Top Runner standards to energy efficiency are difficult to disaggregate from autonomous technological improvement and changes in market demand (e.g., consumers preferring efficient products with lower energy costs), two products clearly demonstrate the Top Runner Programme's impacts.

The best example is residential air conditioners. In Japan, air conditioners began diffusion in the 1970s for cooling, and later for both cooling and heating. Since then, energy efficiency increased because of technological improvements such as inverters and better heat exchangers, but this trend stagnated in the mid 1990s (see Figure 17.3). New Top Runner standards were adopted in 1999. The standards required a coefficient of performance (COP) increase of 66 percent by 2004 compared to the 1997 level (a COP is the ratio between the power input and power output). The adoption of the standards had a significant impact on altering the technological trajectory from increasing heating capacity (in order to expand its market for heating) to improving energy efficiency. Figure 17.3 shows that the efficiency of air conditioners clearly changed in 1998, when the Top Runner Programme began, and in the 2004 target year for the air conditioning standard.

A close look into the energy efficiency trend tells us more about the standard's impact. As shown in Figure 17.4, not only has the Top Runner standard effectively improved the efficiency of high-end products by 50 percent, it also almost doubled the efficiency of low-end products. This shows the Top Runner standards have the power to eliminate low-efficiency products from the market. It is reported that in the 2.8kW class market about seventy models that sold 2 million units in 2003 went out of production to meet the 2004 Top Runner Target (Air Conditioner Evaluation Standard Subcommittee, 2006).

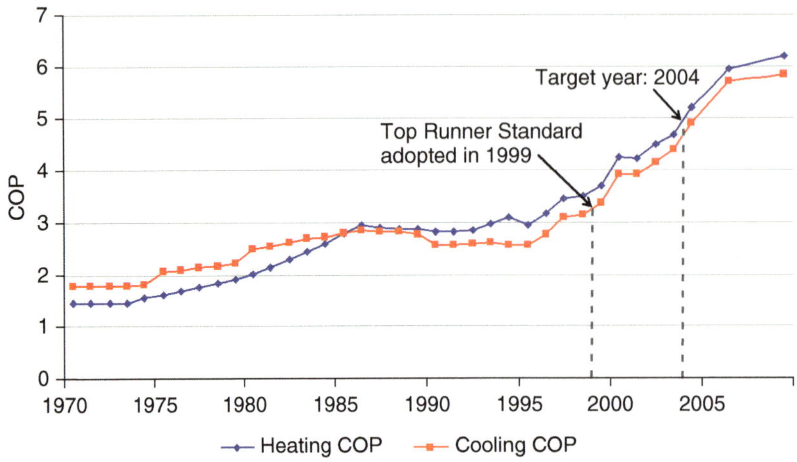

Figure 17.3. Long-term trend of energy efficiency of room air conditioners. *Notes*: COP = coefficient of performance (ratio between power input and power output). *Source*: 1970–2004 data from Murakoshi (2006); 2006 and 2008 data from ECCJ (2006, 2009b) as average of 2.8kW class models.

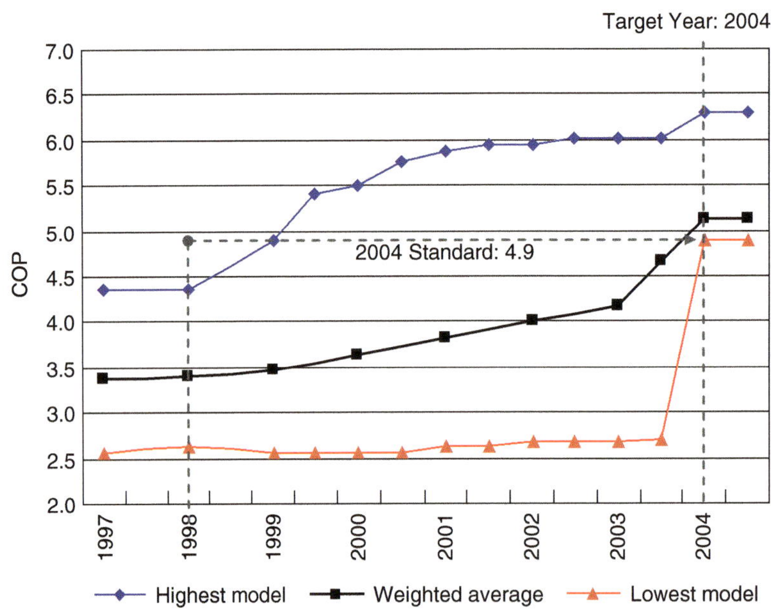

Figure 17.4. Trend of energy efficiency of 2.8kW classroom air conditioners (1997–2004). *Notes*: COP = coefficient of performance (ratio between power input and power output). *Source*: Air Conditioner Evaluation Standard Subcommittee (2006).

Another success story applies to passenger vehicles. While there were important fuel-efficiency improvements in the 1970s and 1980s, efficiency stagnated, and even worsened, in the early 1990s because of the fall of gasoline prices and the increase in vehicle size. But after the introduction of Top Runner standards, energy efficiency improvements clearly accelerated as shown in Figure 17.5. The penetration rate of the Top Runner compatible vehicles in the new vehicle market increased rapidly, from 11 percent in 1997 to more than 90 percent in 2008 (Figure 17.6).

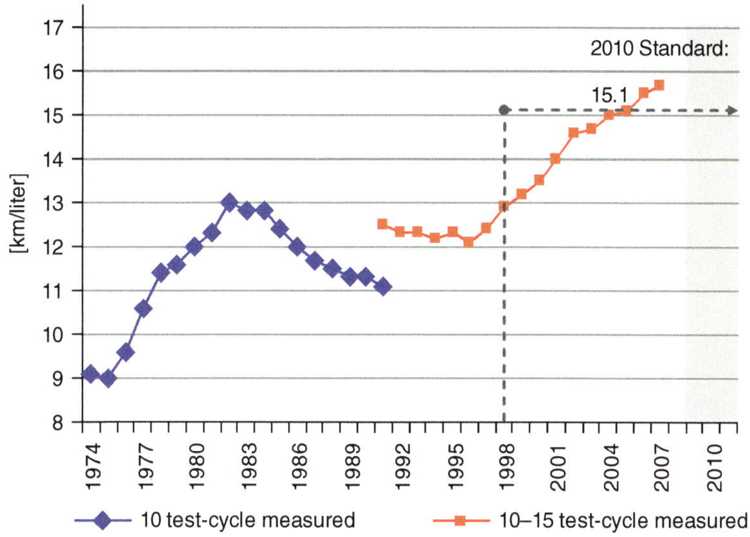

Figure 17.5. Average fuel efficiency of new gasoline-fuelled passenger vehicles sold in Japan.
Notes: Y-axis shows fuel efficiency in km/litre. The Top Runner standard for gasoline-fuelled passenger vehicles of 15.1 km/litre by 2010 was decided in 1998 and was achieved in 2005, five years ahead of the target year.
Sources: 1973–93 data from METI (1996), 1994–2008 data from MLIT (2009).

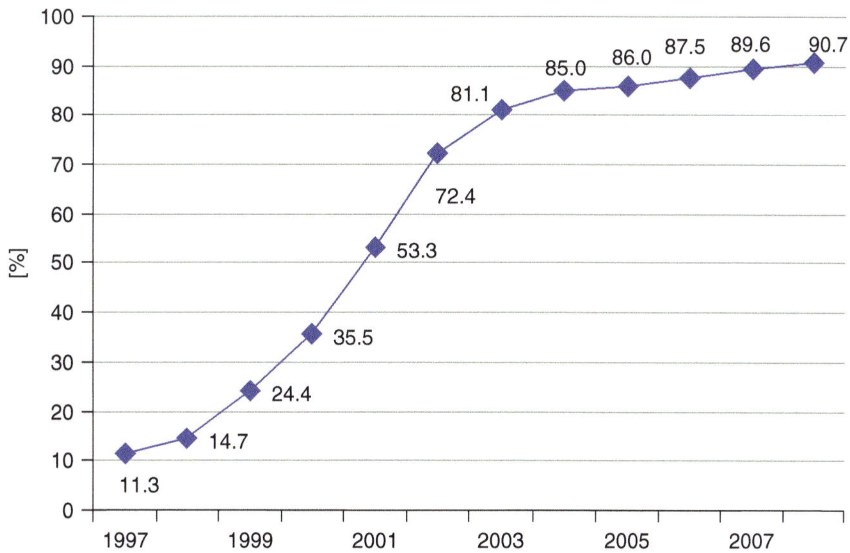

Figure 17.6. Penetration rate of passenger vehicles that are 2010 Top Runner standards compliant.
Source: JAMA (2009).

It should be noted that not all of the fuel efficiency improvement is attributable to Top Runner standards. The fact that fuel efficiency of passenger vehicles started to increase from 1997 (before the 2010 Top Runner standards were decided) implies the effect of other factors such as the rise of environmental consciousness among consumers and the producers' response to the emerging market demand. In 1997, for example, Toyota launched its hybrid vehicle (Prius), which was the culmination of developments started in 1993 under the vision of top management to double fuel efficiency in "vehicles for the 21st century" without any pressure of regulation (Ikari, 1999; see also the case study on hybrid cars). Nevertheless, it is clear the Top Runner standards set a clear market direction towards higher fuel efficiency, reducing the risks faced by firms investing in more efficient vehicles. In interviews conducted with the major car manufacturers in Japan in 2008, they openly recognise that the standards changed their priorities and accelerated their development activities in favour of fuel efficiency improvements.

4 Concerns with the Top Runner Programme

Although the Top Runner Programme has succeeded in driving up the efficiency of a range of end-use products, outstanding issues need addressing.

One problem is the lack of explicit methods to consider consumer impacts. Because Top Runner is based on the most efficient products on the market, any increase of product price associated with improving energy efficiency is not explicitly considered. While it is stipulated that standards should not force consumers to "purchase economically inappropriate high-priced products in the name of energy saving" (ECCJ, 2008: 17), there is no requirement for life cycle or other cost analyses, which is therefore not systematically conducted. This contrasts markedly with explicit prescriptions for life cycle cost analysis in developing minimum energy performance standards, an approach to set energy efficiency standards for appliances adopted in the United States, in the EU, and in other major countries (IEA, 2000).

Lack of life cycle analysis in target setting may lead to price increases so high that the payback period is longer than the lifetime of the product. As an example, Figure 17.7 shows the case of energy-efficient room air conditioners sold in the winter of 2006 in Japan. Many efficient models require more than ten years and sometimes fifteen years to pay back the investment cost. These may also be underestimates as the data is based on the Energy Efficiency Performance Catalogue (ECCJ, 2006), which assumes 3.6 months of cooling operation period and 5.5 months of heating operation period. Those periods are much longer than average operation periods indicated by field surveys (e.g., Nishio and Iwafune, 2009). Many efficient models on the market are sufficiently expensive that they cannot pay back within the typical ten-year to fifteen-year lifetime of room air conditioners.

In addition, cost-effective potentials for efficiency improvement in air conditioner technologies have approached saturation. Motor efficiency of compressors is approximately 95 percent and total heat-insulating efficiency is more than 80 percent with only minimal gains achieved through the 2000s. Efficiency improvements in recent years have instead been achieved mainly by expanding heat exchangers. This option is also facing constraints because of the limited space in Japanese homes. Therefore further tightening of the standards may not be cost-effective.

Another problem is the difficulty in determining a rate of technological improvement that is both challenging yet feasible. In the case of fluorescent lighting, Top Runner standards were set just above the most efficient products already on the market, because the Evaluation Standard Committees made a very conservative assessment on the potential of further technological improvement. Because of this the targets were achieved just after they were established, and they had no impact on efficiency improvement (see Figure 17.8).

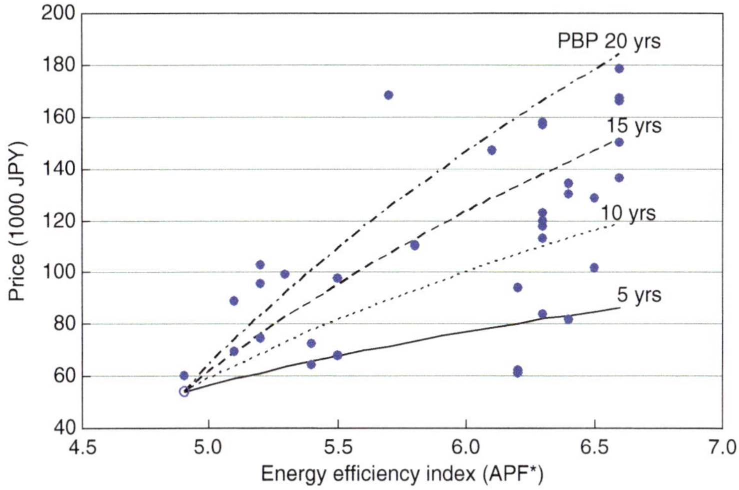

Figure 17.7. Price, energy efficiency, and payback periods of 2.8 kW classroom air conditioners in winter of 2006.
Notes: APF (Annual Performance Factor) is annual cooling and heating demand divided by annual energy consumption to meet the demand. Lines show the highest prices for energy savings to pay back the price increase compared to the least efficient, cheapest model within five-, ten-, fifteen-, and twenty-year payback periods (PBP) based on annual electricity consumption of respective models (ECCJ, 2006) and electricity prices (22 JPY/kWh) in Tokyo. Prices are derived from major online shopping sites for Japanese consumers (+D Shopping and kakaku.com).

The same story applies to liquid crystalline displays, whose targets were achieved almost two years before the target year of 2008. This might partly be a result of successful lobbying by industry, but more fundamentally indicates that for some products the rate of technological improvement is difficult to forecast. For example, one reason for failed target setting of liquid crystalline displays was unexpected advancement in backlight control technology. Similarly, fluorescent lightings with inverters were diffused much faster than expected. This suggests that the Top Runner approach might not be effective for products whose rate of technological development is difficult to forecast.

5 Conclusion

The Japanese Top Runner Programme demonstrates an effective approach for setting mandatory energy efficiency standards based on the most efficient products on the market. After the introduction of Top Runner in 1998, there has been a strong increase in energy efficiency of the targeted products. Although the exact contribution of the Top Runner standards to such efficiency improvement is hard to isolate from autonomous technological improvement and changes in market demand, it has succeeded in accelerating the trend of energy efficiency improvement of some products, such as room air conditioners and passenger vehicles. In those cases the standards provided a clear direction to product development towards higher energy efficiency and eliminated low-efficiency products from the market. This relied on the existence of technological potentials for efficiency improvement.

An important characteristic of the Japanese market that enables the successful functioning of the Top Runner Programme is the market structure dominated by a

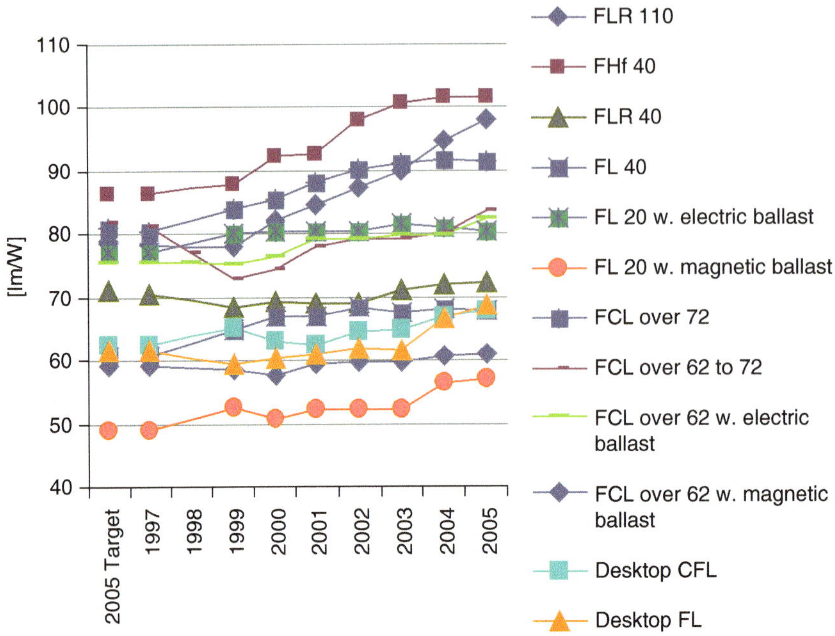

Figure 17.8. 2005 Top Runner standards established in 1997 and actual energy efficiency trend of fluorescent lightings (1997–2005).
Notes: FLR = fluorescent lighting rapid; FHF = fluorescent lighting high frequency; FL = fluorescent lighting; FCL = fluorescent circular lighting; CFL = compact fluorescent lighting. Numbers are lamp size (W).
Source: Fluorescent Lighting Fixtures Criteria Standard Subcommittee (2008).

limited number of domestic producers. They are similar in that they all have high technological competency, could accept strict standards (i.e., no producer is excluded from the market even with strict standards), had incentives to develop energy efficient products to increase competitiveness against foreign producers, and complied with the standards even without strict sanctions (related to Japanese business culture and aversion to public shaming).

In sum, Japanese experience with the Top Runner Programme shows how ambitious policy that matches market conditions as well as technological conditions can work very well to induce remarkable energy efficiency improvements. Because such conditions depend on the country and the phase of technological development of products targeted, careful and adaptive design is required for effective policy making.

FURTHER READING

Tojo (2005) provides a good summary analysis of the Top Runner Programme up to 2005. Siderius and Nakagami (2007) apply the Top Runner insights to a potential European context. Ellis (2007) sets out the broader context of efficiency standards and regulation in an IEA context.

REFERENCES

Air Conditioner Evaluation Standard Subcommittee, 2006. *Final Summary Report by Air Conditioner Evaluation Standard Subcommittee, Energy Efficiency Standards Subcommittee of the Advisory Committee for Natural Resources and Energy*. Tokyo: METI.

ECCJ (Energy Conservation Center, Japan), 2006. *Shouene seinou katarogu 2006 nen huyu ban* [*Energy efficiency performance catalogue*, 2006 winter version], Tokyo: ECCJ (in Japanese).
2008. *Top Runner Program: Developing the World's Best Energy-Efficient Appliances*. Tokyo: ECCJ.
2009a. Final Reports on the Top Runner Target Product Standards, available at: http://www.eccj.or.jp/top_runner/.
2009b. *Shouene seinou katarogu 2009 nen natsu ban* [*Energy efficiency performance catalogue*, 2009 summer version], Tokyo: ECCJ (in Japanese).
2010. *Top Runner Program: Developing the World's Best Energy-Efficient Appliances*, revised edition on March 2010, Tokyo: ECCJ.
Ellis, M., 2007. *Experience with Energy Efficiency Regulations for Electrical Equipment*. IEA Information Paper in Support of the G8 Plan of Action. International Energy Agency (IEA).
Fluorescent Lighting Fixtures Criteria Standard Subcommittee, 2008. Shoumei kigutou no genjou [Current Status of Lighting Equipments], *Energy Efficiency Standards Subcommittee of the Advisory Committee for Natural Resources and Energy*. Tokyo: METI.
IEA (International Energy Agency), 2000. *Energy Labels and Standards*. Paris: IEA.
Ikari, Y., 1999. *Haiburiddo kaa no jidai – sekaihatsu ryosansha Toyota Prius kaihatsu monogatari* [*An Era of Hybrid Cars: History of the Development of Prius by Toyota*]. Tokyo: Kojinsha (in Japanese).
JAMA (Japan Automobile Manufacturers Association), 2009. *Environmental Report 2009*. Tokyo: JAMA.
Jänicke, M., 2008. The Policy Design of Environmental Innovation and Sustainable Production. Bridging the Gap – Responding to Environmental Change – From Words to Deeds. Portoroz, Slovenia, May 14–16.
METI (Ministry of Economy, Trade Industry), 1996. *Shou enerugi souran 1997* [*Encyclopedia of Energy Efficiency & Conservation 1997*]. Tsuusan Shiryou Shuppan Kai, Tokyo (in Japanese).
2004. *Denryoku jukyuu no gaiyou* [*Outline of Electricity Demand and Supply*]. Tokyo: METI (in Japanese).
2007. *Toppu ranna seido no genjou to hyouka ni tsuite* [*Current Status and Assessment of the Top Runner Programme*]. Tokyo: METI (in Japanese).
2011. *Toppu ranna kijun no genjou tou ni tsuite* [*Current Status of the Top Runner Standards*]. Tokyo: METI (in Japanese).
MLIT (Ministry of Land, Infrastructure and Transport), 2009. *Jidousha Nenpi Ichiran* [*Fuel Economy of Automobiles*]. Tokyo: MLIT (in Japanese).
Murakoshi, T., 2006. Minsei bumon ni okeru enerugi shouhi ni tuite [Energy Consumption in the Residential Sector], presented at the Japan Center for Climate Change Actions Seminar, available at: http://www.jccca.org/todohuken (in Japanese).
Nishio, K. & Iwafune, Y., 2009. Measures to Promote CO2 Emission Reduction by Heat Pump Air Conditioners, Report No. Y08026, Tokyo: Central Research Institute of Electric Power Industry (in Japanese).
Nordqvist, J., 2006. Evaluation of Japan's Top Runner Programme, AID-EE Project Working Paper.
Siderius, P. J. S. & Nakagami, H., 2007. Top Runner in Europe? Inspiration from Japan for EU EcoDesign Implementing Measures. Proceedings of ECEEE 2007 Summer Study, pages 1119–26.
Tojo, N., 2005. The Top Runner Program in Japan: Its Effectiveness and Implications for the EU, Report 5515, Swedish Environmental Protection Agency.

ACKNOWLEDGMENTS

The author thanks interviewees from producers and industrial associations for sharing their knowledge. The author also thanks Arnulf Grubler at IIASA for helpful comments on an earlier version of this case study.

18 Solar Innovation and Market Feedback: Solar Photovoltaics in Rural Kenya

Daniel M. Kammen and Arne Jacobson

Authors' Summary

Until the recent worldwide explosion of solar PV deployment, the solar market in Kenya was one of the most dynamic per capita in the world. Up to thirty thousand systems were sold each year, with much of this activity a result of unsubsidized, free market sale of household solar electric systems, which accounted for an estimated 75 percent of solar equipment sales in the country. Solar is the largest source of new electrical connections in rural Kenya.

Despite this undisputed commercial success, product quality has been a significant ongoing concern (Jacobson and Kammen, 2007). Quality problems first became clear with the variations in performance seen in amorphous silicon solar modules that entered the Kenyan market in the early 1990s. This situation created a serious problem in the market, as many potential solar customers were unable to determine which brands performed well and which did not. Through independent testing in 1999, underperforming suppliers and models were identified and consequently withdrawn from the market, but quality problems resurfaced after 2004.

In response, the Kenya Bureau of Standards (KBS) moved to formulate and enforce quality standards for both amorphous and crystalline silicon PV modules. Currently, the KBS requires that import companies secure a certificate that validates their product conforms to the respective Kenyan standards prior to bringing the modules into the country. This certificate of conformity must be issued by an accredited laboratory. No such facility exists in Kenya, so this testing needs to take place in laboratories in Europe, North America, and Asia.

Editors' Guide

Relationship of Case Study to Energy Technology Innovation System Framework

Poor or uncertain product quality undermines the *relative performance advantages* that attract potential end users or *adopters* of new energy technologies like household solar PV systems. The recurrent emergence of quality problems in the Kenyan PV market shows that the supply of high-quality products requires enduring quality control *institutions* that persistently require high performance for all brands on the market. One-time testing efforts or the improvement of individual low-performing brands do not constitute lasting solutions. Product quality is essential as it underpins trust in an innovation. Trust is in turn important not just as a strong characteristic of *adopter preferences* (particularly with end-use technologies) but also because it draws end users into the reinforcement of *shared expectations* for a technology's successful future.

In the case of the solar PV systems in Kenya, product quality was eventually achieved through a combination of *public policy* including standards, and market *institutions* including quality assurance, testing, supplier credibility, and end-user confidence. The standards were drafted by various local market actors forming an effective *advocacy coalition* behind the local market potential. These standards were matched to *adopter preferences* not just for maximum rated power output but also durability and longevity, and drew on international precedents (*knowledge spillovers*).

The necessary institutional capacity to implement the Kenyan PV product quality standards was lacking in the local market. The resulting reliance on foreign testing stations weakened the potential market feedbacks between suppliers and end users of solar PV in Kenya. These opportunities for *exchange and interaction* have become more evident in the recent round of product improvements through discussions coordinated by the Kenyan Renewable Energy Association and involving local industry as well as module importers. This ultimately led to the removal of poor-performing modules from the market. More generally, the role academic and other actors play in the testing and certification process points to the importance of *knowledge generation* in the ongoing deployment of technologies once in the market.

Insights for Energy Technology Innovation Policy

- Standards and other quality assurance mechanisms are an important element to support the diffusion of new energy technologies. This is particularly the case for decentralized technologies like household solar PV whose installation is dispersed among local actors (typically firms) using imported modules. The absence of any single market actor or institution to facilitate learning spillovers associated with scale makes standards all the more important as a market institution.
- Policy instruments such as standards, as well as the mechanisms and institutions to enforce policy, need continuous support. One-off efforts to establish compliance may not have enduring effects.
- The formulation, implementation, and enforcement of standards (and of public policy more generally) also requires appropriate local institutional capacity and the resources associated with building and maintaining this capacity.
- The scale of market demand is linked to the leverage governments have over manufacturers and suppliers to ensure product quality. Growth in the early-adopting markets, as in the Kenyan solar PV case, can provide spillover benefits

to solar markets elsewhere in the region and beyond. As these benefits are not appropriable, this may justify additional international support.

1 Introduction

The solar market in Kenya has been among the largest and most dynamic per capita among developing countries. Cumulative solar sales in Kenya since the mid-1980s are estimated to be in excess of three hundred thousand systems, and annual sales growth has regularly topped 15 percent over the past decade (Acker and Kammen, 1996; Jacobson, 2004). Much of this activity is related to the sale of household solar electric systems, which account for an estimated 75 percent of solar equipment sales in the country (ESDA, 2003). Solar is the largest source of new electrical connections in rural Kenya.

Despite this undisputed commercial success, product quality has been a significant concern in the Kenya market. The debate over solar module quality exploded publicly in May 2004 in the form of a series of sensationalistic newspaper advertisements proclaiming that the low performance of a certain brand of amorphous silicon (a-Si) modules was a "solar panel scandal" (see Figure 18.1). The advertisements sparked a heated debate about quality, consumer rights, and the ethics of negative advertising. Within a short time, though, the focus shifted toward a search for solutions. The results and events presented in this case study represent a positive step toward resolving chronic quality problems that have plagued the Kenya solar market for more than a decade. The issues that emerged in Kenya, far from being an isolated case, were a precursor of what began to take place elsewhere because of the international character of the global solar industry and the degree to which Kenya has for well over a decade been a leader in this rapidly expanding technology sector.

In a series of peer-reviewed publications (Acker and Kammen, 1996; Duke, Jacobson, and Kammen, 2002; Jacobson and Kammen, 2007) and results reported in the local press, we present performance test results for five brands of small a-Si modules that were widely available in the Kenya market during 2004. Our results indicate that while three of the five brands perform at or near their advertised levels, two brands perform far below their nameplate power ratings. Both of the low-performing "brands" were manufactured by the same company, although they were sold under different brand names in Kenya. They can therefore be considered a single line of modules.

The strong performance of the three leading brands supports previous evidence indicating that good-quality a-Si PV modules can be excellent value for the money. At the same time, the very low performance of the two remaining brands is troubling, as many Kenyan consumers lack the information to distinguish among the various competing brands. There is, therefore, a need for market institutions that ensure the quality and performance of solar products sold in the Kenya market.

This conclusion is strongly supported by a series of performance tests that we conducted on a-Si modules in Kenya dating back to 1999 (see Duke et al., 2000; Duke, Jacobson, and Kammen, 2002; Jacobson, 2002; Jacobson et al., 2000a; Jacobson,

Figure 18.1. "Solar Scandal" advertisement in the *Daily Nation* newspaper, May 2, 2004.

Duke, and Kammen, 2000b, 2001; and Jacobson and Kammen, 2004 for results, with papers also available online at http://rael.berkeley.edu/publications and http://www.humboldt.edu/~aej1). The 1999 study played an important role in pressuring

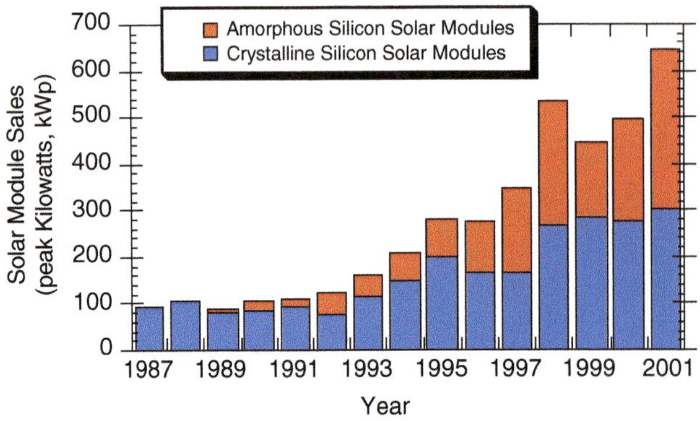

Figure 18.2. Solar module sales from 1987 to 2001.
Sources: Hankins and Bess (1994); Hankins (2000); ESDA (2003).

low-performing brands of a-Si modules to improve or exit the market, while it also verified that high-quality brands of a-Si modules performed well. The result was a stronger solar market and a better-served public. However, within a few years a new line of low-performing a-Si modules entered the market. This confirms the need for vigilant quality monitoring. In other words, the mix of high- and low-performing brands identified in the recent 2004/05 tests is nothing new, and we should not expect performance problems to go away on a permanent basis even with the removal of the current low-performing brands. Persistent problems require persistent solutions, and, in the absence of a vigilant and effective quality assurance program, performance-related problems are likely to reemerge in the future.

2 A Brief History of a-Si Testing in Kenya

Amorphous silicon solar modules entered the Kenya market in 1989. Sales grew rapidly, and by the late 1990s a-Si modules had gained a significant share of the overall solar PV market in Kenya (see Figure 18.2). Although the high sales figures indicate commercial success, a-Si technology has had a mixed reputation for quality in Kenya (e.g., Ochieng, 1999). The first generation of a-Si modules sold in Kenya were made by the Chronar Corporation, which went bankrupt in the mid-1990s in part because of quality-related reputation problems (e.g., Crawford, 1997). Several of Chronar's successors made high-performing a-Si modules, but low-quality versions of the technology remained on the market. This situation created a serious problem in the market, as many potential solar customers were unable to determine which brands performed well and which did not (Duke et al., 2002).

The debate about quality led to a study of a-Si module performance in 1999. This joint study was carried out by researchers (including the authors of this case study) from the University of California, Berkeley and Princeton University in collaboration with Energy Alternatives Africa (EAA) of Nairobi, which has since changed its name to Energy for Sustainable Development, Africa (ESDA). In the context of this case study, we carried out a field study that included performance measurements of 130 a-Si modules and 17 crystalline modules at 145 homes in rural Kenya. We also purchased 14 new a-Si modules in Nairobi that we tested at outdoor testing facilities in the United States and Kenya.

3 Assuring Product Quality: Performance Testing in 1999

3.1 Amorphous Silicon PV Performance Testing and Results – 1999

At the time of the 1999 study, three competing companies manufactured the large majority of the a-Si modules sold in Kenya. The results of the study are shown in Table 18.1 and summarized in Figure 18.3. They indicate that a-Si modules made by two of the three companies (Free Energy Europe and Koncar) performed reasonably well, while modules manufactured by the third company (Intersolar) performed well below the company's advertised levels. Free Energy Europe purchased the factory in France where its a-Si modules are manufactured from Neste Advanced Power Systems (NAPS) in 1998 (Duke et al., 2000). Koncar modules are now sold in Kenya under the brand name Solar Cells. ICP-Solar purchased Intersolar in 2003 (Lane, 2003).

The performance testing showed that the high-performing brands of a-Si modules were an effective, low-cost alternative to crystalline PV. However, the poor performance of the low-performing brand indicated a need for measures to ensure the high quality of all modules sold in the Kenya PV market. For more information about the study, see Jacobson and colleagues (2000a, 2000b), Duke and colleagues (2000, 2002), and Jacobson and Kammen (2007).

3.2 Response to Solar PV Quality in Kenya following the 1999 Study

Intersolar made considerable investments to improve the performance of its a-Si products in the years following the 1999 study. These investments resulted in improved performance over time, as documented in Figure 18.4. These results include tests carried out by Jacobson in 2000 and 2001 at UC Berkeley, as well as performance results from the 2004/05 study that we report in greater detail later in this chapter. The tests from 2000 and 2001 were based on samples of four modules each that were sent by Intersolar to UC Berkeley for testing. The 2004/05 performance was based on a randomly selected sample of modules purchased from retail shops in Kenya.

These results represent a significant success story, as the low-performing Intersolar modules of the 1990s have been replaced with the current generation of high-performing ICP-Solar a-Si modules. However, while the improvements Intersolar and ICP-Solar made are very encouraging, these gains did not, as we explain later, eliminate the quality problems with a-Si technology in the Kenya solar market.

Several years after the 1999 study, a new line of low-performing a-Si modules began to enter the Kenya solar market in significant quantities. The presence of these modules eventually led to the tensions described earlier that came to a head in the first half of 2004. The reemergence of quality problems in the Kenya market confirm that the issue cannot be solved decisively by one-time testing efforts, nor by focusing on the improvement of individual low-performing brands. Rather, institutional solutions that persistently require high performance for all brands are needed to ensure quality.

In the years following the 1999 study, the Kenya Bureau of Standards (KBS) worked in collaboration with the Kenya Renewable Energy Association (KEREA) to draft performance standards for a range of solar products, including but not limited to amorphous silicon solar modules. By 2003, the committee in charge of

Table 18.1. Performance summary for A-Si PV panels tested in Kenya during 1999.

Panel Type	Rated max. power (watts)	Average measured max. power (watts)	Percentage of rated output	Average age of modules (years)	# Modules tested
Koncar (Solar Cells)	12	10.0	83%	2.8	31
NAPS	11	9.7	88%	3.1	31
Free Energy Europe	12	10.6	89%	0.9	32
Intersolar "Phoenix"	11	6.8	61%	2.4	5
Intersolar "Phoenix Gold"	14	7.7	55%	1.5	12
APS	25	22.5	90%	5.0	1
Chronar	10	7.2	72%	5.9	4

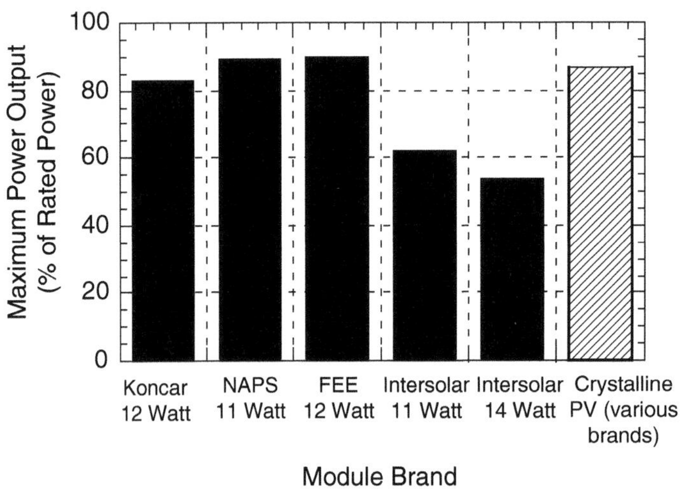

Figure 18.3. Average performance for solar PV modules tested in 1999.

drafting the standards had approved a set of performance standards for both amorphous and crystalline PV modules that drew heavily from the respective codes established by the International Electrotechnical Commission (IEC). The relevant IEC standard for crystalline silicon PV modules is IEC 61215, while the standard for amorphous modules is IEC 61646. The corresponding Kenyan standards are KS 1674 and KS 1675.

The KBS did not, however, have access to the necessary equipment or technical capacity to carry out the specified tests, and the standards, while adopted, remained largely unenforced. Thus, while the move to draft and adopt performance standards may have represented a positive step toward an institutionalized approach to quality assurance, the adoption of unenforced standards did little to protect the interests of Kenyan consumers.

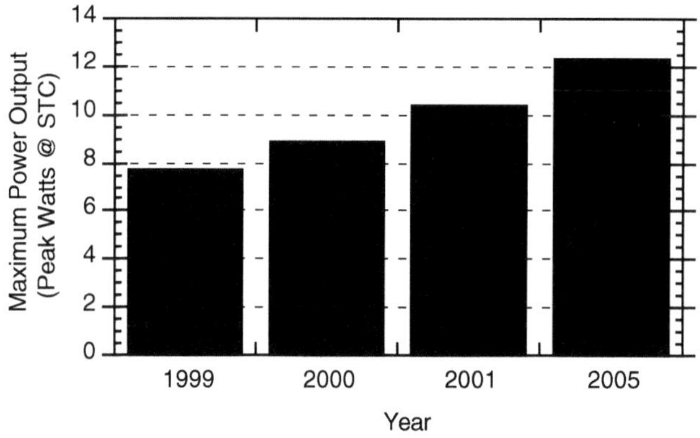

Figure 18.4. Performance trend for fourteen-watt rated Intersolar / ICP-solar modules from 1999 to 2005.
Notes: The performance data for 1999 are from field tests of the twelve intersolar "Phoenix Gold" modules included in the Berkeley-Princeton-EAA study. The 2000 and 2001 results are based on measurements of two separate batches of four "Phoenix Gold" modules by Jacobson at UC Berkeley. The 2005 results are based on the performance of the four ICP-solar module included in the current study.

4 Assuring Product Quality: Performance Testing in 2004/05

4.1 Amorphous Silicon PV Performance Testing – 2004/05

The results presented here are based on tests of five different brands of a-Si solar PV modules sold in Kenya (see Table 18.2 for details). These tests were carried out at Humboldt State University and the University of California at Berkeley between September 2004 and March 2005. In implementing this project, we worked in conjunction with the Kenyan Renewable Energy Association (KEREA), as well as Kenya-based import companies that sell the various brands of amorphous silicon (a-Si) modules included in the study. In particular, KEREA coordinated a series of discussions within the Kenyan solar energy industry about solar module quality, while the solar import companies contributed funds to cover the costs of buying the modules and shipping them to the United States. University researchers coordinated all of the field activities related to selecting and purchasing the solar modules from retail shops in Kenya.

4.2 Performance Test Results

The final, stabilized average maximum power output for each of the brands is given in Figure 18.5. These results indicate that the performance of three of the brands (Free Energy Europe, Solar Cells, and ICP-Solar) is consistent with what one would expect from high-performing twelve-watt solar PV modules. The performance of the remaining two brands (eSolar and SunLink), both purchased from the same manufacturer in China, is well below their advertised level of performance.

It is notable that the final performance of all three of these brands is close to twelve watts despite the differences in their initial performance, as it confirms that the rate of efficiency decline over the first months of operation (so-called Staebler-Wronski degradation) differs from brand to brand. In this study, the average performance of

Table 18.2. Brands of a-Si solar modules included in the 2004/05 study.

a-Si Module brand	Kenyan import company	Country of origin
Free Energy Europe	Chloride Exide, Telesales	France
ICP-Solar	Sollatek	United Kingdom (Wales)
Shenzhen Topray #1 (eSolar)*	Kenital	China
Shenzhen Topray #2 (SunLink)*	Electric Link	China
Solar Cells	Bhatt Electronics	Croatia

* Topray #1 modules were imported by Kenital from the Shenzhen Topray Solar Company and sold in Kenya under the eSolar brand name. Topray #2 modules were imported by Electric Link from the Shenzhen Topray Solar Company and sold in Kenya under at least two brand names in Kenya, including SunLink and SunSolar.

the four ICP-Solar modules dropped by 29 percent prior to stabilization, compared to 21 percent for the Free Energy Europe modules and 20 percent for the Solar Cells modules. This highlights the importance of basing comparisons of relative performance on the final stabilized performance rather than on the initial power output.

It is also important to note that the relative performance of all three of these brands is statistically identical. In other words, although the average performance of the four ICP-Solar modules tested in the study (12.3 watts) was slightly higher than the performance of the Solar Cells (12.1 watts) and the Free Energy Europe (11.8 watts) modules, these differences are well within the margin of error of the measurements methods used in the study. The 95 percent confidence intervals for average stabilized performance for Free Energy Europe modules are 10.7–12.9 watts, for Solar Cells modules are 11.2–13.0 watts, and for ICP-Solar modules are 10.3–14.3 watts. A larger sample size for each module type would be required to determine differences in the relative performance of the respective brands.

One additional significant point is that, while the maximum power output of a solar module under standard test conditions is a key indicator of performance, other performance parameters such as durability and longevity are also important. In this area, Free Energy Europe is noteworthy for the high performance of its "C-version" a-Si module, which has passed the rigorous set of tests required for IEC certification (based on the IEC 61646 standard).

As noted previously, the performance of the two low-performing brands is well below their advertised nameplate ratings. The average stabilized performance for both lines of Shenzhen Topray a-Si modules was approximately six watts, which is well below acceptable levels for fourteen-watt rated modules. The low performance of these modules may be caused by impurities introduced during production and/or other quality control problems in the manufacturing process.

In addition to low power output, we observed problems with module failure for both lines of Shenzhen Topray a-Si modules. In the case of the modules sold under the eSolar brand name, three of the four modules failed completely during their first few months on the test rack. One of the four modules sold under the SunLink brand name also failed during this time period. These failures appear to be caused by water intrusion that led to delamination of the active material of the a-Si modules (see Figure 18.6).

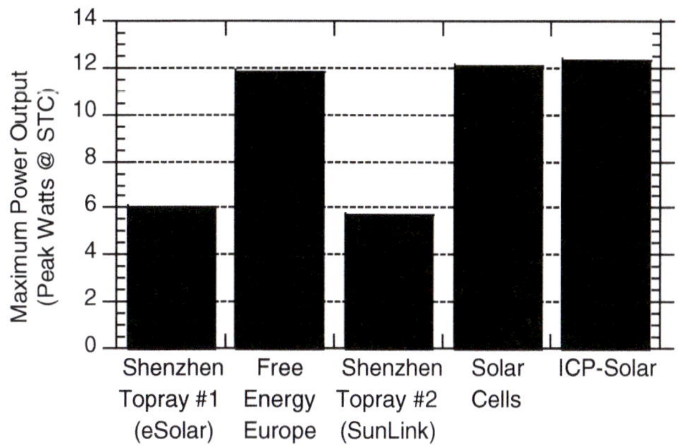

Figure 18.5. Average stabilized maximum power output for five brands of amorphous silicon solar modules sold in Kenya.
Notes: Maximum power at standard test conditions (STC) of 1000 w/m^2 and 25°c.

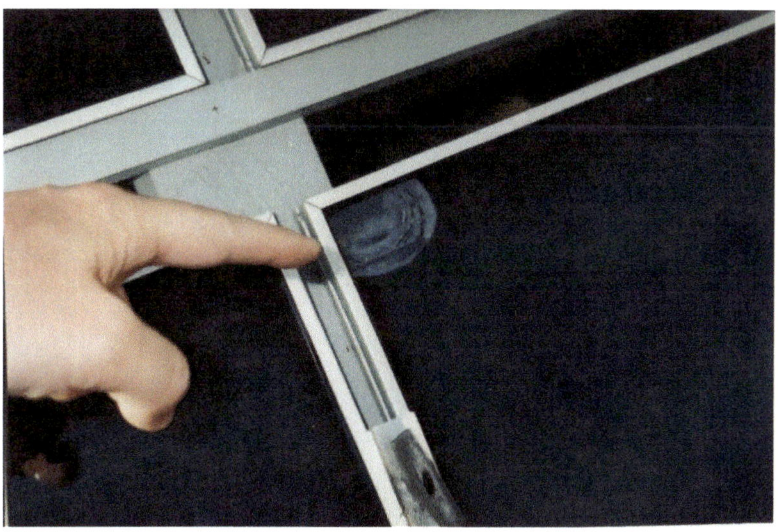

Figure 18.6. Water intrusion related delamination in a Shenzhen Topray a-Si Solar module. Photo: Arne Jacobson.

4.3 Response to Amorphous Silicon Performance Results from the 2004/05 Study

Kenital, the company that marketed the Shenzhen Topray amorphous silicon modules under its eSolar brand name, responded to the results of this study by discontinuing sales of these low-performing modules. Upon learning of the test results, Kenital began importing and distributing the Croatian-made "Solar Cells" a-Si modules.

Electric Link, the other company that had been importing Shenzhen Topray modules, continued to sell the product for a number of months after learning of its low performance. The Kenya Bureau of Standards (KBS) moved to enforce its existing quality standards for both amorphous and crystalline silicon PV modules in the month following the public release of the results of this study in Kenya. Only at this point did Electric Link discontinue sales of the low-performing Shenzhen Topray modules.

5 Discussion and Implications

The history of solar PV in Kenya, focusing on the importance of quality control and the role of performance testing, highlights the importance of an institutional and enforceable approach to quality assurance. Currently, the KBS requires that import companies secure a certificate that validates that their product conforms to the respective Kenyan standards prior to bringing the modules into the country. This certificate of conformity must be issued by an accredited laboratory. No such facility exists in Kenya, so this testing takes place in laboratories in Europe, North America, and Asia (see www.worldbank.org/astae/qpp/PVGAP/App8.pdf for a list of qualified facilities). This means that potential market feedbacks between end users and suppliers of energy technologies may be weakened (with the testing stations only available overseas).

This approach represents a promising step toward an institutionalized approach to quality assurance for the Kenya solar market. While it is perhaps too early to draw conclusions about the long-term significance of these recent events, several solar manufacturers have already sought and obtained IEC certification as the result of this new requirement for participation in the Kenya market. At the same time, the lower-performing brands appear to have been forced from the market.

6 Conclusions

The evolution of renewable energy markets, and their national and international support, has become a topic of international concern (IPCC, 2011). The findings presented here are important not only for Kenya, but also for solar markets elsewhere in Africa and in developing markets worldwide. One of the barriers to the regulatory approach currently implemented in Kenya is a dearth of small, low-cost, certified solar PV modules. In many African countries, the solar markets remain too small to provide the government with sufficient leverage to induce solar manufacturers to pursue certification for the purpose of selling products in that country alone. The Kenya market, however, appears to have grown to a size where it can now successfully provoke such a response. Now that Kenya has taken this step, it should be easier for countries with smaller solar markets to follow suit. This suggests the possibility for significant spillover benefits from Kenya to solar markets elsewhere on the continent and beyond.

On a global stage, the recent growth of on-grid and off-grid renewable energy deployment has raised issues of product quality, testing procedures, and the value of certification and of industry standards. The experience in Kenya is particularly important in efforts to resolve these issues because avenues to address product testing and consumer and industry feedback are all weak in the solar industry generally. These issues are exacerbated in developing nations. More broadly, the Kenya solar story involves a holistic, or "systems" approach to integrating scientific, engineering, industry standards, and adaptive, or "organic" regulation of a fast-growing element of the energy sector.

FURTHER READING

IPCC Special Report on Renewable Energy Sources and Climate Change Mitigation. O. Edenhofer, R. Pichs-Madruga, Y. Sokona, K. Seyboth, P. Matschoss, S. Kadner, T. Zwickel, P. Eickemeier, G. Hansen, S. Schlömer, C. v. Stechow (eds.) Cambridge, UK and New York, NY:.Cambridge University Press.

Jacobson, A. & Kammen, D. M., 2007. Engineering, Institutions, and the Public Interest: Evaluating Product Quality in the Kenyan Solar Photovoltaics Industry. *Energy Policy*, 35, 2960–68.

Kammen, D. M., 2006. The Rise of Renewable Energy. *Scientific American*, September, 82–91.

REFERENCES

Acker, R. & Kammen, D. M., 1996. The Quiet (Energy) Revolution: The Diffusion of Photovoltaic Power Systems in Kenya. *Energy Policy*, 24: 81–111.

Cereghetti, N., Burá, E., Chianese, D., Friesen, G., Realini, A., & Ressonico S., 2003. Power and Energy Production of PV Modules: Statistical Considerations of 10 Years Activity, 3rd *World Conference on Photovoltaic Energy Conversion*, May 11–18, Osaka, Japan.

Crawford, M., 1997. Kiss' New PV Plans: Market Plants, Not Panels. *The Energy Daily*, King Communications Group, Inc.

Duke, R. D., Jacobson, A., & Kammen, D. M., 2002. Product Quality in the Kenyan Solar Home Systems Market. *Energy Policy*, 30: 477–99.

Duke, R. D., Graham, S., Hankins, M., Jacobson, A., Kammen, D. M., Osawa, B., Pulver, S., & Walther, E., 2000. *Field Performance Evaluation of Amorphous Silicon (a-Si) Photovoltaic Systems in Kenya: Methods and Measurements in Support of a Sustainable Commercial Solar Energy Industry*. Washington, DC: World Bank, ESMAP.

Eikelboom, J. A. & Jansen, M. J., 2000. Characterization of PV Modules of New Generations, ECN Report #ECN-C-00–067, Netherlands.

ESDA, 2003. *Study on PV Market Chains in East Africa, Report for the World Bank*, October, 2003, Nairobi: Energy for Sustainable Development Africa (ESDA).

Faiman, D., Bukobza, D., Kabalo, S., Karki, I., Medwed, B., Melnichak, V., de Held, E., & Oldenkamp, H., 2003. *Amorphous, Mono- and Poly-Crystalline Silicon PV Modules: A Comparative Study of Their Relative Efficiencies Under Various Outdoor Conditions*, 3rd World Conference on Photovoltaic Energy Conversion, May 11–18, Osaka, Japan.

Green, M. A., 2005. Silicon Photovoltaic Modules: A Brief History of the First 50 Years. *Progress in Photovoltaics: Research and Applications*, 13: 447–55.

Hankins, M., 2000. A Case Study on Private Provision of Photovoltaic Systems in Kenya. Chapter 11 from *Energy Services for the World's Poor*. Washington, DC: World Bank Energy Sector Management Assistance Program (ESMAP).

Hankins, M. & Bess, M., 1994. Photovoltaic Power to the People: The Kenya Case, report for the Joint UNDP/World Bank Energy Sector Management Assistance Programme (ESMAP), January.

IEC, 1993. IEC – 61215: Crystalline Silicon Terrestrial Photovoltaic Modules – Design Qualification and Type Approval, Geneva: IEC Central Office.

IEC, 1996. IEC – 61646: Thin-Film Terrestrial Photovoltaic (PV) Modules – Design Qualification and Type Approval, Geneva: IEC Central Office.

IPCC 2011. *Special Report on Renewable Energy Sources and Climate Change Mitigation*. O. Edenhofer, R. Pichs-Madruga, Y. Sokona, K. Seyboth, P. Matschoss, S. Kadner, T. Zwickel, P. Eickemeier, G. Hansen, S. Schlömer, & C. V. Stechow (eds.) Cambridge, UK and New York, NY: Cambridge University Press.

Jacobson, A., 2002. AJ's Technical Tips: Designing a Small Solar PV System Part II: Sizing the Solar Panel. *Solarnet*, 4(1).

2004. *Connective Power: Solar Electrification and Social Change in Kenya*. PhD Dissertation, Energy and Resources Group, University of California, Berkeley.

Jacobson, A., Duke, R., Kammen, D. M., & Hankins, M., 2000a. *Field Performance Measurements of Amorphous Silicon Photovoltaic Modules in Kenya, Conference Proceedings*, American Solar Energy Society (ASES), Madison, WI:, June 16–21.

Jacobson, A., Duke, R., & Kammen, D. M., 2000b. Amorphous Silicon PV Panels: Are They a Good Value for the Money? *Solarnet*, 2(2).

2001. Update on the Performance of 14 Watt Phoenix Gold Solar Panels. *Solarnet*, 3(2).

Jacobson, A. & Kammen, D. M., 2004. *Test Results for Five Brands of Amorphous Silicon Solar Modules from Kenya: Initial Maximum Power Output Measurements (Pre-Light Soaking), report delivered to companies in the Kenya solar industry*, September 21, University of California, Berkeley.

2007. Engineering, Institutions, and the Public Interest: Evaluating Product Quality in the Kenyan Solar Photovoltaics Industry. *Energy Policy*, 35: 2960–68.

Kenya Bureau of Standards, 2003a. Kenya Standard: Crystalline Silicon Terrestrial Photovoltaic (PV) modules – Design Qualification and Type Approval, KS1674–2003, Nairobi.

Kenya Bureau of Standards, 2003b. Kenya Standard: Thin-Film Terrestrial Photovoltaic (PV) Modules – Design Qualification and Type Approval, KS1675–2003, Nairobi.

Lane, D., 2003. ICP Solar Technologies Buy UK Plant, *Thin Silicon Newsletter*, number 4, May 2003, http://www.asi-net/newsletter.htm.

Ochieng, F. O., 1999. The Amorphous Question. *Solarnet*, 1(1).

Staebler, D. L. & Wronski, C. R., 1977. Reversible Conductivity Changes in Discharge-Produced Amorphous Si. *Applied Physics Letters*, 31(4): 292–94.

ACKNOWLEDGMENTS

In conducting the field-based testing described in this chapter, we were assisted by Maina Mumbi of Off-Grid Energy Alternative Technologies. This work was supported by the Karsten Family Foundation and the class of 1935 of the University of California, Berkeley.

19 The U.S. Synthetic Fuels Corporation: Policy Consistency, Flexibility, and the Long-Term Consequences of Perceived Failures

Laura Díaz Anadón and Gregory F. Nemet

Authors' Summary

The U.S. government's immediate response to the rapidly rising oil prices of the late 1970s primarily involved demand-side measures such as price deregulation and energy efficiency incentives. Synfuels, however, were promoted as a supply-side response. Subsidies for the demonstration and deployment of synfuels – synthetic liquid or gaseous substitutes for conventional petroleum or natural gas products – received a rare confluence of support from both parties in government (which wanted to show a dramatic response to the public), industry (which saw a potentially profitable new market), and energy experts (Deutch and Lester, 2004). In 1980, the U.S. Congress created the Synthetic Fuels Corporation (SFC) to use the country's extensive coal, heavy oil, and oil shale deposits to ameliorate the energy security and macroeconomic consequences of oil import dependence. With cost estimates of only $60/barrel for the mid-1980s, the development of synfuels production capability was seen as a backstop or "insurance" against seemingly ever-rising oil prices (Deutch, 2005). Five and a half years later, and after expenditures of about $4.5 billion, the long-term legacy of the SFC has been perceived failure.

Editors' Guide

Relationship of Case Study to Energy Technology Innovation System Framework

The history of synfuel programs in the United States, particularly the experience of the Synthetic Fuels Corporation in the 1980s, illustrates many of the innovation system processes relating to *actors and institutions*. Perhaps most salient are the intertwined influences of *shared expectations* and *advocacy coalitions* on both the successes and failures of U.S. synfuel developments. Although various institutions promoted synfuels through most of the early and mid-twentieth century in the United States, the oil crises of the late 1970s established a broad coalition of support, from successive presidents and policy makers (including fiscal conservatives) to technology developers, energy experts, and industrialists. The breadth of this coalition, and the strength of its expectations for the technology, led to an aggressive and well-resourced program to rapidly deploy different synfuel technologies at commercial scales and dramatically increase synfuel production. The goal of substituting more than a quarter of all U.S. oil imports within ten years was testament to the exuberance of these *expectations*, as was the creation of a new form of institution at arm's length from government and designed to respond flexibly and competently to business needs.

But just as these innovation system processes had enabled the technology to move into the demonstration and early deployment stages, so too did they ultimately forestall its further development. Critical here were the strongly context-dependent innovation incentives, particularly the production guarantees whose economic viability were undermined by the collapse in oil prices in the mid-1980s. With additional technical difficulties plaguing the performance and production levels of many of the sponsored projects, the initial *shared expectations* for the synfuels program were recast as overexuberant and unrealistic, undermining political support and related *resource mobilization*, and leading ultimately to the termination of the program. The *legacy of failure* has been long lasting, affecting the U.S. government's appetite for supporting large-scale demonstration projects, and the R&D effort into thermochemical synfuel production. Yet there have also been positive outcomes. In particular, the continuous involvement of major technology developers in addition to that of the U.S. Department of Energy has helped mitigate to some extent the *knowledge depreciation* often associated with boom-bust cycles and has contributed to *knowledge spillovers* into related technologies. Although neither *knowledge generation* nor strengthened *actor networks* were goals of the synfuel program, these outcomes have proved enduring.

Insights for Energy Technology Innovation Policy

- Consistent innovation policies are critical to avoid stop-start or boom-bust cycles in innovation efforts that can have lasting effects on the resources mobilized and expectations needed to support technology development and deployment.
- Codification of knowledge generated through quid pro quo arrangements for public policy support or direct involvement of public entities helps mitigate knowledge depreciation.
- Designing effective innovation policies inevitably entails trade-offs: between managing expectations and building advocacy coalitions; between consistency and flexibility; between commercial-scale projects and granular portfolios.
- Deterministic cost-benefit analysis should not be the only measure used to inform innovation policy if deep uncertainty is associated with market conditions and technical feasibility. Supporting technologies that insure against a risk that does not materialize does not necessarily imply policy failure.
- Rapid commercial scaling up of a technology and/or its output irrespective of market conditions and without sufficient technical information is highly risky. Deciding when to move from applied R&D to demonstration and deployment is difficult, yet crucial. A good management and evaluation system is important to allow for timely decisions about policy goals and activities.

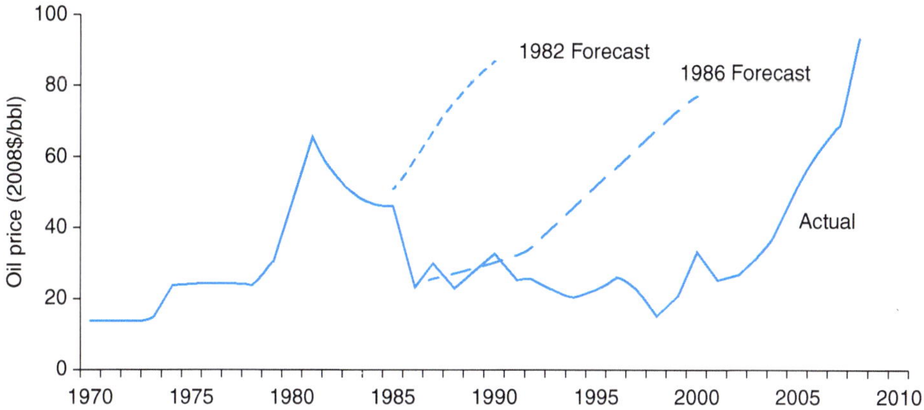

Figure 19.1. Cost of U.S. oil imports and selected world oil price forecasts.
Notes: Oil import costs and forecasts are expressed in 2008 US$ (EIA, 2010a). Oil price forecasts from the early 1980s are by the U.S. Energy Information Administration (EIA, 2010b).

1 Introduction

Following the Iranian Revolution in 1979, oil prices doubled and eventually rose from about $14/barrel to $40/barrel (in 2008$). As shown in Figure 19.1, the U.S. Department of Energy (DOE) projected that oil prices would continue climbing to $80–100/barrel by the mid-1990s (EIA, 2008). And "Project Independence," President Nixon's plan to end oil imports by 1980, was widely acknowledged to have failed.

This was the backdrop for the 1980s synthetic fuels program in the United States. Synthetic fuels – synfuels – are liquid or gaseous substitutes for conventional petroleum or natural gas products, and in this context are derived from fossil fuels such as coal, shale, or tar sands, rather than from biomass. A federally funded R&D program on synfuels dated back to the 1960s and was revitalized during Project Independence with a focus mainly on pilot plants. In 1980, the U.S. Congress passed the Energy Security Act which, among other things, created the Synthetic Fuels Corporation (SFC) and authorized $12.2 billion (in 1980 US$) for its first phase of operation. An expected second phase of operation never took place.

In the 1970s, both fiscal conservatives and environmentalists had opposed the creation of a large-scale government program supporting the commercialization of synfuels because of production cost uncertainties at a commercial scale and environmental impacts of the processes involved. This strong opposition disintegrated in the face of high oil prices in 1979 and the introduction of federal incentives for other alternative sources of energy (including solar and bioenergy).

The SFC's ambitious mandate was to achieve production of 0.5 million barrels per day by 1987, and 2 million barrels per day by 1992, equivalent to the replacement of more than one-quarter of U.S. crude oil and petroleum product imports (EIA, 2010b). To achieve these goals, the SFC was to support private sector projects selected through a competitive bidding process, or by directly negotiating financial contracts for specific projects using production price guarantees and purchase agreements, as well as loan instruments and guarantees (Rothberg, 1983). Commercial and technical viability were the most important selection criteria, but the SFC also took a technology's long-term prospects and environmental impacts into consideration. Contracts issued effectively created small market niches for synfuel project outputs, with the U.S. Department of Defense as a principal purchaser.

The SFC was structured as a quasi-governmental organization unlike other federal agencies: its salaries were higher and operated under a different model, and its budget was not subject to the yearly appropriations process. This structure was in part motivated by the U.S. Congress's lack of faith in the DOE's capability to run large projects. The DOE had already funded synfuel pilot projects that many perceived as unsuccessful (e.g., the Project Gasoline and CO_2 Acceptor pilot projects, see Figure 19.2). In addition, Congress hoped to insulate the operation from volatilities in the annual budget processes and use a model that would appear credible to the private sector (Deutch and Lester, 2004).

Over its four years of operation from 1982–86, the SFC issued nine separate solicitations requesting proposals for financial assistance to aid the construction and/or operation of synfuels projects, and received about 100 distinct proposals (Cohen and Noll, 1991). During its first year of operation, the SFC gave loan guarantees for two shale projects and a heavy oil project, and provided price guarantees for two coal gasification projects (Herrick, 2002). Initially, these projects met project cost and schedule targets. However, once oil prices started to plummet in 1985 (see Figure 19.2), the cost of producing synfuels became considerably greater than their market value and serious discussions began about the continuation of the SFC. Then in 1986, the SFC was terminated.

2 Synthetic Fuel Programs in the United States

2.1 Early History (1925–1978)

The Synthetic Fuels Corporation (SFC) built on a long-standing, federally funded synfuels development program. Between 1925 and 1956, the Bureau of Mines had conducted exploratory research and built small test units. Germany's production of synfuels from coal during World War II motivated Congress to pass the Synthetic Liquid Fuels Act in 1944 for the construction and operation of demonstration plants to produce synthetic liquid fuels from coal, oil shales, and agricultural and forestry products. This early phase of research and experimentation was characterized by technical difficulties and escalating and wide-ranging cost estimates. Two small pilot plants were built but subsequently shut down in the early 1950s as the cost of synfuel development became too high for the government to bear (U.S. DOE, 2010a).

In 1960, the Office of Coal Research was created to find new markets for the politically powerful coal industry and promote the utilization of domestic coal resources in the face of dwindling supplies of oil and gas (Cohen and Noll, 1991). This newly institutionalized program started by supporting research and pilot projects on two processes: coal liquefaction and coal gasification (see Figure 19.2). Coal liquefaction consists of the production of liquids such as gasoline, methanol, and fuel oil from coal either directly (e.g., pyrolysis) or indirectly via coal gasification (producing gas from coal). In both cases, the hydrogen-to-carbon ratio in coal is increased in the liquid synfuels by putting coal in a hydrogen atmosphere, sometimes in the presence of catalysts, and using elevated temperatures and pressures. Coal

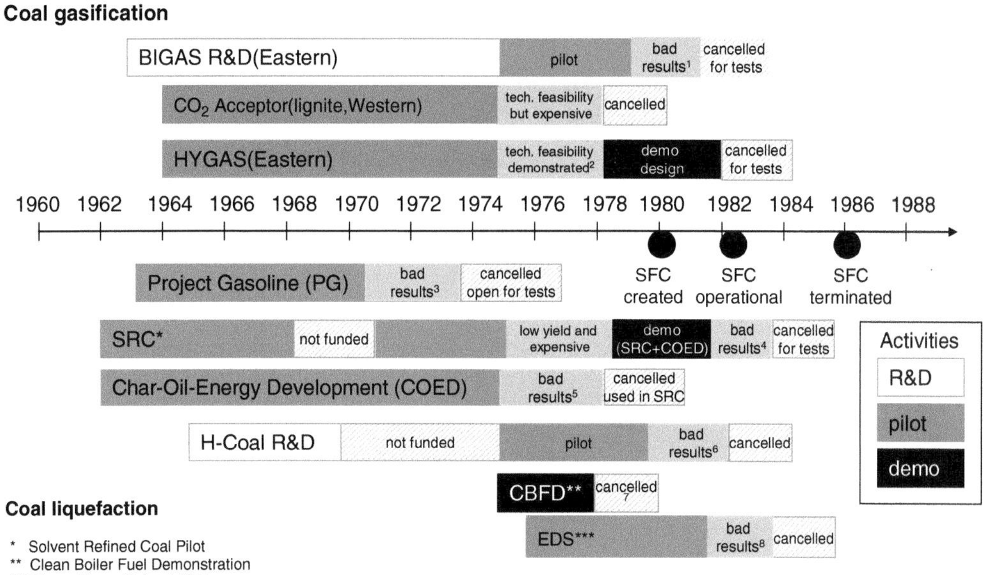

* Solvent Refined Coal Pilot
** Clean Boiler Fuel Demonstration
*** Exxon Donor Solvent Pilot
(1) Design complexities. Pilot operated only briefly in 1977, 1978 and 1979. Pilot scale was 120 tons coal/day.
(2) Costs twice as large as those of deregulated natural gas price. Pilot scale of 75 tons coal/day. Demonstration design was for 7,300 tons coal/day.
(3) Gasoline never produced, problems with all components. Pilot scale plant was 20 tons coal/day.
(4) Major cost overruns (greater than a factor of 2) were found.Performance shortfalls were predicted. The size of the demo plant was going to be 6,700 tons coal/day or about 20,000 barrels per day of oil equivalent. The pilot scale plant was 50 tons coal/day.
(5) Low yield and high sulfur content. Pilot scale was 25 tons coal/day.
(6) Twice as long to design and build, and 66% more expensive. Pilot scale was 250-600 tons coal/day, or 1,400 barrels per day of oil equivalent
(7) Conceptual design was never completed. Caking problems not solved. Demo plant design was 50,000 tons coal/day.
(8) Long delay in design and construction and cost overrun of 50%. Pilot scale was about 250 tons coal/day and 500 barrels per day of oil equivalent.

Figure 19.2. Major coal gasification and liquefaction projects in the United States (1962–86). *Notes*: Projects were supported by the Office of Coal Research (1960–74), the Energy Research and Development Agency (1974–78), and the Department of Energy (1978–80), before the creation of the Synthetic Fuels Corporation in 1980. The timing of the projects is approximate. Data sourced from: Bohn and colleagues (1980); Cohen and Noll (1991); Dravo Corp. (1976); Edgar (1983a, 1983b).

gasification to produce natural gas substitutes rather than liquid fuels was also supported given the importance of increasingly expensive natural gas used in residential and commercial heating as well as industrial-process heat and feedstock.

Attempts to scale up liquefaction and gasification projects from pilot plants to commercial scales faced several challenges, including moving and processing large amounts of coal (solids handling), building components and large reactors to withstand high pressures and temperatures, refining or upgrading raw coal to liquids or to gas products, and dealing with large amounts of waste materials. These challenges were exacerbated by the program's support for unproven direct liquefaction processes over a proven indirect liquefaction process using the Lurgi gasifier. This was possibly due to the latter's incompatibility with Eastern coals.

Cost overruns were endemic, indicating the severe difficulties with projecting costs at commercial scale. The Office of Coal Research (as later with the DOE) persistently underestimated the costs of unproven U.S. gasification and direct liquefaction processes at levels below those of proven European processes (Cohen and Noll, 1991). Figure 19.3 shows that, between 1973 and 1977, the Lurgi fixed bed gasifier, a proven technology in places as varied as Germany, South Africa, the United Kingdom, Pakistan, and Korea (Blazek, Baker, and Tison, 1979), was expected to be

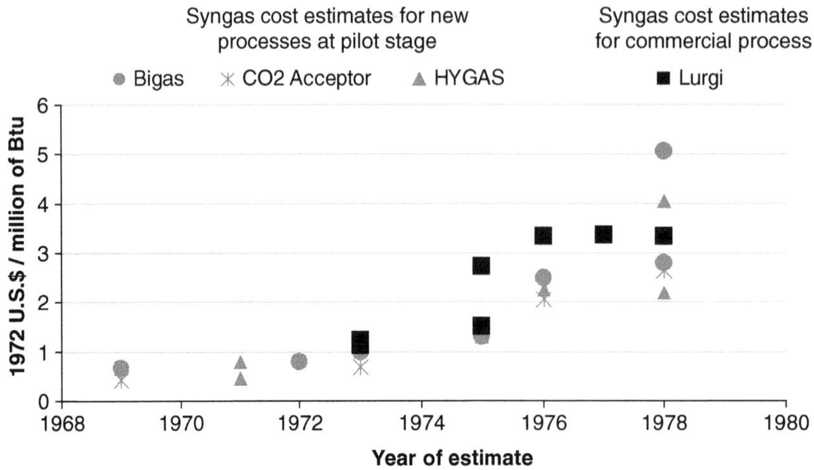

Figure 19.3. Estimates of syngas production costs (1969–78).
Notes: Estimates from the technical literature and government reports of the cost in $ per million BTU of producing syngas from coal using different processes. Symbols in grey represent estimates of new processes only available at the pilot stage. Symbols in black represent estimates (starting in 1973) of the Lurgi process that was already commercial.
Data sourced from: Cohen and Noll (1991, Table 10A-2).

more expensive than newer unproven processes in the United States such as BIGAS, CO_2 Acceptor, and HYGAS. Only in 1978 did the expected cost of deploying the proven Lurgi technology in the United States fall to within the range of these newer processes. The same trend occurred in cost estimates for coal liquefaction. A RAND report attributed the almost fivefold increase in costs between 1971 and 1977 principally to "improved cost estimating capability" based on more detailed engineering designs and more operational experience (Hederman, 1978).

2.2 From Opposition to a Coalition of Support (1973–1980)

The 1973 OPEC oil embargo highlighted the importance of developing synthetic fuels and catalyzed President Nixon's creation of Project Independence, which aimed to make the United States energy self-sufficient by 1980. In 1975, President Ford set a goal to produce 1 million barrels of synthetic fuels per day (Ahrari, 1987) and established the Synfuels Interagency Task Force. The task force carried out a cost-benefit analysis that excluded noneconomic benefits and showed that unless unrealistically low synfuels prices and very high oil prices were used, the program's costs would exceed its expected benefits. Although the need to develop synthetic coal and gas was clear, the cost and development risks were very high, and so without government support they would not happen (Synfuels Interagency Task Force, 1975). In this regard, the task force recommended that loan guarantees be used to enable regulated utilities to access capital markets, and price guarantees be used to remove market risk from future declines in oil prices.

A broad coalition of opponents were against a proposed loan guarantee program on the grounds that it would involve commitment of enormous resources, highly uncertain outcomes, and excessive government involvement in the industry (Ahrari,

1987). Two reports released in 1976 by spending oversight bodies concluded that the loan guarantee policy lacked adequate economic justification and had been insufficiently discussed (Cohen and Noll, 1991). Environmental groups joined fiscal conservatives to oppose commercial synfuels facilities because they anticipated "massive environmental problems, including air pollution, groundwater contamination and disposal of huge amounts of toxic wastes" (League of Conservation Voters, 1987; McCloskey, 1981).

The tide turned in support of the commercialization of synfuels in October 1978, when Iranian oil workers went on strike, followed by protests and the exile of the pro-Western Shah in early 1979. Iranian production dropped from 6 million to 1.2 million barrels of oil per day within two months. In June 1979, a legislative measure was passed to provide loan guarantees, loans, and price contracts to support the domestic production of 0.5 million barrels per day of synfuels. In July 1979, President Carter, who had been criticized for not doing enough, proposed an even bigger ten-year commitment of $88 billion (1980$), supported by an oil windfall profits tax levied on oil producers in the same year. This federal spending would be divided into an initial four-year phase with $20 billion to be disbursed by a proposed new institution, the Synthetic Fuel Corporation (SFC), and a second phase with the remaining $68 billion (Time, 1980).

The SFC's mandate was to achieve production of 0.5 million barrels per day by 1987 and 2 million barrels per day by 1992, which would have replaced one-quarter of crude oil and petroleum imports. With some modifications, this proposal was incorporated into the Energy Security Act and signed into law in 1980 (Ahrari, 1987). The creation of the SFC was supported by technical experts and lawmakers alike (Deutch and Lester, 2004), enshrining their belief that it would be possible to move from liquefaction and gasification plants processing tens or hundreds of tons of coal per day to plants processing thousands and tens of thousands within seven years. The Energy Security Act also included a further $6.8 billion (1980$) of funding for synfuels (through the DOE, through Department of Defense purchase contracts, and for the production of alcohol fuels), as well as $1 billion for solar energy and conservation R&D (Deutch and Lester, 2004).

2.3 The Synthetic Fuels Corporation (1980–1986)

The short history of the SFC was characterized by high turnover in its administration, slowness in the processing of proposals, scandal, and falling support – but relative success in meeting construction deadlines. Delays associated with a change in government meant the SFC did not become operational until February 1982 (Raloff, 1982). Between mid-1980 and early 1982, the DOE received about $6 billion (1980$) to serve as the implementing agency for the SFC (Cohen and Noll, 1991). In 1981, the DOE committed to supporting three projects, two of which were subsequently transferred to the SFC when it became operational (see Table 19.1).

The SFC solicited more projects and by 1985 had received about 100 distinct proposals of which three were funded (Cohen and Noll, 1991). The SFC used several criteria to make decisions about projects to support. Because its objective was to meet its production goals, two major criteria used were commercial and technical viability (i.e., technically realistic, lowest-cost projects), but the SFC also considered the value of having a diversity of technological approaches, of good long-term prospects, of low environmental impacts (including water consumption).

Table 19.1. New synfuel projects in the United States (1980–86).

Project	Project type & design capacity	Contract date	Type of award	Amount in $ billion (current year $)	Brief history
Exxon-TOSCO Oil Shale Project (Colorado) *- transferred to SFC from DOE*	Oil Shale 47,000 barrels per day	Aug 1981	Loan guarantee	$1.15 bn	Costs increased up to 200 percent associated with production at half design capacity and falling oil prices. Terminated by Exxon in May 1982.
Great Plains Coal Gasification Project (North Dakota) *- not managed by SFC*	Coal Gasification (Lurgi gasifier) 137.5 million cubic feet syngas per day (about 16,000 tons of coal per day)	Jan 1982	Loan guarantee	$2.02 bn	First use of Lurgi gasifier in the United States. Completed in 1984 on time and under budget. Operator defaulted on loans in 1985 because of low gas prices. DOE took over plant in 1986 then sold it on to Dakota Gas. Plant still produces syngas and is now being used to demonstrate CO_2 capture.
Union Oil Parachute Creek Shale Oil Project (Colorado) *- transferred to SFC from DOE*	Oil Shale 10,400 barrels per day	Jul 1981 Oct 1985	Price Guarantee ($42.5[a] per barrel) Loan and Price Guarantees	$0.4 bn $0.5 bn	Completed on time. Operated until June 1991 but at half design capacity because of performance problems.
Cool Water Gasification Project (California) *- SFC project*	Coal Gasification (Texaco gasifier with GE107E turbine) 100 MW (660–850 tons coal per day)	Jul 1983	Price Guarantee ($12.5[a] per million Btu)	$0.12 bn	First integrated gasification combined cycle (IGCC) demonstration plant. Completed on time and under budget. Operated successfully until closure in 1998.

Dow Syngas Project (Louisiana) - *SFC project*	Coal Gasification (Dow E-Gas gasifier) 155 MW (1,580 tons of dry sub-bituminous coal per day)	Apr 1984	Price Guarantee ($12.5[a] per million Btu for 10 years)	$0.62 bn	Dow's gasifier at three times larger scale than previous pilot. Operated from 1987 until 1995
Forest Hill Heavy Oil Project (Texas) - *SFC project*	Heavy Oil 1,000–1,400 barrels per day	Sep 1985	Loan & Price Guarantees ($40[a] per barrel for first 50,000 barrels)	$0.06 bn	Completed within budget. Defaulted on loans after two years of operation likely because of price guarantee not covering actual production costs, and filed for bankruptcy.

[a] adjusted for inflation

Notes: All projects managed by the SFC except Great Plains Project.

Data sourced from Bohn and colleagues (1980); Raloff (1986); Andrews (2006); Holt (2005); Priddy (2008); The PACE Company Consultants & Engineers (1985). Additional Information on Exxon-TOSCO project from Bartis and colleagues (2005); Exxon (1982), on Great Plains project from *United States v. Great Plains Gasification Associates* (1987); U.S. DOE (2010b), on Union Oil project from Bartis and colleagues (2005); *O&G Journal* (1991), on Cool Water project from Cohen and Noll (1991), and on Forrest Hill project from Bayrer (1991).

The six projects shown in Table 19.1 differed from those supported during the 1960s and 1970s in that all were completed on time, and in the case of the three coal gasification projects, all utilized Western coal and indirect liquefaction, as opposed to direct liquefaction which had been used previously for Eastern coals.

3 The End of the SFC and the Synfuels Program

Congressional support for the SFC declined because of a combination of falling oil prices, the increasing likelihood that synfuels would not be needed, the failure to meet production goals, and corruption involving the SFC president. By July 1985, SFC projects had only produced 2 percent of the 0.5 million barrels per day goal (Raloff, 1986), and oil prices had fallen to around $20/barrel, a third of the early estimated cost of synfuel production. In addition, the SFC was investigated for nepotism and further criticized for "administrative largesse" – generous salaries and pension benefits for its administrators (Raloff, 1985).

Public support for strong government action on energy seemed restricted to crises. Surveys between 1973 and 1985 found that only in 1974 and 1979, immediately after the two major supply shocks, was energy considered America's "most important problem" by more than one-quarter of the people surveyed (Ahrari, 1987). Environmental organizations mounted a campaign in 1981 to stop the development of synfuels and abolish the SFC (Friends of the Earth, 2010). With all of these factors running against the SFC, the "coalition of opponents" formed again, and the Synthetic Fuels Corporation was abolished in January 1986.

The SFC itself only spent $960 million: 60 percent on the Dow Syngas project, 29 percent on other projects, and 11 percent on administrative expenses (Priddy, 2008). Adding in the $1.5 billion loan default by the Great Plains Coal Gasification project, the total investment in the U.S. synfuels program was $4.5 billion (in 2010$). A further $7.6 billion had been invested in coal liquefaction and gasification R&D and pilot programs through the 1960s and 1970s. And after the authorization of the SFC, an additional $0.6 billion was spent between 1982 and 1984. Nevertheless, the goal that had motivated these investments – producing 2 million barrels of synthetic oil per day by 1992 – was not achieved.

4 Gasification Technology Diffusion and Spillovers

Although the SFC's production targets were not met, the three gasifiers used in the coal gasification projects shown in Table 19.1 are among the most widely used gasification technologies today. These are the Lurgi, GE Energy (previously Texaco), and E-Gas ConocoPhillips (previously Dow) gasifiers. In 1984, there were already 26,000MWth of installed syngas capacity worldwide, three-quarters of which were produced by SASOL, the South African energy company using the Lurgi gasifier, and further processed using Fischer-Tropsch synthesis to produce liquid fuels. By 2007, total gasification capacity globally had more than doubled to 56,000MWth (Tennant, 2010). As shown in Figure 19.4, most of this was produced by six types of gasifiers of which the three noted earlier accounted for over two-thirds. Most of this capacity is used to produce chemicals and liquid fuels.

Several factors may have enabled the diffusion of the GE (Texaco) and E-Gas ConocoPhillips (Dow) designs. First, price guarantees under the SFC created a protected market niche that allowed experience and data to be accumulated on twelve years of operation of the Cool Water project and nine years of operation of the Dow Syngas project. Second, there was relative continuity in the companies engaged in the projects that then commercialized the gasification technologies and the networks that were created.

Constructed in 1984, the Cool Water plant was the first integrated gasification combined cycle (IGCC) demonstration plant in the world. Cool Water was essential to the decision by Tampa Electric company, with Texaco, GE, and other partners (and supported by DOE), to build the Tampa Polk IGCC plant. GE's direct involvement with the Cool Water and Tampa Polk plants allowed it to gain installation and operational experience with the gasification technology and the integration of the IGCC plants. Cool Water also enabled subsequent scale-up of gasification and IGCC plant integration for other plants, including two in the United States and two in Italy. GE claims that Cool Water demonstrated fuel flexibility (e.g., use of coal and fuel oil), and that Tampa Polk proved power and process heat integration.

Similarly, the Dow E-Gas gasifier used in the Dow Syngas project was later used by ConocoPhillips, having acquired the technology, in the 262MW Wabash River IGCC plant under a 50 percent cost-share with DOE. For at least three years ConocoPhillips employees worked at the Dow Syngas facility, creating the opportunity to transfer tacit knowledge (or know-how) in addition to physical infrastructure and technology. The Wabash River plant has also been used to test components, such as ceramic membranes, a warm gas clean up process, syngas cooler coatings, and gasifier temperature measurements (U.S. DOE, 2002).

The DOE was a common partner in the Cool Water and Dow Syngas projects, as well as in the Tampa Polk and Wabash plants, and also served to provide continuity. The Great Plains project was able to provide operational data because once the original operators had defaulted on their loan and plant ownership transferred to the DOE, the DOE had the incentive to recover some of its costs and use it for testing. The Oil Shale Symposium proceedings, the Cameron (later named PACE) Synthetic Fuels reports dating back to 1964, and DOE reports on coal gasification in the early 2000s all contributed to make more technical information available to the wider community (e.g., NETL, 2002).

While lack of a counterfactual makes it hard to determine whether the diffusion of these technologies would not have taken place if the U.S. government had not incentivized commercialization, the actions of the U.S. government likely accelerated diffusion processes. The gasification technologies developed and the operational experience gained may also play an important role if the capture of CO_2 from IGCC plants becomes an important strategy for reducing CO_2 emissions. Spillover benefits can similarly be identified from the \$170 million of support for the two oil shale projects (Parachute Creek and Forest Hill) and the heavy oil project (Exxon-TOSCO) shown in Table 19.1. The Exxon-TOSCO project has provided valuable information regarding the cost, energy consumption, and environmental impacts of oil shale underground mining. The Parachute Creek project's performance problems have uncovered some of the complexities involved in shale oil retorting, which involves heating the rock at high temperature to obtain hydrocarbon liquids for further refining (Bartis et al., 2005).

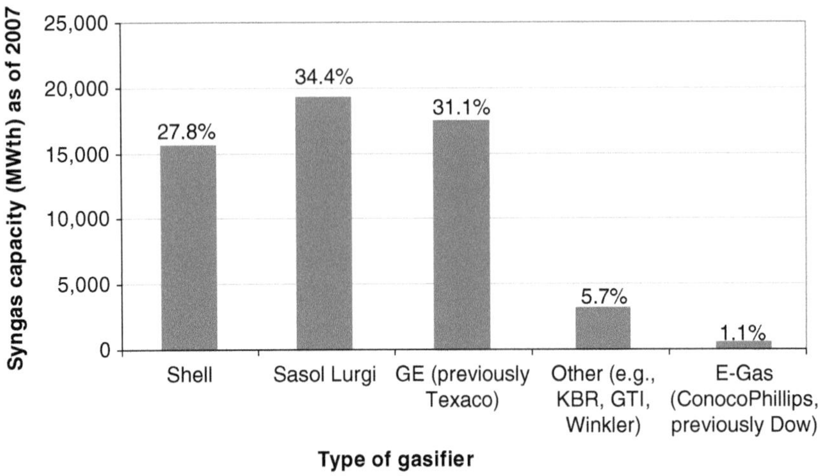

Figure 19.4. Worldwide gasification capacity in 2007 by technology.
Notes: KBR = Kellogg Brown Root. Data sourced from Tennant (2010).

5 Lessons from the Synthetic Fuels Program

Was the SFC a mistake? The simplest interpretation of the SFC is that it represents a policy failure because the projects it funded failed to achieve the program's aggregate production targets. Some conclude that the government's primary mistake lies in its miscalculation of future oil prices or in its decision to promote the demonstration and deployment of synfuels, rather than some other means of addressing high oil prices. However, given the notoriously poor history of forecasting energy prices (Craig, Gadgil, and Koomey, 2002), a program like the SFC is better viewed as insurance against high prices in the future (Deutch, 2005; Schock et al., 1999). Under this interpretation, just because the insurance proves unnecessary does not mean it was a bad idea to buy it.

Another set of interpretations emphasizes the SFC's technical legacy. Projects funded by the SFC, such as the Cool Water project, established the technical foundations for future IGCC plants (PCAST, 1997; U.S. DOE, 2009). The Dow Syngas project laid the foundations for the commercial scale use of the E-Gas gasifier. In addition, the plants served as test beds for new technologies. The Great Plains project, for example, is still used to test CO_2 capture technologies.

In addition, some have suggested that the novel structure of the SFC could serve as a model to support future large demonstration efforts, as it had the flexibility, tools, and competence not present in other U.S. energy-related entities. As an example, the Cool Water project was completed ahead of schedule and under budget despite being the first IGCC demonstration plant in the world (PCAST, 1997). The recently established Advanced Research Projects Agency for Energy, or ARPA-E, which provides financial support for high-risk, high pay-off energy technology projects was modeled after DARPA (an equivalent agency promoting defense-related research) but has elements that are also common to the SFC.

The criticisms of the SFC that remain most relevant relate to the program's overall objective. The imposition of a production goal independent of prevailing energy market prices and without significant experience in full-scale synfuel demonstration projects meant that the SFC program was likely to be much more expensive, risky, and prone to years of backlash than it would have been if its initial goal had been to provide information to the private sector on the technical, environmental, and economic aspects of a synfuels technology (Deutch, 2005). A focus on knowledge generation rather than production may have resulted in selection of different projects, and perhaps a large number of smaller projects – although some of the same projects may have still received funding. On the other hand, it is unclear that a large set of small projects would have generated the same information and spillover benefits as six large ones. And in any case, a knowledge generation approach would not have been consistent with ambitions to dramatically reduce oil import dependence. Expectations for the program were excessively high, with the government presenting the SFC as the solution to energy security concerns. When oil prices collapsed and it became obvious that synthetic fuels were not needed – at least not in the short to medium term – and that the SFC was nowhere near its goal of producing 0.5 million barrels of oil per day, it became politically impossible to maintain the program even though most projects were of technical value. High salaries and corruption exacerbated the SFC's loss of support. This boom-and-bust cycle of excessive hype followed by periods of lower support does not foster a consistent environment for innovation, though it is worth noting the tension between managing expectations and building political support for large (and long-range) programs like the SFC.

The SFC experience also points to the fact that investing large amounts of money to accelerate innovation in a particular area does not guarantee results over a short (or any) period of time. As innovation inputs, time and money are not always interchangeable. Resources at the SFC's disposal were large, but its production goals over the short timeframe set by policy makers was not met.

The SFC experience also indicates that having a good management and independent program evaluation mechanism becomes essential to reduce government expenditures and backlash as soon as there is a high level of certainty that a program will not be cost-effective over the necessary timeframe. This is particularly important in cases like the SFC, because as discussed previously, costs were tied to the market and expectations were set very high. If the program, market, and political developments had been followed more closely, it might have been possible to redefine the SFC goals without shutting down the whole effort before its continuation became politically unsustainable. Redefining program goals might also have reduced the damaging perception of the program as a complete failure. This point also highlights the tension between having consistent initiatives and flexible programs that allow cutting losses.

The ultimate legacies of the SFC program are threefold. First, the perceived failure of the SFC imposed costs that go well beyond the $4.5 billion (in 2010$) spent between 1981 and 1986. The interpretation of technical failure as poor policy has contributed significantly to the U.S. government's aversion to large-scale demonstrations. The first large energy technology demonstration project since the days of the SFC is the controversial FutureGen carbon capture and storage project, which was first announced in 2003 but is not expected to be completed until 2015. Second, the apparent technical infeasibility of direct and indirect liquefaction likely affected the funding and development of biomass to liquid conversion technologies in the United States until today – an effort that has been almost exclusively focused on biochemical as opposed to thermochemical conversion technologies (Williams, 2007). Third, and more favorably, the SFC improved and accelerated the diffusion of liquefaction and gasification technologies that are used in many countries and that have the potential to address current energy problems. And lessons have been learned. Current efforts to produce biologically derived transportation fuels combine a serious commitment to research with a production goal that is much less ambitious than that of the SFC.

FURTHER READING

There are two excellent book chapters on the U.S. synthetic fuels program in books by Cohen and Noll (1991) and Deutch and Lester (2004).

REFERENCES

Ahrari, M. E., 1987. Public Opinion and Synfuels Policy. *Political Science Quarterly*, 102(4): 589–606.

Andrews, A., 2006. *Oil Shale: History, Incentives, and Policy*, Congressional Research Service Report.

Bartis, J. T., LaTourrette, T., Dixon, L., Peterson, D. J., & Cecchine, G., 2005. *Oil Shale Development in the United States: Prospects and Policy Issues*. Santa Monica, CA: Rand Corporation.

Bayrer, R. L., 1991. Appraisal of Current Projects in Synthetic Fuels Technology. *Fuel*, 70(November): 1328.

Blazek, C. F., Baker, N. R., & Tison, R. R., 1979. *High-Btu Coal Gasification Processes*. Chicago, IL: Institute of Gas Technology.

Bohn, E. M., Cowles, J. O., Iyer, R. I., Dadiani, J., & Oyster, J. M., 1980. *Utilization of Synthetic Fuels: An Environmental Perspective*. McLean, VA: U.S. Environmental Protection Agency.

Cohen, L. R. & Noll, R. G., 1991. *The Technology Pork Barrel*. Washington, DC: Brookings.

Craig, P. P., Gadgil, A., & Koomey, J. G., 2002. What Can History Teach Us: A Retrospective Examination of Long-Term Energy Forecasts for the United States. *Annual Review of Energy and Environment*, 27: 83–118.

Deutch, J. M., 2005. *What Should the Government Do to Encourage Technical Change in the Energy Sector?* Cambridge, MA: Massachusetts Institute of Technology.

Deutch, J. M. & Lester, R. K., 2004. *Making Technology Work: Applications in Energy and the Environment*. Cambridge, UK: Cambridge University Press.

Dravo Corp., 1976. *Engineering Support Services Clean Boiler Fuel Demonstration Plant*, Energy Research and Development Administration, Technical Information Center, Chemical Plants Division.

Edgar, T. F., 1983a. Chapter 7 – Coal Gasification. *Coal Processing and Pollution*. Houston, TX: Gulf Publications Company.

1983b. Chapter 8 – Coal Liquefaction. *Coal Processing and Pollution*. Houston, TX: Gulf Publications Company.

EIA, 2008. *Annual Energy Outlook Retrospective Review. Table 4. World Oil Prices, Projected vs. Actual*, Energy Information Administration. U.S. Department of Energy.

EIA, 2010a. *January-last update, Annual Oil Market Chronology* [Online]. Energy Information Administration. U.S. Department of Energy. Available: http://www.eia.doe.gov/emeu/cabs/AOMC/Overview.html. October 12.

EIA, 2010b. *October-last update, Petroleum Navigator* [Online]. Energy Information Administration. U.S. Department of Energy. Available: http://tonto.eia.doe.gov/dnav/pet/hist/wttimus2W.htm. October 16.

Exxon, 1982. Exxon's Abrupt Exit from Shale. *Fortune*, (May): 105–106.

Friends of the Earth, 2010. *Our History* [Online]. Friends of the Earth. Available: http://action.foe.org/content.jsp?key=3653. October 10.

Hederman, W. F., 1978. *Prospects for the Commercialization of High-Btu Coal Gasification*. Santa Monica, CA: Rand Corporation.

Herrick, J. A., 2002. Federal Financing of Green Energy: Developing Green Industry in a Changing Energy Marketplace. *Public Contract Law Journal*, 21(2): 259–60.

Holt, N., 2005. *Gasification and IGCC – Status, Challenges, Design Issues & Opportunities*, Powerpoint Presentation at the GCEP Advanced Coal Workshop, March 15–16, Provo, Utah.

League of Conservation Voters, 1987. *How the 99th Congress Voted on Energy and the Environment* [Online]. League of Conservation Voters Report. Available: http://www.lcv.org/images/client/pdfs/1985–1986_Scorecard.pdf. October 16.

McCloskey, M., 1981. The Environmental Impacts of Synthetic Fuels. *Journal of Energy Law and Policy*, 2: 1–12.

NETL, 2002. *Wabash River Coal Gasification Repowering Project: A DOE Assessment*, National Energy Technology Laboratory, U.S. Department of Energy (US DOE), Report DOE/NETL-2002/1164, January.

O&G Journal, 1991. Unocal to Close Sole U.S. Commercial Oil Shale Plant. *Oil & Gas Journal*, 8(April): 38.

PCAST, 1997. *Federal Energy Research and Development for the Challenges of the Twenty-First Century*, President's Council of Advisors on Science and Technology, Washington, DC: Executive Office of the President.

Priddy, H., 2008. *History of the United States Synthetic Fuels Corporation*, 28th Oil Shale Symposium, Colorado School of Mines, October 13–15.

Raloff, J., 1982. Synfuels Corp. Finally Comes to Life. *Science News*, 121(8).

1985. Washington Deals Synfuels a Big Blow. *Science News*, 128(6).

1986. Congress Kills the U.S. Synfuels Corp. *Science News*, 129(2).

Rothberg, P., 1983. *Synthetic Fuels Corporation and National Synfuels Policy*. Washington, DC: Congressional Research Service, Science Policy Research Division.

Schock, R. N., Fulkerson, W., Brown, M. L., San Martin, D., Greene, R. L., & Edmonds, J., 1999. How Much is Energy Research & Development Worth as Insurance? *Annual Review of Energy and Environment*, 24: 487–512.

Synfuels Interagency Task Force, 1975. *Synthetic Fuels Commercialization Program*, Synfuels Interagency Task Force for the President's Energy Resources Council.

Tennant, J. B., 2010. *Overview of DOE's Gasification Program*. National Energy Technology Laboratory, U.S. Department of Energy.

The PACE Company Consultants & Engineers, 1985. *Pace Synthetic Fuels Report*, The PACE Company Consultants & Engineers.

Time, 1980. Synfuels Success. *Time Magazine*, June 2.

United *States v. Great Plains Gasification Associates*, 1987. 8th Circuit Court of Appeals, 813 F.2d 193, No. 86–5256, United States Court of Appeals.

U.S. DOE, 2002. *Wabash River Coal Gasification Repowering Project: A DOE Assessment*, U.S. Department of Energy (U.S. DOE) National Energy Technology Laboratory report DOE/NETL-2002/1164.

U.S. DOE, 2009. *Pioneering Gasification Plants* [Online]. Washington, DC: U.S. Department of Energy (US DOE). Available: http://www.fossil.energy.gov/programs/powersystems/gasification/gasificationpioneer.html. 1 April 2009.

U.S. DOE, 2010a. *Early Days of Coal Research* [Online]. Washington, DC: U.S. Department of Energy (U.S. DOE). Available: http://www.fe.doe.gov/aboutus/history/syntheticfuels_history.html. November 2.

U.S. DOE, 2010b. *Energy Timeline: from 1971–1980* [Online]. Washington, DC: US Department of Energy (U.S. DOE). Available: http://www.energy.gov/about/timeline1971–1980.htm. October 16.

Williams, R. H., 2007. *Written Testimony of Robert H. Williams. Hearing on the Future of Coal under Carbon Cap and Trade*, U.S. House of Representatives.

ACKNOWLEDGMENTS

This chapter has benefited from contributions by Bob Schock (World Energy Council, UK and Center for Global Security Research, USA), from discussions with Bob Frosch (Harvard), John Deutch (MIT), and from the excellent book chapters listed under "Further Reading."

PART VI

RESOURCES IN THE ENERGY TECHNOLOGY INNOVATION SYSTEM

20 Brazilian Ethanol: Unpacking a Success Story of Energy Technology Innovation

Dustin Meyer, Lynn Mytelka, Rich Press, Evandro Luíz Dall'Oglio, Paulo Teixeira de Sousa Jr., and Arnulf Grubler

Authors' Summary

Brazil's Ethanol Program (ProAlcool), launched in 1975, was a direct response to the dramatic rise in imported petroleum prices in 1973. The military government of the time saw this as a challenge to Brazil's financial stability and energy security, as the country imported 80 percent of the fuel used by its transport sector. Moreover, Brazil had extensive sugar plantations that were facing increased challenges to their exports from European Union trading preferences with African, Caribbean, and Pacific countries and the emergence of corn syrup and other close substitutes for sugar.

ProAlcool was initially a classic import substitution policy. Subsidies were used to expand ethanol production, then in its infancy, and to induce vehicle users to shift to dedicated engines for ethanol that could handle a gasoline blend with more than 5 percent to 10 percent ethanol. When gasoline prices fell a few years later, those who had shifted were left paying the higher costs of ethanol, while the original problem of oil imports remained.

In this changed context, the government decided to invest in the research needed for Brazil to become a more efficient ethanol producer and thus be in a position to eventually eliminate subsidies. At the core of this process was a research partnership that brought together the Brazilian Agricultural Research Corporation and Copersucar, a cooperative of sugar mill and ethanol plant owners. Between 1975 and 2002, ethanol production increased from 0.6 to 12.6 million cubic meters and the price paid to alcohol producers dipped (temporarily) below that of Rotterdam gasoline prices (Goldemberg et al., 2004). By early in the new millennium, increasing yields and reduced processing costs (Van den Wall Bake et al., 2009) had eliminated the need for subsidies. The decision to develop flex-fuel engines, in collaboration with foreign-owned automobile producers, strengthened the domestic auto industry and led to a dramatic shift in consumption habits and practices, thus further building the market. Introduced in 2003, flex-fuel vehicles accounted for 81 percent of light vehicle registrations by 2008 (ANFAVEA, 2008).

EDITORS' GUIDE

Relationship of Case Study to Energy Technology Innovation System Framework

The central narrative of the Brazilian ethanol case study is one of sustained *resource mobilization* and comprehensive *public policy* support to grow, maintain, and diversify a major national industry through periods of relative advantage and through downturns.

The initial successful period of *resource mobilization* for ethanol production and use in Brazil during the 1970s was founded on the institutional history of the sugar and ethanol markets. A relatively small number of *actors and institutions* rooted in this history proved important in the burgeoning innovation system. First, an established cooperative of sugar producers provided a means for *exchange and interaction* between the main market actors. As the market matured, these institutions extended to coordinate the engagement of all the main industry stakeholders, from growers and processors to end-use manufacturers (autos). Second, the potential dilution of this *advocacy coalition* by new entrants was limited both by the effective monopoly position on distribution given to Petrobras, the major Brazilian energy company, and by the inclusion of car manufacturers' interests through the R&D program on flex-fuel engines. Consequently, the interests of potentially disparate actors were aligned behind ethanol innovation efforts. Third, regulatory authority was enshrined in a preexisting market *institution* that not only managed prices and quantities, but could also arbitrage between sugar and ethanol markets (depending on relative prices). Later, the role of clearing house for the subsidies levied on gasoline purchases and redistributed to ethanol producers could also be played by a single actor (Petrobras). In other words, the innovation system had a clear institutional focus for managing the relationship between supply-side and demand-side actors. Fourth, the concentration of the sugar / ethanol market among a few large-scale sugar producers enabled *economies of scale* to be readily captured as production expanded (capital was available, knowledge could readily spill over, and potential market returns were clear). Similarly, the concentration of infrastructure requirements (ethanol-only distribution network and filling pumps) in the hands of a monopoly provider working under clear market incentives tempered the *resistance* to change characteristic of "locked-in" energy systems. Fifth, *knowledge generation* associated with R&D investments was also linked to the particular *institutional* configuration of the innovation system, in this case the public-private partnership between agricultural researchers and sugar mill / ethanol plant owners. These particular institutional characteristics of the Brazilian ethanol history are not generalizable nor directly replicable, yet many alternative versions of this concentrated, aligned configuration of actors and institutions have supported successful innovation outcomes elsewhere.

End users also played a crucial role in the history of Brazilian ethanol. The disadvantages the initial wave of ethanol-only vehicle owners faced in the 1980s placed *adoption and use* on center stage in the renewed innovation efforts of the 2000s. These disadvantages were technical (e.g., inability to cold start) as well as economic (falling relative price of gasoline). In response, concerted innovation investments were made in a new generation of flex-fuel vehicles that removed end users' exposure to uncertain relative prices of ethanol and gasoline. Their *relative advantage* to end users was enhanced by purchase incentives to ensure they were cost competitive and by a preferential tax regime that ensured ethanol remained attractive as a fuel. These incentives also helped overcome the *legacy of failure* and mistrust associated with ethanol-only vehicles, whose market collapsed once oil prices fell.

The phase out of production subsidies led to a period of industry consolidation during which production costs fell significantly through a combination of *knowledge generation* (R&D throughout the value chain), *learning* (yield increases, mechanization), and *economies of scale* (larger distilleries). This illustrates well the interplay between different elements of the innovation system to improve the *relative advantage* of a technology. The cost of ethanol declined by more than two-thirds since the Brazilian ethanol program began. The more recent stabilization of ethanol production costs suggests diminishing returns to learning and scale.

Insights for Energy Technology Innovation Policy

- The singular lesson of the Brazilian ethanol story is the importance of a consistent, continuous, long-term policy framework. This has endured through four turbulent decades and weathered both endogenous difficulties (e.g., technical problems of ethanol-only vehicles, industry consolidation) and exogenous shocks (e.g., oil price volatility, political regime change). This persistence and longevity was underpinned by the coalition of interested stakeholders that was built to support the ethanol program.
- But this came with a cost. A reevaluation of the estimated ethanol production subsidies reveals the magnitude of resource mobilization made necessary not just by the ambition of a national-scale program of industrial capacity expansion and end-use market transformation, but also by its being sustained through periods of clear relative disadvantage (caused by falling oil prices).
- Both the initial ProAlcool Program from the 1970s and the renewed emphasis on ethanol and flex-fuel vehicles in the 2000s combined market-pull measures to provide secure returns on investments in ethanol manufacturing and distribution (e.g., low-interest loans to mills and distilleries, blending mandates, guaranteed prices, preferential tax treatment) with technology-push measures to stimulate innovation (e.g., R&D on ethanol and later flex-fuel engines, research into sugarcane varieties).

1 Introduction

The Brazilian ethanol industry is often held up as an exemplar of successful energy technology innovation and renewable energy policy. Owing to its long growing season and abundance of fertile land, Brazil is naturally suited to agricultural energy production. However, many tropical nations are similarly endowed. What has set

Brazil apart as an ethanol producer is a long-term, consistent policy drawing on multiple sectors of society in support of ethanol production. As the world considers biofuels as a partial solution to a host of energy challenges, and as other nations consider trying to replicate the Brazilian experience, it is important to analyze or unpack the factors behind this innovation history.

The coincidence of the oil price hikes and export challenges faced by Brazil's sugar producers triggered the development of Brazil's ethanol program, ProAlcool, in 1975. This case study analyzes the rationale for the Brazilian decision to develop ethanol production from sugarcane in the 1970s based on distinctive historical experiences as well as established actors. When the price of gasoline fell in the 1980s, the market for ethanol collapsed, but Brazil did not abandon its ethanol pathway. Through a process of policy learning within government, and the development of knowledge networking with and within the private sector, the initial supply-side emphasis was widened to include the demand for ethanol in the transportation sector. Over the next twenty-five years this evolving strategy resulted in a continued market expansion of Brazil's ethanol sector and laid the foundation for the emergence of an innovative biofuels industry.

This case study considers these various sides to the innovation history of Brazilian ethanol. First, we provide a chronology of the main periods in the development of Brazilian ethanol, paying particular attention to the actors and institutions behind ethanol policy as well as to the vehicle technologies that have allowed ethanol to become widely adopted as a transportation fuel in Brazil. Second, we analyze the advances in agricultural and industrial systems that improved the productivity of Brazil's ethanol economy to high levels and that have been associated with the declining production cost profile of ethanol. Third, we unpack the economic dimensions of the ethanol program to provide a more critical picture of its economic success and overall cost than is frequently advanced in the literature.

2 History of Brazilian Ethanol

2.1 Pre-1975: Historical Contingencies

The Brazilian ethanol program was built in the 1970s on three historical foundations: (1) the location and concentration of the Brazilian sugar industry; (2) a long history of ethanol production during periods of crises and as a way of countering sugar price volatilities; (3) the political economy of Brazil in which industrial and energy policies have been shaped strongly by recurring military governments. Throughout these historical developments, important institutions and actor networks were formed that constituted the nexus of the ProAlcool Program introduced in the 1970s.

Sugar has been a defining agricultural commodity for Brazil ever since the establishment of the first Portuguese colony in 1531. Initially located in the northeast, the sugar industry has been concentrated in the southwest of Brazil in the state of São Paulo ever since World War I. Brazil's sugar industry has traditionally been highly concentrated in a geographical region as well as in large plantations and sugar producers. In the 2008/2009 growing season, for example, the 45 largest sugarcane producers in the state of São Paulo accounted for 50 percent of total production, and the 100 largest producers for more than 82 percent (UNICA, 2012). This concentration provided for localized learning and spillover effects in adoption of institutional and technological innovations, further enhanced by the ease of access to knowledge, technology, and capital enjoyed by very large-scale sugar producers. In 1959, a small number of sugar growers and ethanol producers from São Paulo founded a cooperative, Copersucar, to commercialize the sugar and ethanol production of its associates. Its members were also cognizant, however, of the need to develop new sugar varieties and thus improve productivity, which, at about forty-two tons per hectare, was well below the yields of eighty tons per hectare produced by their competitors abroad (Wilkinson and Sorj, 1992). The Copersucar Technology Center (CTC) became a powerful actor and an important center of research in the sugar and ethanol industry.

For much of its early history, ethanol production was used primarily as a means of stabilizing sugar prices or responding to energy crises during wars. The Institute for Sugar and Alcohol (IAA), formed in 1933, was a regulatory agency that regulated prices and set quotas for both sugar and ethanol production, and after 1965, also levied export fees to finance the modernization of Brazil's sugar industry. Ethanol had an important role to play as a fuel, especially during the First and Second World Wars when imported gasoline was scarce. Production levels of 650 million liters from 1941 were only reached again in 1976 (EPE, 2006; Goldemberg, 2009). However, the distilleries controlled by the IAA functioned mainly to remove excess sugar from the market in order to stabilize sugar prices at supported levels (Barzelay, 1986). As a result, the industry was well positioned to rapidly expand ethanol production under the ProAlcool Program.

The historical context of Brazil's political economy also deserves special attention. Two historical milestones in Brazil's ethanol experience were initiated by authoritarian or military governments. The Vargas government instituted the very first mandatory ethanol gasoline blend (5 percent, or E5) in 1931 (Vargas having come to power in a coup d'état in 1930), and President Ernesto Geisel instituted the 1975 ProAlcool Program under a military government. Geisel was deeply familiar with energy issues, having formerly been president of Petrobras, the state-owned oil company created under the second Vargas government in 1945. The political and institutional stage was thus set for launching an ambitious ethanol program in the 1970s as a twin response to the oil price hikes as well as collapsing sugar export markets and prices.

2.2 Phase 1, 1975–1979: Saving the Sugar Industry

Although sugar and alcohol production has a long history in Brazil, the modern ethanol program began when the National Alcohol Program, ProAlcool, was established in 1975. At the time, Brazil was heavily dependent on imported petroleum, and the quadrupling of oil prices in 1973–74 contributed to a foreign debt crisis and brought an end to several years of rapid economic growth that had come to be known as the "Brazilian miracle."

At the same time that oil price shocks were damaging the Brazilian economy, radical changes in world sugar markets were exacerbating the country's economic problems. Brazil had responded to high world sugar prices in the early 1970s by expanding production capacity. In 1975, sugar prices collapsed. In February of that year, a trade and aid agreement between the European Community and seventy-one African, Caribbean, and Pacific countries gave the latter preferential access to the European market for sugar, reducing opportunities for Brazilian sugar exports. In the United States, the other major market for sugar, the food processing industry was beginning its conversion to corn syrup, a close substitute for sugar, in soft drinks. This confluence of factors left Brazil with excess mill capacity and a reduced ability to earn foreign currency.

In response, President Geisel created the Programa Nacional de Álcool, known as ProAlcool. This program was intended to reduce Brazil's dependence on imported oil and to make use of idle capacity in the sugar industry by converting sugar to ethanol. The IAA, Copersucar, and Petrobras were the principal institutions and actors involved in the creation of the ProAlcool program (Bertelli, 2011; BNDES and CGEE, 2008).

ProAlcool was a classic import substitution policy that focused on increasing the sugarcane output of producers at the same time as it maintained low ethanol prices for consumers. Because of Brazil's long history of producing ethanol from sugar, the first phase of the ProAlcool program required few major technical or institutional innovations. Ethanol was initially simply blended with gasoline at rates of 5 percent to 10 percent (E5–E10), which can be used in unmodified gasoline engines and does not require significant modifications at gasoline distribution stations. Ethanol produced and sold in Brazil comes in two qualities: hydrous or containing water (typically some 5 percent) and anhydrous or not containing water (with the water removed chemically after distillation). The introduction of anhydrous ethanol allowed the ethanol/gasoline blend to increase to 20 percent to 25 percent (E20–E25 being currently the market norm in Brazil with pure gasoline not sold at all).

The government incentivized ethanol production with low-interest loans to expand mills and distilleries, guaranteed prices set at the average cost of production, and created demand through blending mandates. At the same time, research into new varieties of sugarcane was carried out by the quasi-governmental agricultural research agency, EMBRAPA; the IAA's research arm, Planalsucar; the Copersucar Technology Center (CTC); and the Campinas Institute of Agronomy (IAC). Petrobras was granted a monopoly on ethanol distribution. Given Petrobras's experience as a gasoline retailer, this helped to speed up the introduction of ethanol and also assured Petrobras's strong integration into the ethanol actor network. During this phase of the ethanol program, sugarcane production increased by 50 percent and ethanol production quintupled to 2.8 billion liters. At the same time, domestic research on (hydrated) ethanol-only engines was stepped up, with the Brazilian government pushing and subsidizing (the initially reluctant) auto companies to develop engines capable of running on 100 percent ethanol. This innovation effort culminated in 1979 with the commercial introduction of the first pure-ethanol vehicle, a Brazilian Fiat 147.

2.3 Phase 2, 1979–1985: Continued Expansion and Innovation

By 1979, the period of easy growth was over. Excess sugar supply capacity had been utilized, and enough ethanol was being produced to provide the nation with an 80/20 gasoline-ethanol mix that was as much ethanol as the conventional gasoline-powered vehicles could accept. To support continued growth, the nation's vehicle fleet would

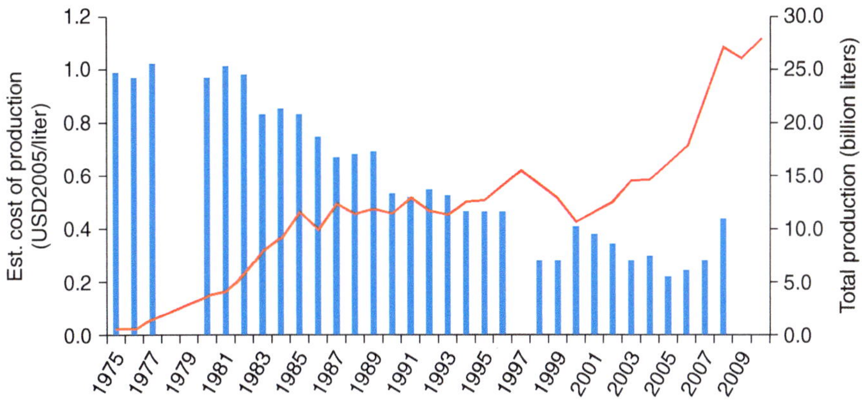

Figure 20.1. Increasing ethanol production (Billion liters) at decreasing cost (US$2005/Liter).
Notes: Brazilian ethanol costs are highly sensitive to the exchange rate in the base year chosen (2005), although the cost dynamics over time invariably show a significant reduction during the period 1975–2000, and a stationary, albeit fluctuating trend thereafter (see text for discussion).
Sources: Production data (1975–2010) from EPE (2011); production cost data (1975–2004) from Van den Wall Bake and colleagues (2009), and (2005–2008) from Valdes (2011).

need to shift toward ethanol-only vehicles, and a new fuel pump infrastructure would need to be built for them. During the first half of 1979, with earlier production targets having been met, ethanol policy was at a standstill.

In the summer of 1979, OPEC announced a 37 percent oil price increase that led to the second oil price shock, and the second phase of the ethanol program began. The government announced that ethanol pump prices would be fixed at 59 percent of the price of gasoline, and that ethanol-only vehicles would be subject to lower sales taxes and licensing fees than conventional vehicles. As shown in Figure 20.2, ethanol-only vehicles were an immediate success, accounting for more than a third of new vehicle sales in 1980, and virtually all of new vehicle sales by 1985 (Furtado, Scandiffio, and Cortez, 2011).

Petrobras built a distribution and pump infrastructure for pure, as opposed to blended, ethanol. Foreign and domestic automakers, under pressure from the government, started manufacturing ethanol-only vehicles. By 1985, virtually all new vehicles sold in Brazil ran on pure ethanol. During this phase of the program, ethanol production quadrupled to 12.8 billion liters (see Figure 20.1). In 1980, domestic ethanol costs were three times higher than international spot market prices or import costs for gasoline (Goldemberg, 2009: 94). Yet dedicated efforts to improve the economics of domestic ethanol production were not deemed necessary, as conventional wisdom at that time assumed ever-escalating oil prices that would ensure domestic ethanol would soon become cost competitive with imported crude oil and gasoline.

2.4 Phase 3, 1985–2003: Uncertainty and Crisis

As the military dictatorship gave way to civilian government in 1985, Brazil entered a period of economic difficulties, including a debt crisis, an IMF-imposed austerity program, and hyperinflation. Struggling to control prices throughout the economy, the government reduced guaranteed prices for ethanol to below the cost of production (Goldemberg et al., 2004), and ethanol output fell for the first time since the program began. When oil prices tumbled in 1986–87, making ethanol increasingly uneconomic, the government banned the construction of new distilleries (Sperling, 1988).

At the same time, world sugar prices recovered, and by 1990 producers were increasing sugar exports even as ethanol output stagnated. Ethanol shortages caused high prices and long waits for drivers of ethanol-only vehicles, and Brazil was forced to import ethanol from overseas. Owners of pure-ethanol vehicles were put at considerable economic disadvantage, a problem compounded by emerging technical performance issues with pure ethanol engines, including the inability to start in cold temperatures and the requirement of more frequent refueling. Consumers lost confidence in the future ethanol supply, and the purchase of ethanol-only vehicles collapsed (see Figure 20.2).

Although ethanol production stayed within the 10–15 billion liter range from 1985 until 2003, reduced subsidies caused the industry to consolidate and increase production efficiency markedly during this period. Goldemberg (2009: 94) reports a doubling of yields from some 3,000 to 6,000 liters ethanol per hectare. Production costs declined significantly through a concerted effort throughout the entire value chain involving research, yield increases, economies of scale in ever-larger distilleries, as well as the utilization of bagasse (a by-product) for electricity generation. Prices were deregulated for anhydrated ethanol used for blending in 1997 and for hydrated ethanol in 1999. Deregulation left blending mandates and relatively high taxes and price controls on gasoline as the major tools with which the government intervened in the ethanol market (Goldemberg et al., 2004).

2.5 Phase 4, 2003–Present: Flex-Fuel Vehicles

Following this period of relative stagnation, standard gasoline engines were reestablished as the norm (see Figure 20.2). The ethanol program was revived in 2003 with the introduction of flex-fuel vehicles (FFVs). The first (full) flex-fuel car commercially introduced in Brazil was the Volkswagen Gol(f) 1.6 Total Flex. Encouraged by the Brazilian government, this innovation was introduced by the Brazilian subsidiaries of major international car companies such as General Motors and Fiat as well as Volkswagen, drawing on indigenous R&D as well as international knowledge and technology spillovers (the Model T Ford produced between 1908 and 1927 had flex-fuel features).

Flex-fuel vehicles (FFVs) were priced competitively with traditional gasoline automobiles (predominantly via government incentives) and allowed consumers to choose the desired blend ratio. FFVs can take pure gasoline, pure ethanol, or any mix. On-board sensors determine the fuel mix and adjust operation accordingly. A significant cost-cutting innovation was the introduction of post-combustion (tailpipe) sensors for establishing the gasoline/ethanol blend of the fuel, as opposed to the previous dominant (and expensive) design of pre-combustion (tank/fuel supply) sensors.

The government incentivized the purchase of FFVs with the same policies it used earlier to increase sales of ethanol-only vehicles, including favorable tax

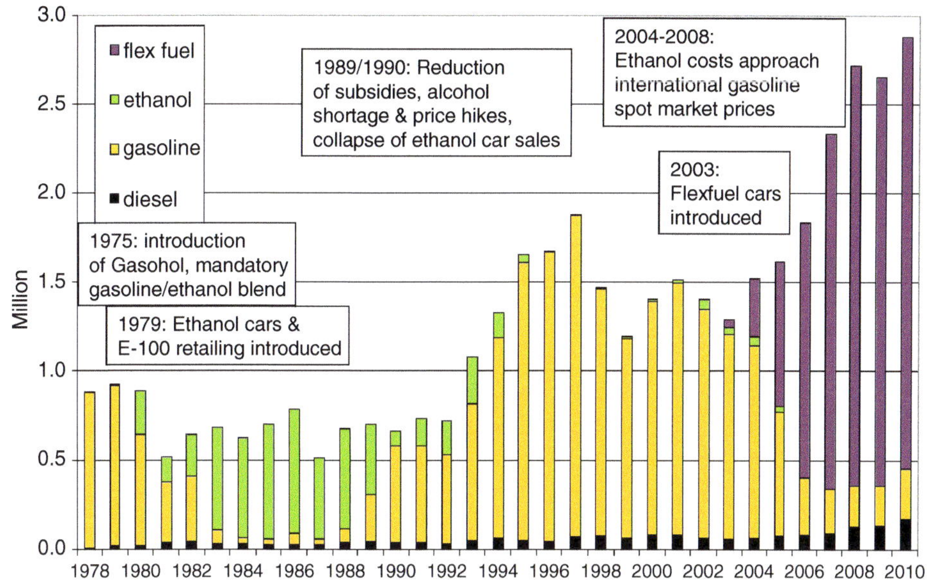

Figure 20.2. Light vehicle registration in Brazil (millions per year) per vehicle fuel type.
Notes: Policy milestones also included.
Source: ANFAVEA (2008, 2012).

treatment at the point of purchase and reduced annual licensing fees. Presently, more than 90 percent of new passenger vehicle sales in Brazil are flex-fuel vehicles (see Figure 20.2), allowing consumers to make decisions, based on price and other factors, about which fuel to use at any given time. FFVs thus resolved the issue that led to a collapse in consumer confidence when oil prices fell in the late 1980s and early '90s. Consumers now have the ability to choose the most economical fuel at any given time, and therefore are no longer as vulnerable as they once were to price spikes or fuel shortages. The rapid proliferation of FFVs in Brazil shown in Figure 20.2 is the most critical development in the recent success of the Brazilian ethanol program. Important, the vehicles were made available at a time of record high petroleum prices, meaning ethanol was frequently the cheaper option of the two fuels.

Because ethanol has a lower energy content than gasoline, ethanol pump prices have to be lower compared to gasoline in order to be competitive. For much of the past decade, the price ratio of ethanol to gasoline has hovered around 60 percent (Valdes, 2011). This price ratio is a result of, on one hand, increased efficiencies in the ethanol value chain, and on the other, gasoline price controls and relatively high gasoline taxes (read: preferential fuel tax treatment of ethanol). In other words, although ethanol prices are no longer regulated or kept artificially low, ethanol is competitive at the pump largely because of a preferential tax regime. A defining characteristic of Brazil's fuel taxation is that it is regionally differentiated (i.e., determined at state levels). The price spread between gasoline (E25 mixture) and pure ethanol varies across states, and generally has been largest in the center of the sugar industry, the state of São Paulo, and lowest in the Amazonian periphery (UNICA, 2012).

Market conditions in 2011 demonstrated the resilience of the transportation fuels sector in Brazil. High world sugar prices that year led producers to divert feedstock into sugar production. At the same time, heavy rains led to poor sugarcane harvests. Because of the flexibility inherent in the FFV fleet, consumers did not suffer shortages or price spikes. Meanwhile, incipient ethanol shortages were managed through decreased blending ratios and corn ethanol imports from the United States (Almeida and Dreibus, 2011).

Whereas ethanol imports in the 1990s signaled a crisis in Brazil's ethanol program, imports in 2011 are perhaps a sign of the program's success. Ethanol has become an important transportation fuel in many countries, in no small part because Brazil has demonstrated its viability. Furthermore, the international market for ethanol is becoming less restricted, as evidenced by the recent elimination of import tariffs in Brazil and the United States. As the international market for ethanol grows, Brazil is well positioned to be a major supplier to the world, although recent ethanol price escalation trends in Brazil and an ever-appreciating currency undermine export potentials and competitiveness, particularly relative to subsidized production like U.S. corn ethanol.

From 1975 to the present, Brazil has maintained policies that have nurtured the ethanol industry by subsidizing supply and creating demand. Whether Brazilian ethanol can compete favorably on price with gasoline depends in large part on exchange rates, among other factors. But economies of scale and technological improvements have brought costs down enough that ethanol is and will likely remain an important transportation fuel, not only in Brazil but also in the United States and Europe, for the foreseeable future.

3 Learning and Cost Reductions

Production costs of ethanol from sugarcane in Brazil have declined by 70 percent over the past thirty-five years (Van den Wall Bake et al., 2009). In the first phase of the ProAlcool program, the largest impacts on cost originated throughout the ethanol production process – from agricultural development, production, and distribution. The majority of cost reductions stemmed from increasing agricultural yields and decreasing industrial costs, and also the use of bagasse (the sugarcane left over after extraction of sugar juice) for industrial heat and electricity generation.

Cost reductions are thus the result of multiple interacting factors, including R&D-based innovation (e.g., in yield increases), economies of scale (e.g., in distillation plants), and the learning by doing effects induced by demand-pull. The innovation history of Brazilian ethanol demonstrates the importance of multiple factors influencing cost and technological innovation whose success lies precisely in their combined effect. Though the initial demand-pull policy efforts were necessary to establish the foundation of a market, the heart of ethanol's success in Brazil were the significant cost reductions that arose from technology innovations at the agriculture stage, distillation/refining stage, and FFV automobile engines. Post-2000 price trends in Brazilian ethanol indicate no further significant cost improvements.

3.1 Increasing Agricultural Yields

Agricultural production of sugarcane has been an important part of the Brazilian economy for centuries. Producers constantly search for methods to increase yields or reduce input costs, but not until the initiation of ProAlcool was substantive agronomical research dedicated to increasing sugarcane yields.

Table 20.1. Comparison of cost structure of brazilian sugarcane in 1976 and 2005.

	Costs [US$(2005)/TC]		Cost reduction (%)
	1976	2005	1976–2005
Land Rent	7.33 (20%)	2.60 (21%)	64%
Soil Preparation	9.78 (27%)	2.33 (19%)	76%
Stock Maintenance	7.47 (21%)	3.31 (27%)	58%
Harvest – Loading	5.97 (17%)	2.47 (20%)	59%
Transport	5.22 (15%)	1.66 (13%)	68%
Total	35.77 (100%)	13.09 (100%)	65%

Notes: Cost structure shown in US$2005 per ton cane (TC) and in percent.
Source: Van den Wall Bake and colleagues (2009).

The result was improvement at every stage of agricultural production (see Table 20.1). New varieties of cane have been developed, many of which were designed via genetic hybrid technology and ideally suited for certain regions or growing conditions. Planting of cane, which utilizes a ratoon system that allows for more than one harvest from a single planting, has been increasingly mechanized throughout the past thirty-five years. Similarly, cane harvesting, long a labor-intensive and dangerous process when done by hand, is transitioning to mechanical harvesting. Fertilizer application, soil preparation, and residue management have been optimized, as has transportation from farm to distillery, minimizing loss of "total reducible sugars." Combined, these efforts have resulted in yield increases of up to 60 percent.

Yield improvements have been bolstered by the development of more sophisticated accounting and systems management tools that cane producers now employ to optimize the timing of operations.

3.2 Decreasing Industrial Costs

The primary objective of the industrial milling process is to extract the maximum amount of sugar from the harvested cane. This process involves extensive mechanical and chemical processes that result in an 80 percent to 90 percent yield efficiency. After treatment, the extracted sugar juice is fermented and subsequently distilled into alcohol. Remaining water is removed via a chemical process in the case of anhydrous alcohol.

The production of bioethanol is hardly different from the production of any alcohol spirit. The general technology for converting sugary liquids to alcohol has existed for centuries: yeast consumes the sugar, converting to alcohol concentrations of approximately 8 percent by volume. This "wine" is then distilled. Despite the maturity of the technical process, collaborative R&D continued to improve the milling process and introduced new innovations, including "open" fermentation, or the extraction of ethanol from the hydrolysis and fermentation of lignocelluloses materials from sugarcane residues. Increasing scale and efficiency and system optimization have been the most important factors in driving down industrial processing costs (see Figure 20.3). The extent of economies of scale present in ethanol distillation plants is remarkable: Van den Wall Bake and colleagues (2009)

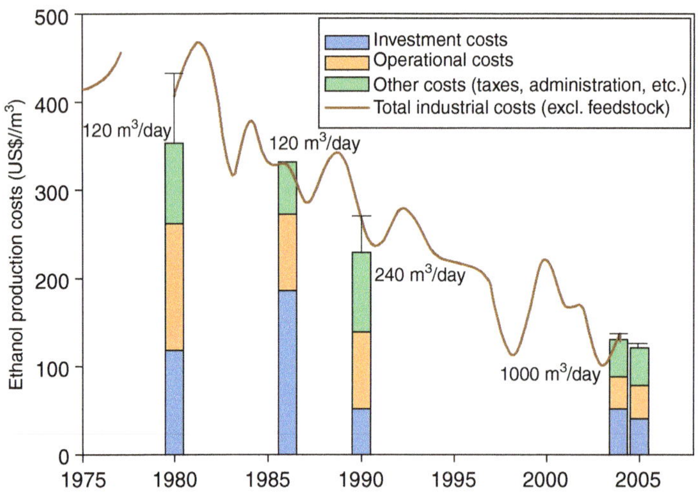

Figure 20.3. Ethanol production costs (excluding sugar feedstock costs) in US$ per m^3, total and per component.
Notes: The average size of new ethanol distilleries in m^3 capacity per day is also shown.
Source: Van den Wall Bake and colleagues (2009).

report typical investment costs of 22–26 million US$ for a new plant built in the 1970s of an average capacity of 120 m3/day compared to 55 million US$ for a plant of some 1000 m3/day built after 2000, that is, an increase in production capacity of more than a factor of 8 only requires an increase in total investments of less than a factor of 2.5 (Grubler et al., 2012).

3.3 Bagasse

Sugarcane bagasse is the cellulosic material left over from the milling and distillation process. Distilleries have long used bagasse as combustion material to generate heat and power for use in the ethanol production process. It has resulted in significant increases in process efficiency and often generates surplus power that is fed into the electricity grid, creating additional revenue (or by-products credit).

4 Subsidies and the Cost of the Ethanol Program

The dominant economic narrative of the Brazilian ethanol program is that after a dedicated, three-decades-long effort it was finally possible to drive down ethanol costs to become cost competitive with fossil fuel alternatives. This version of the innovation history of Brazilian ethanol was supported by an influential paper by Jose Goldemberg published in *Science* (Goldemberg, 2007) that compared Brazilian ethanol prices with Rotterdam gasoline spot market prices over the period 1980 to 2005 using constant US$2004.

The analysis used a US$ reference point as the ethanol program evolved over an extremely turbulent period of economic history in Brazil characterized by prolonged periods of hyperinflation and exchange rate volatility. However, this makes the analysis highly sensitive to the choice of reference year. The US$ – Real exchange rate varied between 0.9 to 3 Real per US$ (IMF, 2012) over the period 1995 to 2011.

The three years with the highest exchange rate (around 3 Real/US$) were in 2002–2004 (compared to a yearly average exchange rate of 1.7 Real/US$ in 2011). The use of 2004 as the base year and for the exchange rate thus can be misleading as it proportionally scales all domestic ethanol prices to the lowest US$ denomination observed historically.

Figure 20.4 compares Goldemberg's original 2007 data with more recent Brazilian price trends and the Rotterdam spot market price, expressed in US$2004 (top panel) as well as Brazilian Real2010 (bottom panel). The original price data in Goldemberg (2007) were expressed on a volume basis. Given the lower heating value of ethanol compared to gasoline (21.2 versus 32.2 MJ/liter), the original price data have been adjusted and expressed on a better comparable per unit energy content (GJ) basis.

As Figure 20.4 shows, the relative economics of Brazilian ethanol and the inferred levels of subsidies are critically dependent on the exchange rate used in the comparison. Using a 2004 base year and exchange rate basically confirms Goldemberg's original conclusion: since 2004, Brazilian ethanol is cheaper than imported gasoline. This suggests an economic success story. Conversely, using a 2010 base year and exchange rate and denominating in (constant) Brazilian Real2010 yields a quite different picture: only in three out of a total of thirty-six years spanned by the time series (2004, 2007, and 2008) is ethanol at par with imported gasoline prices. It is significantly more costly in thirty-three out of thirty-six years, including the post-2008 period.

Both economic comparison metrics agree that prior to 2004 Brazilian ethanol prices were significantly above those of imported gasoline. The difference between domestic ethanol and international gasoline prices (times ethanol sales volumes) yields a proxy, even if imperfect indicator of the total subsidy implicit in the program (see Figure 20.4).

Subsidization of the ethanol industry has taken many forms since 1975. Supply incentives included low and negative interest rate loans to mills, and regulated demand at fixed prices. Consumer incentives included capped low prices for ethanol, and reduced taxes and licensing fees on ethanol-only, and later flex-fuel vehicles.

Gasoline consumers paid the largest of these subsidies. Gasoline prices were kept at a fixed interval above ethanol until ethanol prices were deregulated in 1997 (for anhydrated) and 1999 (for hydrated) ethanol. Petrobras, which had a monopoly on both ethanol and gasoline distribution, effected the transfer of funds from gasoline consumers to ethanol producers through pricing. It collected the subsidy when selling gasoline to distributors and delivered it to ethanol producers when buying ethanol at fixed prices (Moreira and Goldemberg, 1999). This intra-sectorial subsidy ended with price deregulation, though the government continues to support ethanol by taxing it favorably relative to gasoline.

This subsidy has been a significant, if not the dominant, portion of total subsidies to ethanol producers. It can be estimated by considering prices from the position of the refiner/blender. Whatever amount of ethanol was injected into the system downstream of the blender, an energetically equivalent amount of gasoline could have been injected at lower (or greater) cost. The difference between what was spent on ethanol and what, hypothetically, would have been spent on an energetically equivalent amount of gasoline is an estimate of the subsidy.

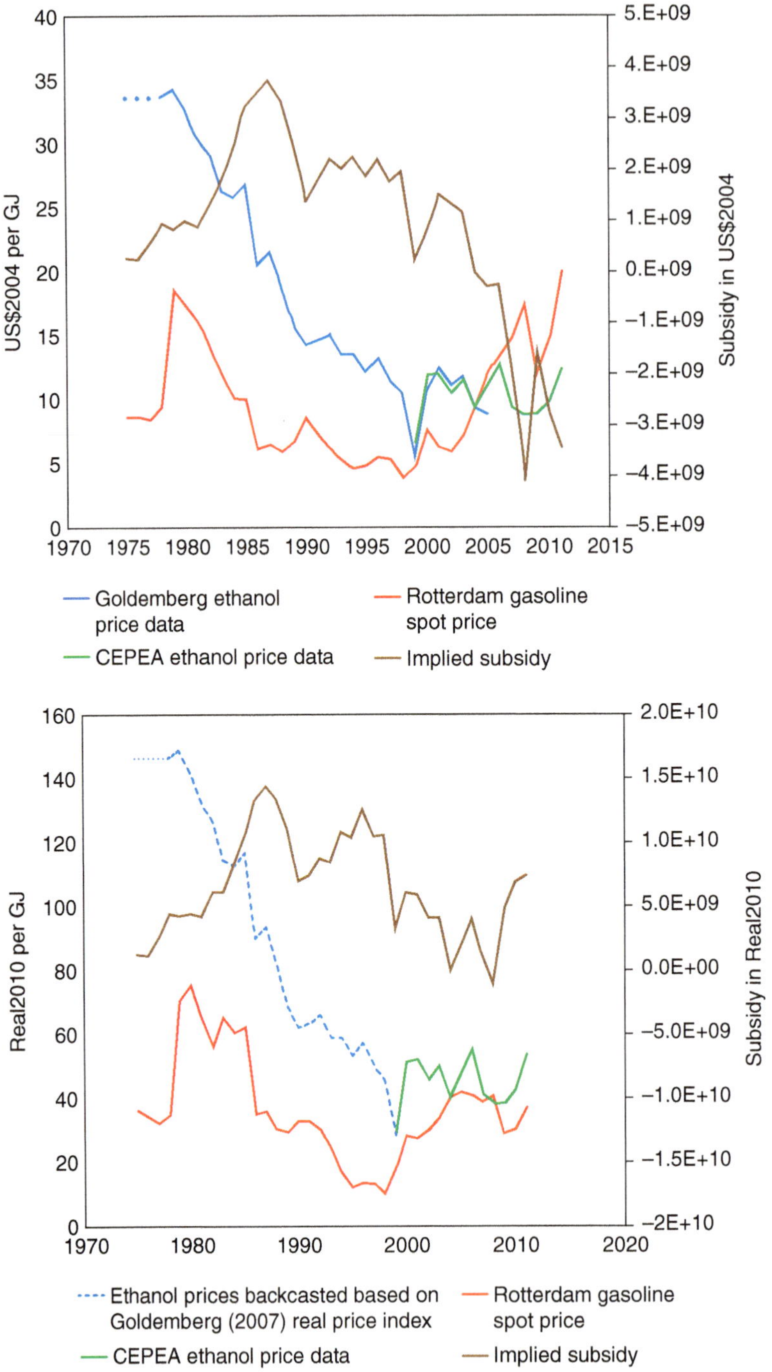

Figure 20.4. Brazilian ethanol prices vs Rotterdam gasoline spot market prices, with inferred subsidy.

Notes: Top panel shows prices in US$2004 per GJ; bottom panel shows prices in Real2010 per GJ. Inferred subsidy (US$2004 and Real2010) shown on right y-axis estimated as price differential times ethanol sales volume, with positive values indicating ethanol subsidy and negative values indicating macroeconomic gains).

Sources: Goldemberg (2007) for ethanol prices to 2005 adjusted from volume to energy content basis; CEPEA (2012) for ethanol prices from 2000–11; IEA (2012) for nominal gasoline prices converted to constant US$ using the U.S. GDP deflator.

Using 2004 base year and exchange rates and the US$ as denominator yields a cumulative net subsidy of 35 billion US$2004 (cumulative subsidy of 45 billion US$2004 1975–2000 minus negative subsidies or macroeconomic savings of 10 billion US$ in the 2001–11 period). Using 2010 as a base year, exchange rates and the Brazilian Real as denominator yields a cumulative net subsidy of 240 billion Real2010 over the period 1975–2011, or of some 136 billion US$2010 (with the 2010 exchange rate being 1.76 R/US$).

The large range of estimated subsidy between 35 to 136 billion US$ illustrates the ambiguity of economic comparisons under widely varying exchange rates as well as the significant levels of resource mobilization necessary by such an ambitious, national-scale, long-term program of energy technology innovation. Cumulative subsidies prior to 2000, often referred to as the "learning investment" necessary to drive down long-term costs and prices, range between 45 to 114 billion US$.

These price differential–derived subsidies are only a crude and macro-level perspective and in all likelihood represent a conservative, lower bound estimate. To illustrate this point, consider the situation for the year 2008, which both indicators agree was the year with the most favorable price ratio of domestic ethanol to imported gasoline. The price difference between domestic ethanol and gasoline spot market prices ranged between 8.6 US$2004/GJ to 2.3 Real2010/GJ in favor of ethanol. In the last week of October 2008, for example, average retail prices differed between 1.513 and 2.511 Real per liter for E100 and E25 respectively (ANP, 2008). Considering the different heat values, this translates into prices of 71.4 and 85.3 Real/GJ for E100 and E25 respectively, that is, a price spread of 13.9 Real/GJ, mostly accounted for by the preferential (lower) tax treatment of ethanol. Multiplying by total ethanol sales of 460 PJ in 2008 (EPE, 2011) and assuming that the entire price spread on an energy content basis is attributable to tax rate differences gives an estimated foregone tax revenue of some 6.4 billion Real in 2008 (3.5 billion US$ in 2008), the year when domestic ethanol had a decisive (wholesale) price advantage over imported gasoline.

5 Conclusion and Lessons Learned

In 2010 Brazil produced roughly 28 billion liters of ethanol (EPE, 2011). This is about fifty times as much as was produced when the ProAlcool program was created in 1975, and about 30 percent of the world total (EPE, 2011; RFA, 2012) Furthermore, ethanol accounts for roughly 20 percent of the highway transportation fuel mix in Brazil, with gasoline comprising 28 percent, and diesel fuel (predominantly used in trucks) making up the balance (EPE, 2011).

Industry projections indicate continued expansion as Brazil strives to satisfy not only its internal market, but overseas demand as well. The U.S. Environmental Protection Agency, for instance, has classified Brazilian ethanol as an "advanced biofuel" that satisfies renewable fuel requirements. This drives up demand for Brazilian ethanol and has led to a curious situation: tankers loaded with Brazilian ethanol destined for the United States now cross paths on the high seas with tankers loaded with U.S. corn ethanol destined for Brazil. The two fuels are chemically equivalent, but Brazilian ethanol is perceived as having environmental benefits that increase its value and that will continue to drive the industry's expansion.

Many lessons can be taken from the story of Brazilian ethanol. Investments in the form of direct and indirect subsidies were very substantial relative to the size of the Brazilian economy. Perhaps just as significant as the subsidies themselves is the fact that they were partially wound down, especially following the collapse of oil prices in 1986–87. The partial withdrawal of subsidies forced the industry to consolidate and to innovate, modernize, and improve ethanol economics. Also important was the coordinated engagement of multiple sectors and stakeholders of Brazilian society, including not only the growers and processors that supplied ethanol but also the auto industry that developed the vehicles to burn it.

Perhaps the foremost lesson of all from the Brazilian ethanol experience is the importance of a consistent, long-term policy framework. The Brazilian ethanol project began in earnest in 1975. In the nearly four decades since, oil prices have soared and collapsed, Brazil's political and economic systems have come apart and been remade, and the ethanol program itself has faced several periods of crisis. Yet policies to support the Brazilian ethanol industry, and the research and investment that underlie its performance, have never been abandoned. This decades-long continuity in policy has made the innovation history of Brazilian ethanol unique.

FURTHER READING

For further, more detailed treatments of the Brazilian ethanol story, see Van den Wall Bake and colleagues (2009); Valdes (2011).

REFERENCES

Almeida, I. & Dreibus, T., 2011. Brazil to Become Net Importer of US Ethanol. *Bloomberg News*. Nov. 11, 2011 ed.

ANFAVEA, 2008. *Brazilian Automotive Industry Yearbook*, São Paulo and Brasilia: ANFAVEA (Brazilian Automotive Industry Association).

2012. *Brazilian Automotive Industry Yearbook*, São Paulo and Brasilia: ANEFAVEA (Brazilian Automotive Industry Association).

ANP, 2008. *Síntese dos Preços Praticados – Brasil. RESUMO III – Gasolina R$/l (in Portuguese)*, Brazil, October 2008, Agência Nacional do Petróleo (ANP).

Barzelay, M., 1986. *The Politicized Market Economy: Alcohol in Brazil's Energy Strategy*. Berkeley: University of California Press.

Bertelli, L. G., 2011. A Verdadeira Historia do ProAlcool. Available: www.biodieselbr.com/proalcool/historia/proalcool-historia-verdadeira.htm.

BNDES & CGEE, 2008. *Sugar-Cane-Based Bioethanol Energy for Sustainable Development*. Rio de Janeiro.

CEPEA, 2012. *Ethanol Price Index Data Series* [Online]. CEPEA (Center for Advanced Studies on Applied Economics). Available: http://www.cepea.esalq.usp.br/english/ethanol/?id_page=243.

EPE, 2006. *Brazilian Energy Balance*. Rio de Janeiro: Empresa de Pesquisa Energética (EPE) and Brazilian Ministry of Mines and Energy.

2011. *Brazilian Energy Balance*. Rio de Janeiro,: Empresa de Pesquisa Energética (EPE) and Brazilian Ministry of Mines and Energy.

Furtado, A. T., Scandiffio, M. I. G., & Cortez, L. A. B., 2011. The Brazilian Sugarcane Innovation System. *Energy Policy*, 39(1): 156–66.

Goldemberg, J., 2007. Ethanol for a Sustainable Energy Future. *Science*, 315(5813): 808–10.

2009. The Brazilian Experience with Biofuels (Innovations Case Narrative). *Innovations: Technology, Governance, Globalization*, Fall 2009, 4(4): 91–107.

Goldemberg, J., Coelho, S. T., Nastari, P. M., & Lucon, O., 2004. Ethanol Learning Curve – The Brazilian Experience. *Biomass and Bioenergy*, 26(3): 301–304.
Grubler, A., Aguayo, F., Gallagher, K., Hekkert, M., Jiang, K., Mytelka, L., Neij, L., Nemet, G., & Wilson, C., 2012. Chapter 24 – Policies for the Energy Technology Innovation System (ETIS). *Global Energy Assessment – Toward a Sustainable Future*. Cambridge, UK and New York, NY: Cambridge University Press and Laxenburg, Austria: The International Institute for Applied Systems Analysis.
IEA, 2012. *Energy Prices and Taxes*. IEA (International Energy Agency). online data available at OECD eLibrary.
IMF, 2012. International Financial Statistics. *IMF eLibrary online data base*. IMF (International Monetary Fund).
Moreira, J. R. & Goldemberg, J., 1999. The Alcohol Program. *Energy Policy*, 27(4): 229–45.
RFA, 2012. *World Fuel Ethanol Production* [Online]. Renewable Fuels Association (RFA). Available: http://ethanolrfa.org/pages/World-Fuel-Ethanol-Production.
Sperling, D., 1988. *New Transportation Fuels: A Strategic Approach to Technological Change*. Berkeley: University of California Press.
UNICA, 2012. *Quotes and Statistics 2012* [Online]. UNICA (Brazil Sugarcane Industry Association). Available: http://english.unica.com.br/dadosCotacao/estatistica/.
Valdes, C., 2011. *Brazil's Ethanol Industry: Looking Forward*, USDA Economic Research Service.
Van den Wall Bake, J. D., Junginger, M., Faaij, A., Poot, T., & Walter, A., 2009. Explaining the Experience Curve: Cost Reductions of Brazilian Ethanol from Sugarcane. *Biomass and Bioenergy*, 33(4): 644–58.
Wilkinson, J. & Sorj, B., 1992. *Structural Adjustment and the Institutional Dimensions of Agricultural Research and Development in Brazil: Soybeans, Wheat and Sugar Cane, Working Paper No. 76*. Paris: OECD Development Centre.

21 Global R&D, Market Formation, and Diffusion Investments in Energy Technology Innovation

Arnulf Grubler, Laura Díaz Anadón, Kelly Sims Gallagher, Ruud Kempener, Anastasia O'Rourke, and Charlie Wilson

Authors' Summary

This case study provides a quantitative overview of the financial resources mobilised globally as fundamental inputs into the energy technology innovation system. Available data from a diverse range of sources are compiled and summarised throughout the innovation life cycle from research and development (R&D) through market formation to widespread market diffusion. Attention is paid to the less well-reported types of financial resource inputs including private (as well as public) R&D and niche market investments by risk-taking venture capital. The resulting synthesis provides a first-order snapshot of relevant financial flows for energy technology innovation, and provides a framework to situate additional detailed case studies on research, development, and demonstration (RD&D) expenditures outside the IEA member countries, and commercial or "diffusion" investments into end-use technologies. Taken together, these case studies paint a comprehensive picture of financial resources mobilised for energy technology innovation. The key elements of this picture are: (1) the scale of resource mobilisation increases dramatically across successive innovation stages; (2) the structure of current energy investments is highly asymmetrical, with end-use technologies dominating diffusion investments but energy supply technologies, particularly fossil fuels and nuclear energy, dominating early stage innovation investments; (3) six major emerging energy economies now account for a significant proportion of global energy innovation investments. These findings point to the need for new institutional arrangements for the comprehensive and systematic gathering and assessment of energy innovation investments to inform decision making.

Editors' Guide

Relationship of Case Study to Energy Technology Innovation System Framework

Financial investment and expenditure data are the principal metrics used to characterise *resource mobilisation* into the energy technology innovation system.

Financial resource inputs create intangible technological knowledge (e.g., via R&D and *knowledge generation*) as well as tangible knowledge (e.g., via investments in the *adoption and use* of capital stock). Financial resources are frequently used to characterise particular innovation stages (e.g., R&D investments) or particular innovation processes (e.g., *learning* investments). The principal contribution of this case study is to compile and synthesise financial resource mobilisation data across all *stages of the innovation life cycle* within the energy technology innovation system, as well as across all components of the energy system from energy supply, through conversion and distribution, to energy end use. This systemic picture allows meaningful analysis of the relative magnitude of resource inputs into different parts of the innovation system, and what this implies for the effective functioning of the system as a whole.

Insights for Energy Technology Innovation Policy

- Both the magnitude and structure of the US$1–5 trillion resource mobilisation into the ETIS annually are influenced by market incentives and related public policies. The alignment and consistency of these incentives across a wide range of policies is therefore critical.
- The increasingly international dimension of resource mobilisation into the ETIS is evident. Emerging economies have become major players in energy technology R&D, niche markets, and diffusion investments alike. New institutions and mechanisms for international cooperation and coordination are needed, as well as inclusive membership in international agencies such as the International Energy Agency to technology development and information exchange.
- A pronounced asymmetry in innovation efforts in favour of energy supply technologies and to the detriment of energy end use suggests the need to rebalance public and private sector ETIS portfolios with a greater emphasis on energy end-use technologies and efficiency.

1 Introduction

In this case study, we collate and summarise available data on the financial resources mobilised to support and induce energy technology innovation throughout the life cycle from research and development (R&D) through market formation to widespread market diffusion. We provide what – to our knowledge – is a first attempt at a comprehensive quantitative overview of financial investment inputs throughout the energy technology innovation system. We attend to the less well-reported types of financial resource inputs including private (as well as public) research and development (R&D) and niche market investments by risk-taking venture capital (see box for definitions). Novel empirical data on two particularly underreported types of financial investments are treated in more detail in companion case studies on RD&D expenditures in the major developing countries, and on commercial or "diffusion" investments into end-use technologies. This case study provides the synthetic overview necessary to piece together these various facets of the resource

Table 21.1. Synthesis of global investments (2005$ billion PPP) in energy technology innovation.

Technology	Innovation Stage					
Class	RD&D		Market Formation		Diffusion	
End use & efficiency	>>8	[1]	5	[8]	300–3500	[15]
Fossil fuel supply	>12	[2]	>>2	[9]	200–550	[16]
Nuclear	>10	[3]	0	[10]	3–8	[17]
Renewables	<12	[4]	~20	[11]	>20	[18]
Electricity (Gen+T&D)	>>1	[5]	~100	[12]	450–520	[16]
Other* and unspecified	>>4	[6]	<15	[13]	n.a.	
Total	>50	[7]	<150	[14]	1000-<5000	[19]

Notes: First order estimates of investment data by innovation stage (columns) and technology class (rows).

* Other includes hydrogen, fuel cells, other power and storage technologies, and basic energy research.

RD&D investment data from IEA (2009b); WEC (2001); Kempener and colleagues (2010); see also the case study on emerging economies: [1] public + private RD&D of 1.8 billion + >>6 billion respectively; [2] public + private RD&D of 2 billion + >10 billion respectively; [3] public + private RD&D of >6.2 billion + >3.4 billion respectively; [4] public + private RD&D of 2–5 billion (range from IEA (2009b); Kempener and colleagues (2010) versus NEF/SEFI (2009)) + 7 billion respectively (private RD&D data includes renewable electricity); [5] and [6] public RD&D only; [7] rounded lower bound.

Market formation investment data from (NEF/SEFI, 2009): [8] includes 2 billion asset finance + estimated 2 billion venture capital; [9] estimated 2 billion from venture capital only; [10] mature technology at diffusion stage; [11] estimated 2.4 billion venture capital + 24.8 billion biomass/biofuels – 8 billion Brazilian ethanol considered mature technology at diffusion stage; [12] estimated 90 billion asset finance + 8 billion venture capital; [13] unaccounted for technology categories; [14] rounded upper bound.

Diffusion investment data: [15] first order estimate from Wilson and Grubler (2011), see also the case study on end-use investments, with range spanning end-use technologies (upper bound) to just energy-using components of end-use technologies (lower bound); [16] and [18] see Table 21.3 for details; [17] estimate based on 2–3 GW reactor completions per year (IAEA-PRIS, 2010) at $1500–2500/kW; [19] rounded range.

mobilisation story. Taken together, they paint a comprehensive picture of financial investments in energy innovation.

Evidently, money is not the only resource that needs to be mobilised: the development of knowledge, skills, supporting institutional settings, and so on are all also important. Financial resources should not be interpreted as a single metric through which the system for energy technology innovation can be comprehensively described. Yet financial investments are an important and observable form of innovation input, a core element of any embodiment of technological change, and data – albeit far from comprehensive – are more readily available than other input, output, or outcome metrics. (We discuss this at length in the case study on assessment metrics.) In addition, financial investment data offer a commensurable metric across technologies, across innovation stages and processes, and across different sectors and markets. Finally, financial data are useful for policy makers and industry alike as both operate under budget constraints.

Table 21.1 provides an overview of the investment estimates that are discussed in detail in this case study. Investment and expenditure data are expressed in international $ using purchasing power parity (PPP) to improve comparability between economies (although a drawback of PPP is that simple *ex post* adjustments to a common base year are not possible for international, cross-country statistics).

We classify our investment estimates by broad technology class and by innovation stage. The first innovation stage is characterised by research, development, and demonstration (RD&D) expenditures. These are a well-defined expenditure category in macroeconomic and corporate accounts. Thereafter, definitional boundaries become unavoidably blurred. The market formation stage is described by investments that either rely on risk-taking private finance including venture capital or private equity funds, or that require policy inducements and market incentives such as feed-in tariffs or production tax credits. The final widespread or mass market diffusion stage represents the commercialisation of mature technologies that no longer require dedicated policy incentives. This distinction from the market formation stage is inevitably somewhat arbitrary as the broader market environment with its relative prices, tax regimes, rules and standards, and so on, is strongly influenced by policy. But in the diffusion stage, these are less dedicated to supporting specific technologies.

2 RD&D Investments

National and international statistical agencies routinely collect research, development, and demonstration (RD&D) expenditures at the macroeconomic level (see OECD, 2007). The data are usually differentiated by funding source (public or private sectors), by RD&D-performing institution (government laboratories, universities, or private firms), and by economic sector. Methodologies, data collection, and compilations are well established (OECD, 2002). But energy-related or technology-specific RD&D data are not reported separately in these macroeconomic statistical frameworks. This creates data challenges for assessing energy technology innovation (Dooley, 2000). The exception is for public sector RD&D expenditures in member countries of the International Energy Agency or IEA (IEA, 2009a). Comparable information for non-IEA countries (including Brazil, China, and Russia) is sparse and for private sector RD&D is extremely fragmented. Yet our estimates suggest they account for around three-quarters of all energy-related RD&D globally, and rising. Comprehensive data to separate out demonstration investments from R&D expenditures are also few and far between.

2.1 Public Sector Energy RD&D

Figure 21.1 summarises the trends in public sector energy RD&D since 1974 in IEA member countries and contrasts it with total public RD&D expenditures. Two defining characteristics of energy RD&D are its comparably small magnitude (5 percent of total government RD&D) and its cyclical "boom and bust" profile with rapid expansion in the wake of the oil crises of the 1970s, subsequent collapse, and only gradual recovery after the year 2000. This stands in stark contrast to the continually expanding total R&D budget in OECD countries. These trends are extensively discussed in the literature (e.g., Dooley and Runci, 2000; Doornbosch and Upton, 2006) and used to argue for a persistent "R&D underinvestment" in

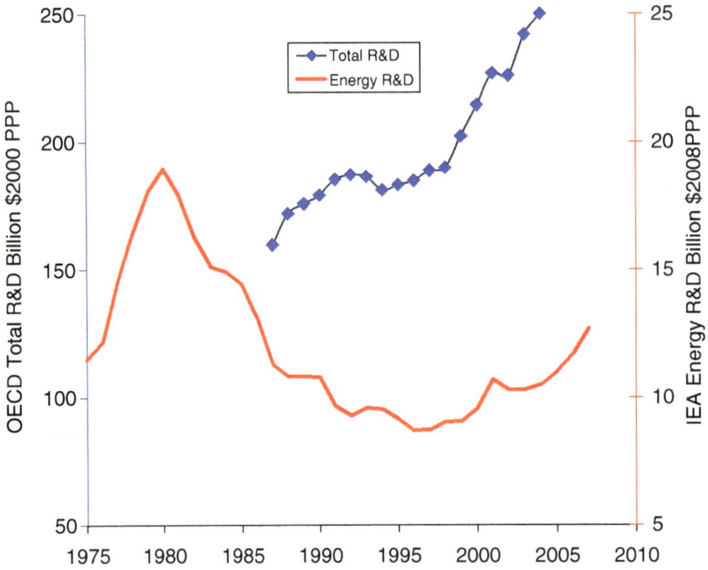

Figure 21.1. OECD public R&D expenditures (2000$ PPP) vs public energy RD&D in IEA countries (2008$ PPP).

Notes: Y-axes expenditure scales differ by a factor of ten. The period 1985–2000 shows opposing time trends.

Sources: OECD data from Doornbosch and Upton (2006); IEA data from IEA (2009b).

energy innovation by researchers (e.g., Nemet and Kammen, 2007) and business executives (e.g., AEIC, 2010).

The contribution of demonstration investments to the totals shown in Figure 21.1 is unclear. The IEA (2009a) reports a total of some $0.55 billion for seven countries for which demonstration-specific data are available, with the United States accounting for more than $0.44 billion. This total equates to some 4 percent of total public energy technology RD&D in all IEA member countries in 2008, pointing to the dominance of R&D over technology demonstration proper.

Figure 21.2 summarises the historical evolution of IEA member country energy RD&D by broad technology class, illustrating a third defining characteristic: asymmetries in public RD&D portfolios. Total public RD&D in IEA member countries in 2008 amounted to some $12.7 billion. Close to $5 billion was spent on nuclear (fission and fusion), $3 billion on "other" energy technologies (hydrogen, electricity transport and distribution, basic energy research), and about $1.5 billion apiece on fossil fuels and energy efficiency. Fossil fuel and nuclear technologies receive the lion's share of RD&D funding in IEA member countries and in the major developing economies.

As noted, comparable RD&D statistics for non-IEA member countries are lacking, giving rise to the incorrect perception that energy RD&D and technology development is primarily performed in developed economies. The companion case study by Kempener and colleagues provides a survey of available RD&D data in non-IEA major energy economies or "BRIMCS" countries (Brazil, Russia, India, Mexico, China, South Africa). This survey reveals that public energy RD&D in the six BRIMCS countries amounted to some $2.7 billion compared with US$4.4 billion in the United States in 2008 (IEA, 2009a).

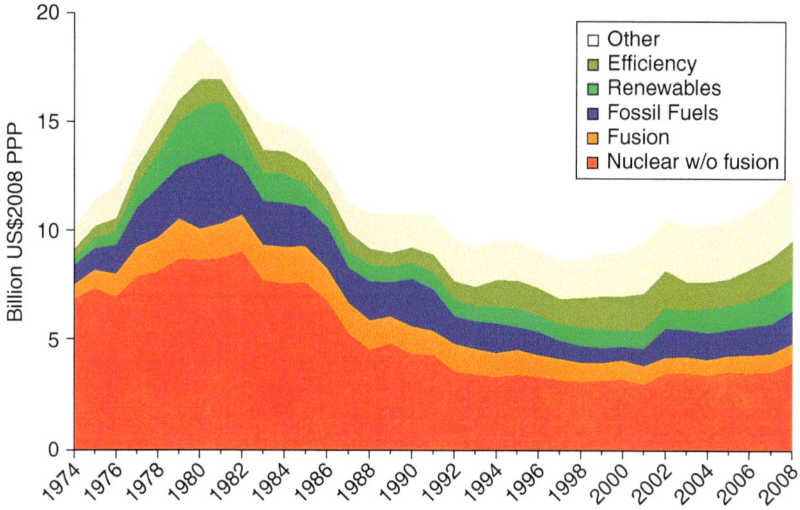

Figure 21.2. Public energy RD&D in (2000$ million PPP) in IEA countries by technology class (1974–2008).
Source: based on IEA (2009b).

In the BRIMCS countries, traditional distinctions between public and private sources of RD&D funding become increasingly blurred by the dominant position of wholly or partially state-owned enterprises. RD&D expenditures in state-owned oil, gas, and electricity companies are strongly determined by government policies. Including these raises the six BRIMCS countries' total current energy RD&D expenditures to some $15 billion, which is equivalent to the entire $13 billion public energy RD&D budget of the twenty-eight IEA member countries and about half the estimated $25 billion combined public and private energy RD&D in OECD countries.

2.2 Private Sector Energy RD&D

Macroeconomic RD&D statistics by economic sector are available for a sample of nineteen OECD countries in the OECD Structural Analysis database (OECD, 2009). However, the latest year reported is 2002. Data for other countries are not collected and available in any internationally comparable form. As the directorate general for research of the European Commission noted: "Despite the growing need, statistics on energy R&D expenditures in the private sector remain a problem" (EC, 2005).

In 2002, business enterprises excluding extractive industries (coal mining, oil, and gas extraction) performed R&D equivalent to $433 billion (PPP in 2000$) in the nineteen OECD countries sampled. The only sectoral breakdown relevant for energy innovation are "coke, refined petroleum products and nuclear fuel" ($2.7 billion) and "electricity, gas, and water supply" ($2 billion), with an OECD total of energy-related private sector RD&D of less than $5 billion. Of course, RD&D performed in the manufacturing sector has a bearing on energy use, such as in electrical machinery ($13 billion), motor vehicles ($50 billion), or aircraft ($20 billion; see NSF, 2010). How much of the RD&D performed in these sectors is energy related remains unknown. The six automobile manufacturers listed among the top twenty-five global

Table 21.2. Private energy RD&D (in 2001$ billion) for selected OECD countries.

RD&D Expenditure (2001$ Billion)	U.S.	U.S.	Japan	Korea	Sweden	France	Denmark	Spain	Total
reference year	*2000*	*2004*	*1997*	*1998*	*1997*	*1998*	*1993*	*1998*	*~2000*
Efficiency			6.10						6.1
Fossil fuels	0.81	1.04	0.84	0.03		0.56			2.2
Nuclear	0.03	0.03	1.00			1.17			2.2
Renewables			0.29						?
Other or nonspecified	0.36	1.21				0.05			?
Total	1.20	2.28	8.6	0.06	0.1	1.78	0.08	0.08	11.9

Notes: Latest available data for the United States shown for comparison (NSF, 2009).
Source: WEC (2001).

corporations in 2006 spent between $4.6 billion (Honda) and $7.5 billion (Toyota) on R&D, with a total of some $39 billion. Other corporations whose R&D is likely to include an important energy component are Siemens ($6.6 billion), Samsung ($5.9 billion), Matsushita ($4.9 billion), Sony ($4.6 billion), and Bosch ($4.4 billion) (NSF, 2010). The consumer good and appliance products of these companies also suggest that private energy-related RD&D may be dominated by energy end-use technologies.

The only available survey of private sector RD&D specific to the energy sector is the study conducted by the World Energy Council (WEC, 2001), covering the period 1997–2000 for a sample of seven OECD countries and expressed in 2001$ billion (PPP). Total private energy RD&D amounted to some $12 billion annually. The Kempener, Anadón, and Condor (2010) survey on BRIMCS countries (see also the case study on end-use investments) puts these OECD numbers into perspective, with private sector and state-owned enterprises also investing around $12 billion (in 2008 PPP) albeit based on more recent data.

The WEC data are summarised in Table 21.2 and although dated, usefully illustrate the magnitude and technology allocation of private energy RD&D. It is also noteworthy that, with the exception of Japan, private RD&D on energy efficiency is either unrecorded or subsumed within the "other or nonspecified" category. Private sector energy RD&D thus appears to track the public RD&D portfolios' emphasis on energy supply technologies, particularly nuclear and fossil fuels.

Time series data from the United States show that private energy RD&D follows comparable trends to public energy RD&D (see Figure 21.3). Both are influenced by the rise and fall of oil prices. Nemet and Kammen (2007) argue that "the signal of commitment that a large government initiative sends to private investors outweighs any crowding-out effects associated with competition over funding or retention of scientists and engineers." In other words, "shared expectations" – an important institution within the energy technology innovation system – may help explain the trends seen in the interrelationship depicted in Figure 21.3. And expectations may wane as well as wax: a coupled decline in RD&D budgets is seen after 1985.

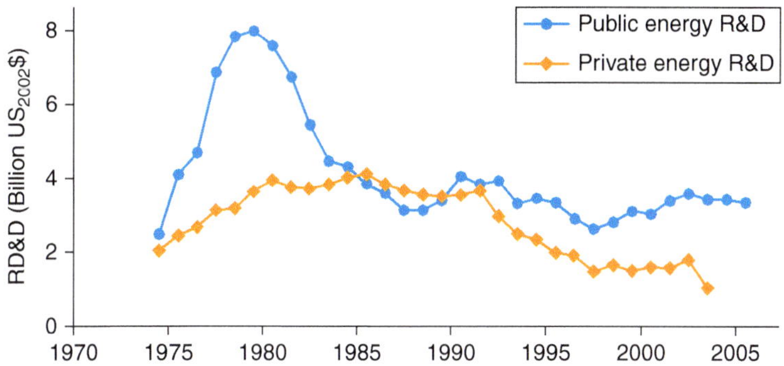

Figure 21.3. Trends in U.S. public and private energy RD&D (US2002$ billion).
Source: Nemet and Kammen (2007).

2.3 Total Energy RD&D

Based on the limited data available, the order of magnitude estimate of global energy RD&D amounts to some $50 billion per year with some $15 billion in public sector RD&D and up to $35 billion invested by the private sector (see Table 21.1). About half of all energy RD&D is spent on fossil fuels and nuclear technologies.

An alternative estimate by the Sustainable Energy Finance Initiative (SEFI) of global sustainable energy RD&D amounts to some $12.4 billion in 2005 (UNEP/SEFI/NEF, 2009). This breaks down as $5.6 billion public and $6.8 billion private RD&D, although insufficient detail is provided to allow a more in-depth comparison.

3 Market Formation Investments

Market formation investments include public and private investments in the initial phase of diffusion during which technologies are often in "niche markets." These are characterised by end users' high willingness to pay for novel performance advantages, and some degree of insulation from the cost pressures of a competitive market. Niche markets are often created or protected by policy incentives or regulation.

For the purposes of tracking investments against conventionally reported categories, we define market formation investments as investments in or by companies that have solved the core technological problems out of R&D and that are now seeking to commercialise their products into relatively small, specific markets, at least initially. These investments are not, on their own, commercially viable, and so are too early for mass market adoption (and so "diffusion" investments). We thus emphasise private investments that take advantage of markets created by government policies, such as renewable electricity portfolio standards or feed-in tariffs, and renewable fuel regulations. Even with this definition, market formation investments in the energy sector are difficult to estimate because many transactions are unreported, and there is no accepted measurement approach.

3.1 Estimating Market Formation Investments

Investments in most renewable energies can be classified as market formation investments given their reliance on policy incentives. A recent UNEP/SEFI/NEF

(2009) report analysed investments into "sustainable energy," which it defined as solar, wind, biofuels, biomass and waste to energy, marine and small-hydro, geothermal, efficiency, and other low-carbon technologies/services. It excludes large-scale hydro (>50 MW) and all nuclear power. The UNEP/SEFI/NEF study reported investments into sustainable energy had risen to US$223 billion in 2008, although this includes $67 billion of purely financial transactions including mergers and acquisitions, refinancings, and buyouts (UNEP/SEFI/NEF, 2009). These transactions represent changes in ownership rather than investments into the formation of markets for energy innovations (see box for definitions of investment types).

Box. Definition of Types of Market Formation Investment. Source: UNEP (2011)

- *Venture capital and private equity (VC/PE)*: all money invested by venture capital and private equity funds in the equity of companies developing energy technologies. Similar investment in companies setting up generating capacity through special purpose vehicles is counted in the asset financing data.
- *Public markets*: all money invested in the equity of publicly quoted companies developing energy technologies and power generation. Investment in companies setting up generating capacity is included in the asset financing data.
- *Asset financing*: all money invested in energy production or electricity-generation projects, whether from internal company balance sheets, from debt finance, or from equity finance. This excludes refinancings.
- *Mergers and acquisitions (M&A)*: the value of existing equity purchased by new corporate buyers in companies developing energy technologies or operating energy projects.

Of the US$160 billion technology-specific investments reported by UNEP/SEFI/NEF (2009), US$103 billion comprised various types of risk-taking private finance (venture capital, private equity, asset finance, and new equity issuances – see box for definitions), which we consider as clearly constituting market formation investments. A further US$18 billion was on public and private RD&D, and the remaining US$35 billion on small and residential projects and the Brazilian ethanol programme could be considered diffusion investments.

A breakdown by asset class of private market formation investments in sustainable energy technologies is shown in Figure 21.4. Asset financing comprises the dominant share. Asset finance includes the building of new assets (projects, plants, etc.) using either project finance that ring fences the asset in a "special purpose vehicle" or balance sheet financing. Asset finance is more common for investments in large and more mature technologies, but in this context are considered market formation investments given their exposure to and reliance on government subsidies and other market incentives.

New listings on public markets grew sharply from 2006 to 2007, but have settled back to broad equivalence in magnitude to venture capital (VC) and private equity (PE) investments. The compound annual growth rate across the three asset classes

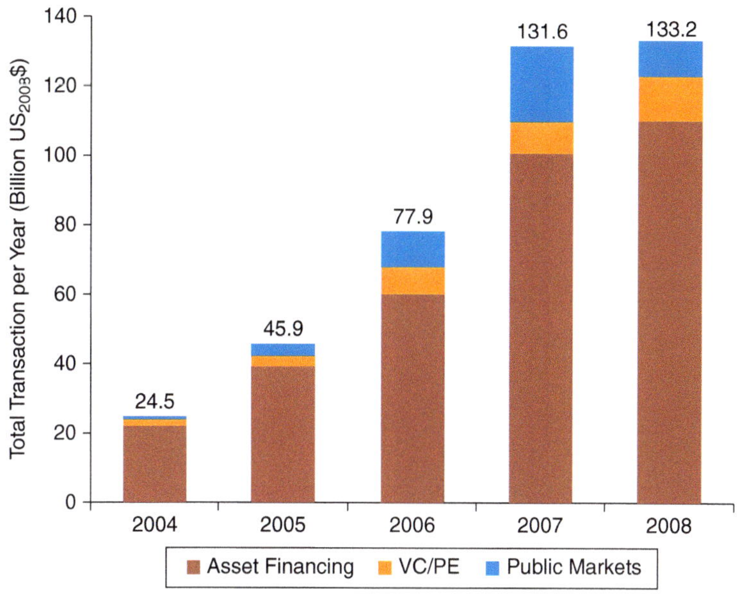

Figure 21.4. Annual investments (US2008$ billion) in market formation for sustainable energy technologies (2004–2008).
Notes: 2008 data exclude US$8 billion in Brazilian ethanol.
Sources: UNEP/SEFI/NEF (2009); O'Rourke (2009); Bloomberg New Energy Finance database, courtesy of ERD3 Project Harvard (UNEP/SEFI/NEF, 2009).

from 2004 to 2008 was 63 percent. The principal technology class invested in was wind power; investments from 2004 to 2008 grew annually at 51 percent. Investments in energy efficiency were small (only 2 percent of total investments from 2004 to 2008). The principal regions invested in were OECD countries, notably Europe (45 percent), and North America (30 percent).

3.2 Venture Capital and Private Equity (VC/PE) Investments

Although more risk-taking than conventional debt and equity investors, venture capital (VC) and private equity (PE) investors typically only commit to a technology once much of the innovation risk has been mitigated, when markets are at least somewhat defined, and when the innovating company is formed and functioning. VC/PE investors aim to profit from the rapid scaling up of the technology in early markets demonstrating commercial potential to allow them to sell their equity stake at a high multiple of the initial investment. VC/PE investors' core role is to bring not just capital but also commercial skills and expertise to aid and accelerate this transition from demonstration to diffusion. In the energy sector, this is particularly relevant for the less mature sustainable energy technologies.

The UNEP/SEFI/NEF (2009) data combined with data on fossil fuel investments in the Thomson/Reuters VentureXpert database (Thomson Financial, 2009) show dramatic growth in VC/PE investments into energy – specifically into clean energy technologies – since the mid to late 2000s. From 2002 to 2008, the total amount of energy (fossil and nonfossil) investments made by VC/PE investors worldwide grew from US$1.1 billion to some US$14.6 billion with a 53 percent compound annual

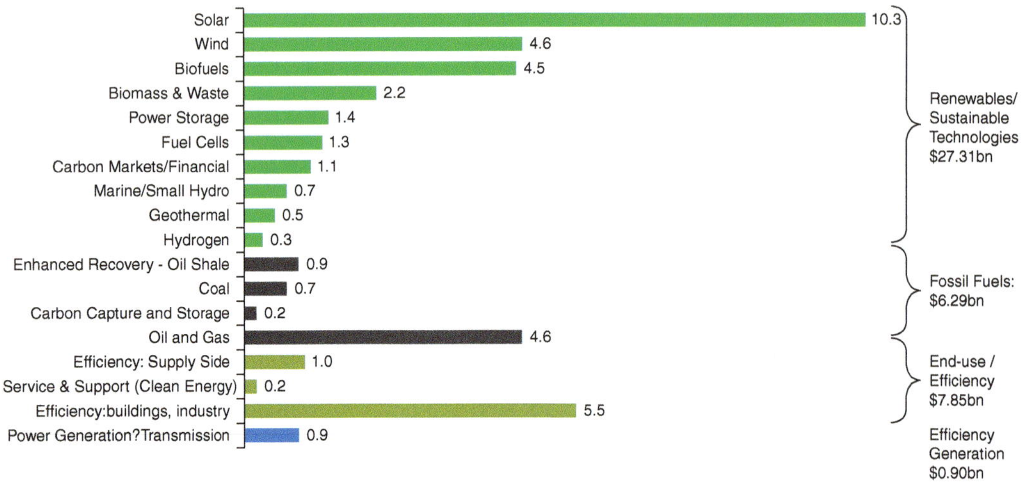

Figure 21.5. Venture capital / private equity investments (US2008$ billion) in energy technologies (2002–2008).
Notes: "Biofuels" excludes investments in Brazilian ethanol companies.
Sources: O'Rourke (2009); Bloomberg New Energy Finance database, courtesy of ERD3 Project Harvard (NEF/SEFI, 2009); Thomson Reuters VentureXpert Data BaseThomson Financial (2009).

growth rate (see Figure 21.5). This is likely an underestimate as VC/PE investments are often not publicly reported, investments made by individual investors are usually not reported, and data outside the North American and European dominance of professional VC/PE firms are limited. Over the 2002–2008 period as a whole, at least US$40.9 billion was invested in energy technology firms in some 2,375 transactions. The number of investment rounds (or transactions) grew at 26 percent per year. The bulk of investments were into North American and European companies with nonfossil-based energy technologies.

Figure 21.5 shows that the majority of energy investments made by VC/PE investors are in sustainable energy (particularly renewable power generation) and end-use efficiency technologies. Solar is the dominant technology class, capturing some 30 percent of the global total over the 2002–2008 period and growing particularly rapidly from 2005 in terms of both numbers of transactions and total amounts invested. End-use efficiency, including smart energy metering in buildings, demand response software systems, high-efficiency lighting, and so on, also grew markedly in this period.

4 Diffusion Investments

"Diffusion" investments into mature commercial energy technologies are not collected systematically. Modelling studies, as well as limited survey data, allow estimates of capital investments in energy supply technologies, but analogous data for energy end-use investments are almost entirely lacking. A companion case study on end-use investments aims to provide a plausible first order estimate to address this data gap and to provide a comparator for the estimates of energy supply investments reviewed here.

4.1 Diffusion Investments in Energy Supply Technologies

Data on energy supply investments are typically based on limited surveys or model estimates that combine statistical data and/or estimates of capacity additions with average technology-specific investment costs. Modelling studies have been available since the mid-1990s from the research community (e.g., Nakicenovic, Grubler, and McDonald, 1998; Nakicenovic and Rogner, 1996; Riahi, Grubler, and Nakicenovic, 2007), as well as in the work of the IEA, particularly the World Energy Investment Outlook (IEA, 2003), but also in the Energy Technology Perspectives series (IEA, 2006, 2008a) and the World Energy Outlook series (e.g., IEA, 2008b, 2009b). These latter reports include unique survey data on energy supply investments, particularly in the oil and gas industry.

A common feature and drawback of almost all modelling studies is that energy supply investments are not reported for their reference or base year values, but instead as cumulative totals over the model projection horizon (typically thirty years). This makes an assessment of current investment levels and allocations difficult, and a robust comparison among different modelling studies almost impossible. Here we draw heavily on the only modelling study that has disclosed its base year energy investment values (Riahi et al., 2007), and compare this with the surveys reported in the IEA's World Energy Outlook series (IEA, 2006, 2008b, 2009b). Riahi and colleagues' (2007) estimates are based on capacity additions and price levels in 2000 as the reference year, expressed in US1990$ and converted here to a common US2005$ basis using the U.S. GDP deflator although price levels remain in year 2000 terms. (Energy technology price escalation after 2000 is a feature observed for almost all energy supply technologies, but technology-specific price deflators are not available, hence reported prices and inferred investment numbers refer to year 2000 levels.) As the energy sector has seen significant price escalation since 2004 (particularly for oil and gas), Riahi and colleagues' estimates can be considered a lower bound and the IEA estimates an upper bound (see Table 21.3).

Despite differences in estimated investments per technology class, the available data shown in Table 21.3 suggest a likely order of magnitude range of annual energy supply investments from around US$700 billion (2005$) up to some US$840 billion in 2007/2008. Investments are dominated by electricity generation and transport and distribution (T&D), which total around US$500 billion. Fossil fuel supply investments, particularly upstream (i.e., exploration and production), account for US$250–400 billion, mostly for oil and gas. The renewable energy technologies that featured prominently in market formation investments are thus minor players under the market conditions characterising current diffusion investments. Liquid and gaseous biofuels account for US$20 billion (included in the lower bound of fuel investments), including US$8 billion for Brazilian ethanol (UNEP/SEFI/NEF, 2009). Large-scale hydropower makes up less than 17 percent of current energy supply investments (<US$100 billion for annual capacity additions of between 25–30 GW).

Major uncertainties include the accounting for oil and gas exploration activities (at some US$40 billion). Some companies categorise these as RD&D activity for future oil/gas reserves. On this basis, oil and gas exploration would represent the single largest type of RD&D spending across the energy sector as a whole.

Uncertainties also exist for electricity transport and distribution (T&D) infrastructure investments for which only modelling study data are available and

Table 21.3. Estimates of diffusion investments (US2005$ Billion) in energy supply technologies.

Investments (US$2005 Billion) In energy supply technologies		Low estimate (prices & activity in 2000)	High estimate (prices & activity in 2005–2007)
FUELS – UPSTREAM:			
	Exploration fossil fuels	n.a.	40
	Extraction fossil fuels	180	180–360
FUELS – DOWNSTREAM: [1]		n.a.	100–140
	Synfuels, fossil	1	7
	Biofuels	20	n.a.
	Other	20	n.a.
TOTAL FUELS		>220	300–550 [2]
POWER GENERATION:			
	Fossil	110	n.a.
	Nonfossil	100	n.a.
	Total (Fossil + nonfossil)	210	220–300
	T&D	>>70	?-230-?
TOTAL POWER GENERATION		>500	450–520 [3]
TOTAL ENERGY SUPPLY		>720	750–840 [3]

[1] Includes refining, pipelines, and other infrastructure. [2] Minimum excludes exploration; maximum includes exploration. [3] Minimum-maxima range is not additive from sub-component min-max range.
Notes: T&D = Transport and Distribution of electricity.
Sources: Low estimates from Riahi and colleagues (2007); high estimates from IEA (2006, 2008b, 2009b). See text for details.

estimates differ by about a factor of three. The IEA (IEA, 2008b) projection of average annual electricity T&D infrastructure investments of US$230 billion over the period 2007–15 appears extremely high, and is comparable to current investments in electricity generation capacity. It is also interesting to note that no studies report actual data for current investments in nuclear energy (even though nuclear capacity additions feature prominently in many projections). According to the IEA (2002; 2009b), installed nuclear capacity expanded by 20 GW between 1999 and 2007. The IEA (2008b) reports an increase from 358 to 376 GW between 2000 and 2006, which yields an average annual net increase in nuclear capacity of between 2–3 GW, mostly in Asia where investment costs are comparatively modest at an estimated 1500–2500 $/kW. This suggests current investments of between US$3–7.5 billion/year for nuclear reactors, which makes this the only technology in which RD&D investments *exceed* diffusion investments.

Evidence regarding the time trend of energy supply investments is scarce. An intriguing empirical finding from the United States shows a significant decline in energy supply investments as a share of sector revenues for electricity generation in the second half of the twentieth century (see Figure 21.6). The declining share of revenues reinvested in capital stock in the U.S. electricity sector suggests a substantial thinning of resources available for capital turnover and diffusion of new technologies as a twin result of slowing demand growth and energy sector deregulation and

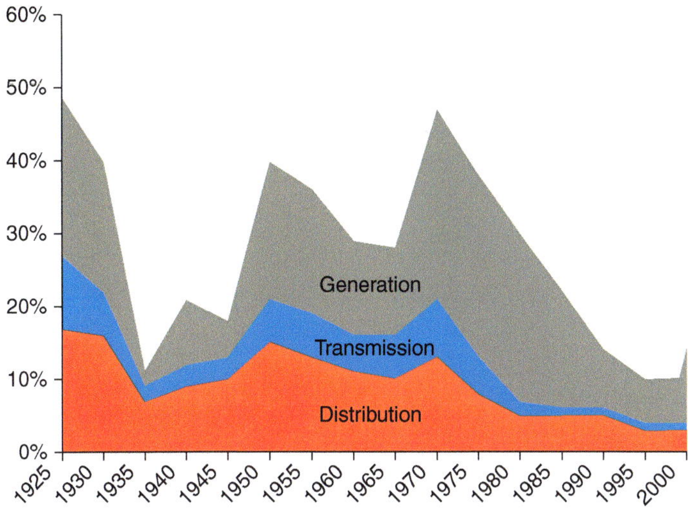

Figure 21.6. Declining investments as a share of revenues in the U.S. electricity sector (1925–2000).
Source: modified from EPRI (2003).

liberalisation. It remains unclear if this trend is a specific phenomenon of the United States or IEA countries, but it certainly emphasises the importance of comprehensive time series data on energy sector investments to support decision making.

This assessment of diffusion investments has focussed on the global level for the simple reason that regionally disaggregated investment data are lacking. Modelling studies suggest that energy supply investments in the year 2000 were distributed about 60:40 between developed and developing economies (using the Annex I and non-Annex I definitions of the United Nations Framework Convention on Climate Change). Short-term projections (e.g., to 2030 by IEA, 2009b) suggest roughly a 50:50 split in the energy supply investment needs of Annex I and non-Annex I countries, for a global total of cumulative energy supply investments from 2008 to 2030 of some US$25 trillion (2008$).

4.2 Diffusion Investments in Energy End-Use Technologies

The distributed, decentralised nature of investments in end-use technologies by households and firms, and their conventional classification as consumer expenditures rather than investments, explains the absence of energy end-use investment estimates in the literature. Compounding these data challenges are definitional problems. Even defining what constitutes an energy end-use technology is difficult. In the case of a new building, for example, the system boundary might be drawn around the heating and air conditioning systems, or more broadly to include the building envelope and structures that contribute to the demand for heat or cooling. Indeed, the entire building may be considered an end-use technology. Some modelling studies include incremental energy end-use technology investments associated with energy efficiency improvements over a reference scenario (e.g., IEA, 2009b). Apart from introducing additional definitional ambiguities (i.e., what constitutes incremental investments), the modelling is invariably partial, omitting certain technologies and treating others inconsistently. More commonly, modelling studies deal with energy end-use investments by "assuming away" missing data through exogenous (policy-independent) energy efficiency trends.

These data issues and challenges are discussed in more detail in a companion case study that presents a global, bottom-up estimate of total investment costs in energy end-use technologies. This case study uses volume data (production, delivery, sales, and installations) and cost estimates to approximate total investment costs in 2005 both for end-use technologies (broadly defined) and their specific energy-using components (narrowly defined). Here, we summarise these estimates to provide a comparator for the energy supply investments presented earlier. We recommend the reader to the companion case study for all details of method and data (see also Wilson and Grubler, 2011).

Investments in 2005 in end-use technologies (broadly defined) are estimated to be in the order of US$1–3.5 trillion. Investments in the energy-using components of these end-use technologies (narrowly defined) is in the order of US$0.1–0.7 trillion. These investment ranges, particularly at the upper end, should be treated as underestimates, as the analysis omitted many end-use technologies. We therefore suggest a range of annual end-use investments to be conservatively in the order of US$0.3–4.0 trillion. This compares with the range of annual energy supply investments in the order of US$0.7 trillion.

Although the two ranges span the same orders of magnitude, the upper bound of end-use investments is four times higher than its energy supply equivalent. Interesting, this result aligns with the IEA's estimation that demand-side investment needs exceed supply-side investment needs by a factor of four to five in climate policy scenarios (IEA, 2008a). Disaggregating the data by region shows that approximately two-thirds of the end-use investments in 2005 are in Annex I countries; the remaining one-third are in developing economies.

5 Conclusions

In this case study, we have compiled disparate data on RD&D and market formation investments and on diffusion investments in energy supply technologies, and have integrated these with novel datasets on RD&D investments in BRIMCS countries, and on diffusion investments in end-use technologies. The result is a synthetic snapshot of the financial resources mobilised globally as investments into the energy technology innovation system, distinguishing both the stage of the innovation life cycle (RD&D, market formation, diffusion) and the type of technology (upstream, downstream, power generation, T&D, end use).

From this picture, summarised in Table 21.1, certain key findings emerge. First, the scale of resource mobilisation increases across successive innovation stages, from RD&D (~$50 billion), to market formation (~$150 billion), to the dominant diffusion stage (>$1000 billion). If large-scale technological change is on the agenda, changes in the diffusion environment and associated incentives for technology adoption and diffusion – for example, through changes in relative prices – will be essential *in addition to* the resources invested in developing improved technologies in the earlier innovation stages.

Second, the structure of current energy investments is highly asymmetrical. End-use technologies dominate diffusion investments but are strongly underrepresented in earlier innovation investments (based on available data). The comparatively large support for energy supply technologies such as fossil fuels and nuclear energy in RD&D is therefore brought into sharper relief, and is mismatched with the challenges facing the energy system from energy access to energy security and climate change mitigation, all of which call for vastly improved energy end-use efficiency.

Third, six major emerging economies – Brazil, the Russian Federation, India, Mexico, China, and South Africa, known collectively as "BRIMCS" countries – now account for a significant proportion of global energy innovation investments. However, significant regional imbalances, particularly in the support for energy RD&D, persist. The increasing globalisation of energy innovation in general, and of energy technology RD&D in particular, suggests that new mechanisms for international technology cooperation and coordination might be needed, including appropriate institutional (re)designs beyond the limited existing membership and scope of the IEA.

Finally, it is clear that formidable data problems are associated with assessing energy technology innovation, pointing to important areas of future research and renewed initiatives to provide better technology-specific data. Institutions to collect and share these data at both national and international levels are also needed.

FURTHER READING

Riahi and colleagues' (2007) study is noteworthy for its reporting of base year investment data, and should be read as a contrasting approach to the IEA World Energy Outlook studies (IEA, 2008b; 2009b).

REFERENCES

AEIC, 2010. *A Business Plan for America's Future*. American Energy Innovation Council.

Dooley, J. J., 2000. *A Short Primer on Collecting and Analyzing Energy R&D Statistics*. Batelle.

Dooley, J. J. & Runci, P. J., 2000. Developing Nations, Energy R&D, and the Provision of a Planetary Public Good: A Long-Term Strategy for Addressing Climate Change. *Journal of Environment & Development*, 9(3): 215–39.

Doornbosch, R. & Upton, S., 2006. *Do We Have the Right R&D Priorities and Programmes to Support the Energy Technologies of the Future?* Paris: OECD.

EC, 2005. *Energy R&D Statistics in the European Research Area. Brussels:* European Commission.

EPRI, 2003. *Electricity Technology Roadmap – Meeting the Critical Challenges of the 21st Century*. Palo Alto, CA: Electric Power Research Institute (EPRI).

IEA, 2002. *World Energy Outlook*. Paris: International Energy Agency.

2003. *World Energy Investment Outlook*. Paris: International Energy Agency.

2006. *World Energy Outlook*. Paris: International Energy Agency.

2008a. *Energy Technology Perspectives 2008 – Scenarios and Strategies to 2050*. Paris: Organization for Economic Cooperation & Development.

2008b. *World Energy Outlook*. Paris: International Energy Agency.

2009a. *Energy Technology R&D Statistics* [Online]. International Energy Agency (IEA). Available: http://www.iea.org/stats/rd.asp. March 2010.

2009b. *World Energy Outlook*. Paris: International Energy Agency.

Kempener, R., Anadón, L. D., & Condor, J., 2010. *Governmental Energy Innovation Investments and Policies in the Major Emerging Economies: Brazil, Russia, India, Mexico, China, and South Africa*. Cambridge, MA: Energy Research, Development, Demonstration and

Deployment Project, Energy Technology Innovation Policy Group, Belfer Center of Science and International Affairs.

Nakicenovic, N., Grubler, A., & McDonald, A. (eds.) 1998. *Global Energy: Perspectives*. Cambridge: Cambridge University Press.

Nakicenovic, N. & Rogner, H. H., 1996. Financing Global Energy Perspectives to 2050. *OPEC Review*, 20(1): 1–23.

NEF/SEFI, 2009. *Analysis of Trend and Issues in the Financing of Renewable Energy and Energy Efficiency. Basel, Switzerland:* UNEP & Basel Agency for Sustainable Energy, New Energy Finance (NEF) / Sustainable Energy Finance Initiative (SEFI).

Nemet, G. F. & Kammen, D. M., 2007. U.S. Energy Research and Development: Declining Investment, Increasing Need, and the Feasibility of Expansion. *Energy Policy*, 35(1): 746–55.

NSF, 2009. *Research and Development in Industry: 2004. Arlington, VA:* National Science Foundation.

2010. *Science and Engineering Indicators 2010. Chapter 4: Research and Development: National Trends and International Linkages. Arlington, VA:* National Science Foundation.

O'Rourke, A. R., 2009. *The Emergence of Cleantech*. New Haven, CT: Yale University.

OECD, 2002. *Frascati Manual: Proposed Standard Practice for Surveys on Research and Experimental Development. Paris:* Organisation for Economic Co-Operation and Development.

2007. Science, Technology and Innovation Indicators in a Changing World: Responding to Policy Needs. Paper discussed at the OECD Blue Sky II Forum, September 25–27, 2007 Ottawa, Canada. Organisation for Economic Co-operation and Development.

2009. *R&D Expenditure in Industry. Paris:* OECD STAN (Structural Analysis Database).

Riahi, K., Grubler, A., & Nakicenovic, N., 2007. Scenarios of Long-Term Socio-Economic and Environmental Development under Climate Stabilization. *Technological Forecasting and Social Change*, 74(7): 887–935.

Thomson Financial, 2009. *Thomson Reuters VentureXpert Data Base* [Online]. Thomson Reuters. Available: http://vx.thomsonib.com/NASApp/VxComponent/NewMain.htm. July 15, 2010.

UNEP, 2011. *Global Trends in Sustainable Energy Investment 2011: Analysis of Trends and Issues in the Financing of Renewable Energy and Energy Efficiency*, United Nations Environment Programme (UNEP), Bloomberg New Energy Finance (NEF), Frankfurt School for Climate & Sustainable Energy Finance.

UNEP/SEFI/NEF, 2009. *Global Trends in Sustainable Energy Investments 2009*, Nairobi: UN Environment Programme; Sustainable Energy Finance Initiative; New Energy Finance.

WEC, 2001. *Energy Technologies for the 21st Century: Energy Research, Development and Demonstration Expenditure 1985–2000: An International Comparison*. London.

Wilson, C. & Grubler, A., 2011. *A Comparative Analysis of Annual Market Investments in Energy Supply and End-Use Technologies*. Laxenburg, Austria: International institute for Applied Systems Analysis.

ACKNOWLEDGEMENTS

This case study benefitted from the many discussions between the lead and contributing authors of chapter 24 of the *Global Energy Assessment*.

22 Energy RD&D Investments in the Major Emerging Economies and the United States

Ruud Kempener, Laura Díaz Anadón, Kelly Sims Gallagher, and Kejun Jiang

Authors' Summary

This case study provides an overview of energy research, development, and demonstration (RD&D) investments in six major emerging economies: Brazil, the Russian Federation, India, Mexico, China, and South Africa (referred to as "BRIMCS" countries). RD&D expenditures are differentiated by technology category (fossil fuel, nuclear, electricity, renewables, efficiency, and other) and by funding source (government, state-owned enterprises, private industry, nonprofit organisations), based on the most recent published data available in each country. For comparison purposes, corresponding data for the United States are included. This data synthesis draws on a wide array of sources that have not been systematically compiled and compared to date. Various key findings emerge. First, energy RD&D in BRIMCS countries is substantial, amounting to some $14 billion (in PPP terms), which is slightly greater than the entirety of the public energy RD&D budget of all IEA countries combined ($12.7 billion). However, the relative RD&D effort in BRIMCS compared to IEA countries is sensitive to the comparison metric (e.g., GDP, energy use, energy production). Second, the significance of energy RD&D expenditure in BRIMCS countries challenges the traditional view that new energy technologies are predominantly developed within IEA countries and highlights the need to include BRIMCS countries in a comprehensive global strategy to accelerate energy technology innovation. Third, the highest level of RD&D support in BRIMCS countries is for fossil fuel and nuclear energy. Renewables and energy efficiency are strongly underrepresented in expenditures as well as in statistical reporting.

Editors' Guide

Relationship of Case Study to Energy Technology Innovation System Framework

Research, development, and demonstration (RD&D) investments are widely used as a metric to characterise *resource mobilisation* for *knowledge-generation* activities within the energy technology innovation system. Resulting *knowledge stocks* can be *codified* (e.g., as patents or research publications), intangible (e.g., as know-how in engineers or designers), or embodied (e.g., as prototypes or demonstration projects). Private *actors* (firms, industries, enterprises) play increasingly important roles in the RD&D process, although this depends on the policy and market environments in specific countries and for specific technologies. State-owned enterprises and government mandates for revenue allocation to RD&D are more prevalent in the major developing country RD&D efforts.

More broadly, resource mobilisation is strongly linked to the particular *institutional configuration* of national innovation systems. In the United States, the federal government and the network of national laboratories are the principal *actors*, with the more recently established ARPA-E specifically tasked with *entrepreneurial risk taking* in search of breakthrough technologies. In Mexico, three publicly funded research centres dominate the energy RD&D landscape. In Russia, the major national energy companies that enjoy significant *economies of scale* in their respective markets play a lead role in R&D programmes organised in technology streams. In India, an expansive hierarchy of government departments and agencies is involved in organising and directing RD&D investments, in particular to partially state-owned enterprises. The Chinese innovation system is also complex in terms of the multiplicity of actors and institutions involved, from government ministries, research laboratories, higher education institutes, and state-owned as well as private enterprises.

Insights for Energy Technology Innovation Policy

- The significance of energy RD&D expenditure in BRIMCS countries challenges the traditional view that developed economies and IEA members lead on energy innovation. BRIMCS country governments control larger amounts of energy RD&D funding than IEA country governments. BRIMCS countries should be included in a comprehensive global strategy to accelerate energy technology innovation.
- RD&D investments are strongly weighted towards fossil fuels and nuclear power. Both renewables and end-use efficiency are strongly underrepresented in financial terms as well as in statistical reporting. This weighting of resource inputs imbalances the overall innovation system unless a similar skew in innovation outputs is intended.
- A more comprehensive and systematic effort is needed to collect information on energy RD&D investments by central governments, other institutions, and private firms. Efforts to synthesise RD&D data clearly show the inconsistency and patchiness with which data are reported or otherwise available. This is particularly the case for private sector RD&D investments and for disaggregated investments by technology types. This undermines the capacity for policy makers to make informed decisions.
- Comparative data on RD&D between countries and across technology types are necessary to prioritise opportunities for international RD&D cooperation and for the coordination of national policies and innovation activities.

1 Introduction

In the 1970s, concerns about the vulnerability of countries to shocks in energy supply, primarily oil, prompted the largest energy-consuming countries to create the International Energy Agency (IEA). Besides working on operational measures such as sharing oil reserves, the IEA has the long-term objective of reducing oil dependence by implementing energy conservation programmes, enlarging available fossil energy resources through exploration and development, and developing new forms of energy (Lantzke, 1975). These have been pursued through joint energy research, development and demonstration (RD&D) projects, and regular reporting and analysis of national energy technology policies and research activities (Keohane, 1978).

Thirty-five years later, the global energy system has changed. Emerging and some large non-IEA countries like Brazil, Russia, India, Mexico, China, and South Africa (referred to as the "BRIMCS" countries) have become important players in energy production, conversion, and use. Yet scant information is available about energy RD&D activities outside the IEA members. The IEA does provide energy statistics for nonmembers, but the collection of data on energy technology policies is restricted to its twenty-eight member countries and does not include BRIMCS countries. Although other international institutions have been established to promote information exchange on energy policy issues, including the International Renewable Energy Agency and the Clean Energy Ministerial, no institution has the mandate to systematically collect information on energy RD&D investments across different energy technologies and across all countries.

In this chapter, we collate and analyse data on energy RD&D investments in the BRIMCS countries and compare them with energy RD&D investments in the United States, a major IEA member country. Our aim is to deepen understanding of energy RD&D efforts outside developed countries in the IEA and to identify and prioritise opportunities for RD&D cooperation, coordination of national policies, and innovation activities.

2 Methodology

No international effort exists to systematically collect information on energy RD&D investments in the BRIMCS countries. To fill this gap we collected data on public and private investments in energy RD&D for the period from 2000 to 2008 from a wide range of data sources including annual reports of ministries and state-owned enterprises (SOEs), databases provided by regulators or statistical organisations, and interviews with local experts. Data were collected in local currencies, but for comparative purposes are reported in million 2008 purchasing power parity (PPP) adjusted international dollars (World Bank, 2010). For details on the compiled database, see Kempener, Anadón, and colleagues (2010).

RD&D investment data were categorised according to funding source and technology. Funding sources distinguish central government from other sources (Kempener et al., 2010). Central government funding is defined as: (1) funding directly provided by a central government to universities, industry, national laboratories, or other national or international organisations for energy RD&D

Table 22.1. Definition of energy technology categories.

Categories	Technologies
Fossil energy	Oil and gas (excluding exploration), both conventional and non-conventional
	Coal (including carbon capture and storage or CCS)
Nuclear energy	Fusion
	Fission
Renewable energy	Solar energy (heating, PV, and solar thermal applications)
	Wind energy
	Ocean energy
	Biomass and waste energy
	Geothermal Energy
	Hydropower
Energy efficiency (all end-use technologies)	Industry
	Appliances
	Residential and commercial buildings
	Transportation (including vehicles)
	Agriculture
Transmission, distribution, and storage	Transmission and distribution (including pipelines, smart grids)
	Stationary energy storage
	Hydrogen
General energy technologies	Unspecified and/or insufficient information to disaggregate RD&D data by technology

activities; (2) funding for energy RD&D from state-owned enterprises (SOEs) that are 100 percent owned by the central government; or (3) funding for energy RD&D resulting from government mandates that require SOEs to spend a particular percentage or amount of their revenues on energy R&D or energy RD&D. Funds specifically targeted to support RD&D projects on energy technologies are also considered energy RD&D investments. Energy RD&D funding categorised as "other" includes funding from state or local governments, from SOEs that operate independently, from industry, or from nonprofit organisations.

Five categories of energy technology are used to differentiate the targets of RD&D investment (see Table 22.1). A sixth category of "general energy technologies" is included as a residual to capture investments that could not be linked to a particular energy technology. For further details, see Gallagher, Anadón, and colleagues (2010). RD&D investments that target other research areas (such as information and communications technologies, nanotechnology, or behavioural studies) that might have indirect impacts on energy resources, infrastructure, or use are not included unless funded through an energy RD&D programme.

Some of the available data sources provide only aggregate values for energy RD&D funding. As a result, in some cases it was not possible to determine how much support was directed towards different types of energy technologies. Conversely, some data sources only contain information on specific energy RD&D projects or programmes, and do not provide an overview of all sources of funding. Consequently, we should keep three important considerations in mind when interpreting the data. First, aggregate energy RD&D investments reported separately from energy RD&D investments in specific energy technologies are included under the sixth "general energy technologies" category. Second, years and energy technology categories for which no data on energy RD&D investments are available are left blank. Third, available data on energy RD&D investments may be incomplete. Totals should therefore be treated as lower bounds.

3 RD&D Institutions and Expenditures by Country

The next seven sections provide brief descriptions of the main funding agencies and funding allocation mechanisms for energy RD&D in the United States and the BRIMCS countries, followed by an overview of energy RD&D investments in those countries between 2000 and 2008.

3.1 United States

3.1.1 RD&D Institutions

The U.S. Department of Energy (DOE) is a major funder for government energy RD&D and is also responsible for a large range of national laboratories conducting energy RD&D. The DOE Office of Science conducts basic energy RD&D in areas including high-energy physics, nuclear physics, and basic energy sciences. Other DOE offices covering energy efficiency and renewable energy, electricity delivery and energy reliability, nuclear energy, and fossil energy are responsible for promoting innovation in a range of energy technologies in association with the network of eleven national laboratories. Most of these laboratories also receive funding from other offices.

The DOE is also responsible for the loan guarantee programme that issues loan guarantees to reduce the borrowing rate of qualifying energy technology projects, and the Advanced Research Projects Agency – Energy (ARPA-E). ARPA-E was created in 2007, and in April 2009 received US$400 million for two years to fund high-risk, high-payoff transformational energy research to bridge the gap between basic energy research and industrial innovation. Because the loan guarantee programme focuses on deployment and ARPA-E started after 2008, neither programme is included in this synthesis.

3.1.2 RD&D Investments

Table 22.2 provides an overview of U.S. energy RD&D investments. Technology categories are explained in Table 22.1. The data on federal government energy RD&D are based on a database developed by Gallagher and Anadón (2010). Industry data are based on an industry R&D study by the National Science Foundation (NSF, 2008). Government fossil energy RD&D includes coal, gas, petroleum, and carbon capture R&D projects, and a carbon capture and storage project (FutureGen). Government renewable energy RD&D includes international renewable energy programmes, energy efficiency RD&D includes policy and management issues, and general energy technologies (unspecified) includes the budget for basic energy sciences. The "other" category of funding source includes industry funding.

Table 22.2. Available data on energy RD&D investments in the United States between 2000 and 2008.

U.S. Energy RD&D Expenditure (in mln 2008 PPP $Int)		2000	2001	2002	2003	2004	2005	2006	2007	2008
fossil energy (incl. CCS)	government	308	633	713	633	610	420	577	583	659
	other	986	867	863	1067	1162				
nuclear energy (incl. fusion)	government	354	365	466	461	480	529	583	678	770
	other		27	25		34				
renewable energy sources	government	346	410	415	326	292	306	272	579	699
	other									
energy efficiency	government	624	671	549	567	535	486	467	442	525
	other									
transmission, distribution, and storage	government	30	32	89	202	252	286	314	286	319
	other									
energy technologies (unspecified)	government	924	1163	1150	1152	863	924	980	1120	1160
	other	285	535	784	728	1350				
total	**government**	**2586**	**3274**	**3382**	**3341**	**3031**	**2951**	**3193**	**3688**	**4132**
	other	**1271**	**1430**	**1671**	**1794**	**2545**				

Table 22.2 shows that energy RD&D investments in fossil energy, nuclear energy, and renewable energy have doubled between 2000 and 2008 in real terms. The fastest growth rate in investments between 2000 and 2008 has been in the area of transmission, distribution, and storage, while energy efficiency investments have slowly declined.

3.2 Brazil

3.2.1 RD&D Institutions

Energy RD&D funding in Brazil comes from mandatory contributions by the SOEs in the oil and gas industry and the electricity sector. In 2008, the Brazilian government owned 56 percent of Petrobras, the state oil company. Between 2002 and 2008, the government owned 52 percent to 58 percent of Eletrobras, the state electricity company.

Energy RD&D in the electricity sector is monitored, evaluated, and allocated by the electricity regulator. Annual energy RD&D funding is based on a minimum of 1 percent of utilities' net operating income. This funding is directed towards energy R&D or energy efficiency demonstration projects in the market with the exact allocation depending on whether the utilities are generators, transmission companies, or distribution companies. Some RD&D funds support national R&D programmes, such as the Brazilian scientific and technological development fund and the sectoral CT-Energy programme that funds research on end-use energy efficiency in industry. Remaining funds are used for in-house R&D projects assessed and evaluated according to guidelines by the regulator (Soares et al., 2008).

Similar regulation is in place to collect energy R&D funds from Brazil's state-owned oil company, Petrobras. In 2008, Petrobras was mandated to redistribute 25 percent of royalties that exceed 5 percent of oil and natural gas production to fund energy R&D in the CT-Energy and CT-Petro programmes. CT-Petro supports innovation in the production chain of oil and natural gas.

Brazil's Ministries of Science and Technology, Mines and Energy, Education, and External Affairs all fund nuclear energy RD&D. The National Nuclear Energy Commission within the Ministry of Science and Technology is Brazil's nuclear energy regulator and also the most important organisation supporting nuclear technology and innovation with an R&D budget supporting health, agriculture, environment, and industry applications of nuclear energy (Obadia, 2010).

3.2.2 RD&D Investments

Table 22.3 provides an overview of Brazil's energy RD&D investments. Technology categories are explained in Table 22.1. Data come from RD&D investments from CT-Energy, CT-Petro, regulator-mandated R&D investments in energy efficiency, and the annual reports of Petrobras and Eletrobras. Central government funding for fossil energy RD&D includes funds in CT-Petro (which supports exploration, production, supply, gas/energy, management, and centres). "Other" funding includes Petrobras's R&D investments in exploration and production, supply, distribution, and gas. Central government funding for nuclear energy RD&D includes a proportion of CT-Energy funds (15 percent in 2006), which is assumed to remain constant over time. According to ERAWatch (2010a), in 2008 the Nuclear Energy

Table 22.3. Available data on energy RD&D investments in Brazil between 2000 and 2008.

Brazil energy RD&D expenditure (in mln 2008 PPP $Int)		2000	2001	2002	2003	2004	2005	2006	2007	2008
fossil energy (incl. CCS)	government	170	129	79	76	64	68	89	101	79
	other	352	361	453	571	595	745	1192	1240	1167
nuclear energy (incl. fusion)	government		9	4	9	9	9	10	7	8
	other									
renewable energy sources	government	7	40	36	50	46	44	66	46	15
	other									
energy efficiency	government	9	35	40	47	41	40	67	46	3
	other									
transmission, distribution, and storage	government	22	94	104	125	111	106	175	122	14
	other									
energy technologies (unspecified)	government		13	6	14	14	14	16	11	12
	other		224	203	178	165	199	218	209	184
total	**government**	**208**	**319**	**270**	**321**	**286**	**281**	**424**	**333**	**131**
	other	**352**	**585**	**656**	**750**	**760**	**945**	**1410**	**1449**	**1351**

Commission and the Brazilian Nuclear Industries also had R&D budgets of R$ 511.8 million (2008 PPP $Int 350 million) and R$ 432.8 million (2008 PPP $Int 296 million) respectively, but these are not included in this synthesis as further details are unknown. Central government funding for renewable energy RD&D includes a proportion of CT-Energy funds as well as regulator-mandated funding. The data available on energy efficiency RD&D includes CT-Energy's funding for R&D on basic industry technologies and regulator-mandated R&D investments in energy efficiency, thermoelectric generation, and environment. Central government transmission, distribution, and storage RD&D includes CT-Energy's support for RD&D on electricity and hydrogen and regulator-mandated R&D on power system security, planning, control, optimisation, reliability, and measurement. RD&D funding from Eletrobras is reported under "other" funding for the residual category of unspecified energy technologies because it includes funding for a range of areas from distribution management to wave energy to conductivity.

Table 22.3 shows that the majority of Brazil's energy RD&D investment is focussed on fossil energy and transmission, distribution, and storage. These funds are tied to the revenues generated by the oil and gas and electricity SOEs. However, Table 22.3 does not include the Nuclear Energy Commission's budget as the proportion allocated to energy RD&D is not known. Because Table 22.3 assumes constant proportions of RD&D within CT-Energy and regulator-mandated R&D budgets, it is difficult to determine whether the government's energy RD&D funding priorities have changed, although it is clear that both budgets overall have remained fairly stable.

3.3 Russia

3.3.1 RD&D Institutions

The Federal Agency for Science and Innovation manages a significant part of the civil R&D budget and is therefore one of the most important institutions in Russian R&D and innovation (MES, 2006). The current applied science programme (2007–12) has five priority areas of which one is "Power Engineering and Energy Efficiency" (Khlunov, 2007). The RD&D budget for this area increased from RUB 6,960 million (2008 PPP $Int 704 million) in 2004 to RUB 23,567 million (2008 PPP $Int 1324 million) in 2008, with more than 80 percent invested in the government's "federal targeted programs" (Aksenova, 2009). Research grants are allocated to scientific research organisations and industry on the basis of tenders (ERAWATCH, 2010b). Energy R&D is also supported through materials research including projects on gas flaring, solar panels, efficient lighting systems, and insulation materials (Aksenova, 2009). In July 2009, the government implemented a new federal programme on nuclear power (Government Directive, 2009), but this falls after the 2008 cutoff for this data synthesis.

Little information is available on energy RD&D funding from industry, although Gokhberg (2003) estimates that in 2001 about 33 percent of Russia's gross domestic investments in R&D came from the private sector. Table 22.4 includes the available R&D budgets from five Russian companies: Gazprom in the natural gas industry; Rosneft, Lukoil, and Surgutneftegas in the oil industry; and Unified Energy System in the electricity sector.

3.3.2 RD&D Investments

Table 22.4 provides an overview of Russia's energy RD&D investments. Technology categories are explained in Table 22.1. Data draws on statistical overviews provided by Gokhberg (2010b) and augmented by Reutov (2007) and by annual reports of five major national energy companies. Federal fossil energy RD&D funding is based on R&D programmes for CCS, co-generation, oil and gas recovery, extraction of nontraditional resources (deep wells, bitumen sands, and arctic shelf), and processing of gas including liquefied natural gas (LNG). "Other" funding of fossil energy RD&D are Gazprom and Lukoil's R&D investments, which provide data only for selected years. Federal renewable energy RD&D funding represents biofuels and biomass programmes, federal energy efficiency RD&D represents programmes to improve lighting efficiency and system management programmes, and federal energy RD&D on transmission, distribution, and storage is based on a hydrogen programme that includes work on storage and fuel cells. In this latter technology category, "other" funding is from Unified Energy System's R&D investments in 2000 and 2001. Data on other federal energy RD&D programmes are reported under the residual unspecified energy technologies category (Klimenko, 2008). "Other" funding in this category reports the difference between the data points presented previously and Gokhberg's biannual data (2010a) on Russia's gross energy RD&D investments from all central government, industry, and nongovernmental sources.

Table 22.4 shows that there is little information about Russia's allocation of energy RD&D to the different energy technologies, although a qualitative analysis suggests that most energy RD&D is directed towards nuclear energy, fossil energy, and transmission, distribution, and storage. A list of "critical priority technologies" released in 2006 (Klimenko, 2008) sheds some light on these areas of focus by specifying nuclear power technologies; hydrogen; new and renewable sources; fossil fuel power generation and fuel production; energy-saving systems for transportation, distribution, and consumption of heat and electricity; and efficient engines and propulsion devices for transport systems. Two of five technological development priorities cited in a 2009 presidential address are energy (including efficiency, conservation, and new fuels) and nuclear technology (Smith, 2010).

3.4 India

3.4.1 RD&D Institutions

Gathering information about the allocation of energy RD&D in India is difficult because the central government has several layers, all active in energy RD&D. The Planning Commission proposes R&D initiatives, but the ministries and departments have to make decisions and implement programmes. Each ministry includes SOEs and institutions responsible for funding and allocating energy RD&D budgets. The Ministries of Power, Coal, Petroleum and Natural Gas, New and Renewable Energy, as well as the Department of Atomic Energy and the Ministry of Science and Technology through the Departments of Science and Technology, and Scientific and Industrial Research, are all involved in supporting or running energy RD&D programmes.

3.4.2 RD&D Investments

Table 22.5 provides an overview of India's energy RD&D investments. Technology categories are explained in Table 22.1. Data on RD&D investments from each of these

Table 22.4. Available data on energy RD&D investments in Russia between 2000 and 2008.

Russia energy RD&D expenditure (in mln 2008 PPP $Int)		2000	2001	2002	2003	2004	2005	2006	2007	2008
fossil energy (incl. CCS)	government								23	20
	other	339		263	256	280	399	152	261	411
nuclear energy (incl. fusion)	government									
	other									
renewable energy sources	government								16	14
	other									
energy efficiency	government								29	25
	other									
transmission, distribution, and storage	government						1	1	26	22
	other	34	64							
energy technologies (unspecified)	government			28	25	20	16	14	52	45
	other	183		378		398		587		508
total	**government**			**28**	**25**	**20**	**17**	**15**	**145**	**126**
	other	**555**	**64**	**642**	**256**	**677**	**399**	**739**	**261**	**918**

Table 22.5. Available data on energy RD&D investments in India between 2000 and 2008.

India energy RD&D expenditure (in mln 2008 PPP $Int)		2000	2001	2002	2003	2004	2005	2006	2007	2008
fossil energy (incl. CCS)	government	53	58	72	224	146	96	307	186	106
	other			1	3320	2389	1564	559	1378	694
nuclear energy (incl. fusion)	government	207	289	298	296	295	738	866	987	965
	other									
renewable energy sources	government	38	52	43	43	38	18	31	45	57
	other									
energy efficiency	government									
	other									
transmission, distribution, and storage	government	12	12	19	8	11	12	27	30	35
	other									
energy technologies (unspecified)	government									
	other									
total	**government**	**311**	**410**	**431**	**570**	**491**	**865**	**1231**	**1248**	**1163**
	other			**1**	**3320**	**2389**	**1564**	**559**	**1378**	**694**

ministries are included, augmented with data on R&D investments in SOEs. "Other" funding for fossil energy RD&D includes R&D budgets of SOEs that are only partially owned by the government and the loans for R&D activities that are provided to them by the Oil Industry Development Board. Central government funding for nuclear energy RD&D is based on the total R&D budgets of the Bhabha Atomic Research Centre and the Indira Gandhi Centre for Atomic Research, although it is possible that a large fraction of their budgets is used to investigate military applications. Central government funding for renewable energy includes ministry budgets and programmes as well as the R&D budget of the National Institute of Renewable Energy. Central government funding for transmission, distribution, and storage includes the budget of the Central Power Research Institute and the R&D budgets of two SOEs.

Table 22.5 shows that India's energy RD&D investments are mostly directed towards fossil and nuclear energy technologies. Nuclear RD&D investments more than quadrupled between 2000 and 2008 (although it is not possible to determine the proportion allocated to military applications). The RD&D budgets for renewable energy and for transmission, distribution, and storage are substantially smaller, but have received a relatively stable level of support.

3.5 Mexico

3.5.1 RD&D Institutions

Two organisations are central to funding and directing energy RD&D in Mexico. The first is the Ministry of Energy, which finances three public research centres (OECD 2009): the Electrical Research Institute; the Mexican Institute of Petroleum; and the Institute for Nuclear Research. The second is the National Council on Science and Technology (CONACYT), which has three sectoral funds for energy research (19 percent on energy sustainability, 29 percent on hydrocarbons, and 52 percent on research and technological development) and also supports postgraduate and regional research centres. Part of the funding for these programmes comes from PEMEX, Mexico's state-owned oil and gas company. In 2006–2007, PEMEX was mandated to allocate 0.05 percent of the total value of crude oil and natural gas production to RD&D programmes. In 2008, this tripled to 0.15 percent, of which the majority went to hydrocarbon RD&D (Pemex, 2008).

3.5.2 RD&D Investments

Table 22.6 provides an overview of Mexico's energy RD&D investments. Technology categories are explained in Table 22.1. Data draws on government and industry investments provided by CONACYT (2008) and PEMEX's annual reports. Federal government funding for fossil energy, nuclear energy, and transmission, distribution, and storage RD&D are based on the RD&D budgets of their corresponding research institutes. Fossil energy R&D also includes R&D funds provided to PEMEX. "Other" funding for fossil energy R&D includes PEMEX's budget for scientific and technological research, and funding for energy efficiency includes R&D investments by motor vehicle and other transport industries. "Other" funding for unspecified energy technologies is based on R&D investments in two sectors reported by CONACYT (2008): (1) coal, oil derivatives, and nuclear energy; and (2) electricity, gas, and water supply.

Table 22.6. Data on energy RD&D investments in Mexico between 2000 and 2008.

Mexico energy RD&D expenditure (in mln 2008PPP $Int)		2000	2001	2002	2003	2004	2005	2006	2007	2008
fossil energy (incl. CCS)	government	230	183	203	218	97	78	62	140	
	other							0.1	0.1	0.2
nuclear energy (incl. fusion)	government	54	45	40	22	33	35	33	32	
	other									
renewable energy sources	government									
	other									
energy efficiency	government									
	other	213	0	662	566	219	263			
transmission, distribution, and storage	government	55	56	83	74	79	82	73	79	
	other									
energy technologies (unspecified)	government									
	other	26	24	21	23	18	19			
total	**government**	**339**	**284**	**326**	**314**	**209**	**194**	**167**	**252**	
	other	**239**	**24**	**684**	**589**	**237**	**282**			

Table 22.6 shows broadly stable energy RD&D budgets for fossil energy, nuclear energy, and transmission, distribution, and storage, through their respective government-funded research institutes. No information is available about energy RD&D investments on renewable energy sources or energy efficiency. From the data available, the extent to which Mexico's car manufacturing industry R&D investments relate to energy efficiency is also unclear.

3.6 China

3.6.1 RD&D Institutions

China's energy innovation system is very complex. Central government ministries, commissions, and other agencies often have overlapping areas of responsibility and action. The Ministry of Finance distributes funds for energy RD&D to the Ministry of Science and Technology, the National Development Reform Commission, the National Natural Science Foundation of China, and the Chinese Academy of Sciences. These institutions then allocate funding according to their respective missions to different research institutes (including national laboratories), higher education institutions, and enterprises (enterprises of all kinds, not only state-owned enterprises) (Gao, 2010).

Furthermore, industry data are broken down into funding for energy-related science and technology activities by SOEs and by "other" enterprises. The China Statistical Yearbook does not provide any breakdown of what percentage of funds for science and technology activities in industry is spent on R&D, nor on the respective allocations to "new" versus "commercially available" technologies. Here we assume that all funding for science and technology activities in enterprises relates to "new" energy technologies and can therefore be included within China's energy RD&D investments.

3.6.2 RD&D Investments

Table 22.7 provides an overview of China's energy RD&D investments. Technology categories are explained in Table 22.1. Funding for science and technology activities by SOEs is included under government investments. Table 22.7 shows that government investments have increased rapidly since 2004. Most of this growth has taken place through increased R&D investments in SOEs, while government funding for energy-related science and technology activities in "other" enterprises in these industries declined over the same period.

Table 22.7 also shows that 30 percent to 50 percent of China's total energy RD&D funding is for fossil energy. There is hardly any breakdown of the data for other energy technologies except for some data on "new fuel vehicles" projects (Siegler, 2009), on nuclear energy science and technology activities in institutions of higher education, and on large R&D investments in the transmission network (Delman and Chen, 2008).

3.7 South Africa

3.7.1 RD&D Institutions

South Africa has limited government funding for energy RD&D. The Department of Science and Technology has a small budget for energy resources and supply, and

Table 22.7. Available data on energy RD&D investments in China between 2000 and 2008.

China energy RD&D expenditure (in mln 2008 PPP $Int)		2000	2001	2002	2003	2004	2005	2006	2007	2008
fossil energy (incl. CCS)	government	1004	684	999	563	3499	3760	4586	5541	6755
	other	767	1101	1028	1873	40	55	46	105	289
nuclear energy (incl. fusion)	government		24	6	5	19	23	25	41	12
	other		17	4	3	14	16	18	27	7
renewable energy sources	government									
	other									
energy efficiency	government		63	75	86	87	75	114	144	136
	other			4	3	17	15	12	6	26
transmission, distribution, and storage	government									
	other									
energy technologies (unspecified)	government	1631	1140	913	976	2609	2637	3257	3320	4900
	other	554	845	780	938	745	575	704	725	985
total	**government**	**2634**	**1911**	**1992**	**1629**	**6214**	**6496**	**7983**	**9045**	**11803**
	other	**1321**	**1963**	**1816**	**2818**	**816**	**661**	**779**	**863**	**1307**

Table 22.8. Available data on energy RD&D investment in South Africa between 2000 and 2008.

South Africa Energy RD&D Expenditure (in mln 2008 PPP $Int)		2000	2001	2002	2003	2004	2005	2006	2007	2008
fossil energy (incl. CCS)	government									
	other	125	96	117	117	116	63	65	165	164
nuclear energy (incl. fusion)	government	146	126	105	451	217	209	211	263	133
	other	26	25	22	118	43	41	38	35	31
renewable energy sources	government									
	other	8	11	12	19	7	12	8	11	7
energy efficiency	government									
	other									
transmission, distribution, and storage	government									
	other	63	91	94	154	59	61	38	38	26
energy technologies (unspecified)	government			0.3		0.5	0.5	0.5	10	9
	other									
total	**government**	**146**	**126**	**106**	**451**	**217**	**210**	**211**	**273**	**142**
	other	**223**	**223**	**245**	**409**	**225**	**178**	**148**	**248**	**229**

some energy-related R&D funds for natural resources, transportation, and natural sciences, technology, and engineering. The South African Department of Energy also funds energy RD&D indirectly and largely through a subsidiary research institute that is involved in the search for technologies to meet the future energy needs of South Africa, both fossil and renewable. The main research organisation under the umbrella of CEF Ltd. is South Africa's National Energy Research Institute (SANERI), which in 2008 merged with the National Energy Efficiency Agency to form SANEDI. The majority of government support for energy RD&D in South Africa was either carried out or funded by SANERI (in 2007 SANERI received 40 million ZAR, while the budget of the DST for energy research was 2 million ZAR). The South African government funds nuclear energy R&D indirectly through the state-owned Nuclear Energy Corporation and through a 69 percent ownership share in a company that develops nuclear pebble bed modular reactors (the government's ownership share ended in 2010). In addition, two large companies conduct energy R&D in South Africa: Eskom, which is involved in nuclear, renewables, and transmission, distribution, and storage; and Sasol, which produces liquid fuels.

3.7.2 RD&D Investments

Table 22.8 provides an overview of South Africa's energy RD&D investments. Technology categories are explained in Table 22.1. "Other" funding for fossil energy RD&D includes Sasol, and funding for nuclear, renewables, and transmission, distribution, and storage RD&D includes Eskom. Federal government funding for unspecified energy technologies is based on the Department of Science and Technology's annual R&D contributions to programmes on "energy resources" and "energy supply."

Table 22.8 shows that South Africa's government has limited involvement in directing energy RD&D investments towards particular energy technologies, except for its indirect involvement in R&D activities in the nuclear energy industry. Furthermore, the energy RD&D budget of the national energy research institute is limited in comparison to the indirect funds provided to nuclear energy RD&D. Most energy RD&D in South Africa takes place in industry, dominated by the electricity-related activities of Eskom and the coal, oil, and gas activities of Sasol.

4 Comparative Country Analysis

The previous sections provided a panel data set of energy RD&D investments in the BRIMCS countries and the United States. Table 22.9 presents a comparative synthesis of energy RD&D investments in each of these countries for the most recent year for which data was available. The results should be interpreted with caution as this synthesis is based on official data available (with the extent to which data is available differing between countries) which are subject to missing data and underestimation.

Two main conclusions can be drawn from this synthesis. First, fossil and nuclear energy receive the highest level of RD&D funding in both the United States and the BRIMCS countries. Even though support for renewable energy RD&D programmes increased substantially between 2000 and 2008 in the United States, Brazil, and India, absolute levels of funding remain small compared to fossil and nuclear energy. Energy RD&D funding in transmission, distribution, and storage has

Table 22.9. Comparison of latest energy RD&D data in the United States and the BRIMCS countries.

In Million 2008 PPP $Int*	Fossil Energy (incl. CCS)	Nuclear Energy (incl. fusion)	Renewable Energy Sources	Energy Efficiency	Transmission, Distribution, and Storage	Energy Technologies (unspecified)	Total
United States-Gov't	659	770	699	525	319	1160	4132
United States-Other~	1162	34	no data	no data	no data	1350	2545
Brazil-Gov't	79	8	46	46	122	12	313
Brazil-Other	1167	no data	no data	no data	no data	184	1351
Russia-Gov't	20	no data	14	25	22	45	126
Russia-Other	411	no data	no data	no data	no data	508	918
India-Gov't	106	965	57	no data	35	no data	1163
India-Other	694	no data	no data	no data	no data	no data	694
Mexico-Gov't	140	32	no data	no data	79	no data	252
Mexico-Other	0.1[1]	no data	no data	263[3]	no data	19[4]	282
China-Gov't	6755	12	no data	136	no data	4900	11803
China-Other	289	7	no data	26	no data	985	1307
South Africa-Gov't	no data	133	no data	no data	no data	9	142
South Africa-Other	164	31[2]	7	no data	26	no data	229
BRIMCS-Gov't	**7100**	**1149**	**>117**	**>208**	**>259**	**>4966**	**>13799**
BRIMCS-Other	**2724**	**>>38**	**>>7**	**>>289**	**>>26**	**>1696**	**>4781**
BRIMCS-GRAND TOTAL	**9824**	**>1187**	**>124**	**>497**	**>285**	**>6662**	**>18580**

* Data from the United States, Brazil, Russia, India, China, and South Africa mainly based on 2008, Mexico on 2007.

"Other" includes (whenever available) funding from state and local governments, partially state-owned enterprises, NGOs, and industry.

~U.S. data on industry expenditure is from 2004 (NSF 2008).

[1] Based on PEMEX's fund for scientific and technological research on energy.

[2] Based on total nongovernmental investments into PBMR Ltd.

[3] Based on 2005 R&D expenditure in car manufacturing industry (CONACYT, 2008).

[4] Based on 2005 R&D expenditure in utilities sector (CONACYT, 2008).

>These cumulative values are based on data from only three to four BRIMCS countries, so actual expenditures are expected to be higher.

>>These cumulative values are based on data from two or fewer BRIMCS countries, so actual expenditures are expected to be much higher.

gained substantial support in China, Brazil, and Mexico, which can be explained by the important role of SOEs in the electricity industry in these countries.

Second, governments and 100 percent government-owned SOEs in the BRIMCS countries invested a minimum of US$ 13.8 billion in energy RD&D in 2008, with around 90 percent of these funds coming from SOEs (see Table 22.9). Governments in the BRIMCS countries may thus control larger amounts of energy RD&D funding than the governments in IEA member countries, whose total energy RD&D investments were US$12.7 billion in the same year. This finding is subject to data limitations, particularly the lack of systematic reporting, and definitions of what constitutes RD&D varying widely between different data sources.

Table 22.10 provides a comparison of energy RD&D investments in the United States and the BRIMCS countries relative to their GDP, energy use, and energy production. The comparison is based on data from the most recent year for which the most complete aggregate data was available in each country. The results show significant differences between countries, with the relative amounts of energy RD&D investments highly dependent on the comparison metric. China's government investments in energy RD&D are higher than any other country based on any metric, but funding from "other" sources is lower than any other country. Conversely, "other" funding in Brazil is relatively large because of the R&D expenditure by Petrobras and Eletrobras. In Russia, government investments are relatively low if measured in terms of energy use and production because of its abundant energy resources. South Africa's government energy RD&D investments relative to GDP are higher than in most other countries, although their absolute value is lowest (see Table 22.9). Similarly, the size of energy RD&D investments from India's government relative to India's GDP, energy production, or energy use is higher than the United States, whereas their absolute value is less than 30 percent that of the United States. In general, therefore, energy RD&D investments of "governments" or "others" cannot be explained as a simple function of GDP, energy consumption, or energy production.

5 Conclusion

This chapter provides an overview of energy research, development, and demonstration (RD&D) investments in six major emerging economies: Brazil, Russia, India, Mexico, China, and South Africa (the "BRIMCS" countries). Energy RD&D investments are categorised by funding source (government and other) and by technology category.

Analysis of available information shows that fossil fuel and nuclear energy received the highest level of RD&D funding between 2000 and 2007 in the BRIMCS countries as well as in the United States. In 2008, renewable energy RD&D funding in the United States exceeded fossil energy RD&D funding, but in the BRIMCS countries RD&D investments in fossil and nuclear energy RD&D remain the largest. The significant investments in energy RD&D made by the BRIMCS countries in absolute terms points to the need to include the BRIMCS countries in a comprehensive global effort to accelerate energy technology innovation and meet the world's energy challenges. The relative efforts of the different BRIMCS countries and the United States highly depend on whether GDP, energy use, or energy production are used as measures for comparison.

Table 22.10. Energy RD&D investments relative to GDP, energy use, and energy production.

Energy RD&D investments (% of GDP $*10^2$)							
	United States	Brazil	Russia	India	Mexico	China	South Africa
government	2.9	1.8	0.5	3.5	1.7	14.9	5.7
other	2.0	7.7	3.9	2.1	2.0	1.7	5.2
Energy RD&D investments relative to energy use (mln 2008 PPP $int /mln tonnes oil equivalent)[a,b]							
	United States	Brazil	Russia	India	Mexico	China	South Africa
government	1.8	1.4	0.2	2.0	1.4	6.0	2.0
other	1.1	6.1	1.4	1.2	1.6	0.7	1.8
Energy RD&D investments relative to energy production (mln 2008 PPP $int /mln tonnes oil equivalent)[a,b]							
	United States	Brazil	Russia	India	Mexico	China	South Africa
government	2.5	1.5	0.1	2.6	1.0	6.5	1.7
other	1.6	6.7	0.7	1.5	1.1	0.7	1.6

[a] U.S. gov.='08, U.S other ='04, Brazil gov. ='07, Brazil other ='07, Russia gov. ='08, Russia other ='08, India gov. ='08, India other ='08, Mexico gov. ='07, Mexico other ='05, China gov. ='04, China other ='04, South Africa gov. ='07, South Africa other ='07.

[b] Data on energy production and energy use from the World Bank WDI (2010). Data for the year 2008 is not available, so energy RD&D investments in 2008 are relative to energy production and energy use in 2007.

Finally, this overview emphasises that a more comprehensive and systematic effort is needed to collect information on energy RD&D investments by central governments and other institutions. Currently, governments in the BRIMCS countries still have limited information on their energy RD&D investments and to the extent information is available, it is often not categorised into different priority areas. This impedes effective policy making as well as the search for mutually beneficial international energy RD&D activities.

FURTHER READING

IEA, 2011. Clean Energy Progress Report, IEA Input into the Clean Energy Ministerial, June 2011.

Anadón, L., 2011. Transforming U.S. Energy Innovation, Energy Research, Development, Demonstration and Deployment Project, Energy Technology Innovation Policy Group, Belfer Center of Science and International Affairs, November 2011.

REFERENCES

Aksenova, I., 2009. Personal Communication: Energy R&D funding in Russia, email correspondence with M. Bunn, November 2, 2009.

Anonymous, 2008. Statement of Katharine A. Frederiksen Acting Assistant Secretary Office of Policy and International Affairs U.S. Department of Energy, U.S.-China Economic and Security Review Commission, Washington, DC: 10.

CONACYT, 2008. Informe General del Estado de la Ciencia y la Tecnología (in Spanish), Consejo Nacional de Ciencia y Tecnologia, accessed December 2009 at: http://www.conacyt.mx/.

2010. Difusion del Fondo CONACYT-SENER (antecedentes), CONACYT; SENER, accessed August 2010 at: http://www.conacyt.gob.mx/Fondos/Sectoriales/SENER/Sustentabilidad-Energetica/2010/Difusion-del-fondo.ppt.

Delman, J. & Chen, Y., 2008. Nordic Collaboration with China in Energy Research and Development, Nordic Institute of Asian Studies, commissioned by Nordic Energy Research, 127, accessed October 10 at: http://www.nordicenergy.net/section.cfm?id=3–0&path=142.

ERAWATCH, 2010a. National Profiles: Brazil, ERAWATCH, accessed August 2010 at: http://cordis.europa.eu/erawatch/index.cfm?fuseaction=ri.content&topicID=329&countryCode=BR.

2010b. Russian Federation; Federal Agency for Science and Innovation, ERAWATCH, accessed August 2010 at: http://cordis.europa.eu/erawatch/index.cfm?fuseaction=org.document&uuid=E45D3E10–043B-5E51–3526AADB9C2B6169.

Gallagher, K. S. & Anadón, L. D., 2010. DoE Budget Authority for Energy Research, Development, and Demonstration Database, Energy Technology Innovation Policy Research Group, John F. Kennedy School of Government, Harvard University, accessed at: http://belfercenter.ksg.harvard.edu/publication/20013/doe_budget_authority_for_Energy_research_development_demonstration_database.html.

Gallagher, K. S., Anadón, L. D., et al., 2010. Trends in Investments in Global Energy RD&D, Wiley Interdisciplinary Reviews: Energy Innovation, R. A. Pielke, Wiley.

Gallagher, K. S., Holdren, J. P., et al., 2006. Energy-Technology Innovation. *Annual Review of Environment and Resources*, 31(1): 193–237.

Gao, C., 2010. Personal Communication, email correspondence with R. Kempener, June 19, 2010.

Gokhberg, L., 2003. S&T and Innovation in the New Economy: Search for Systemic Solutions, The First Globelcis Conference Innovation Systems and Development Strategies for the Third Millennium, Rio de Janeiro, accessed at: redesist.ie.ufrj.br/globelics/slides/Leonid%20Gokhberg.pdf.

2010a. Personal Communication: Gross Domestic Expenditure (GERD) on Energy R&D, email correspondence with A. Grubler, February 19, 2010.

2010b. Personal Communication: Gross Domestic Expenditure (GERD) on Energy R&D, email correspondence with A. Grubler, February 19, 2010.

Government Directive, 2009. Nuclear Power Technologies of the New Generation for 2010–15 and prospectively to 2020, R. F. Government Directive Russia, Federal Targeted Programme.

Kempener, R., Anadón, L. D., et al., 2010. Governmental Energy Innovation Investments and Policies in the Major Emerging Economies: Brazil, Russia, India, Mexico, China, and South Africa, Energy Research, Development, Demonstration and Deployment Project, Energy Technology Innovation Policy Group, Belfer Center of Science and International Affairs, 1–61, accessed November 2010 at: http://belfercenter.ksg.harvard.edu/publication/20517/.

Keohane, R. O., 1978. The International Energy Agency: State Influence and Transgovernmental Politics. *International Organization*, 32(4): 929–51.

Khlunov, A., 2007. EU-Russia: New Perspectives for S&T Cooperation for 2007–2013, International Conference EU-Russia Cooperation: Priorities for Science and Technology 2007–2013, Moscow, accessed at: http://hse.ru/temp/2007/files/02_22_konf/Khlunov.pdf.

Klimenko, A., 2008. Russian R&D Policy in the Field of Energy Technology, FASI-IEA NEET Workshop "Cooperation in the Field of Energy Technologies," Moscow, accessed at: http://pt21.ru/content/view/11/14/lang,en/.

Lantzke, U., 1975. The OECD and Its International Energy Agency. *Daedalus*, 104(4): 217–27.

MES, 2006. Strategy Development of Science and Innovation in the Russian Federation for the Period up to 2015, Ministry of Education and Science of the Russian Federation, 123, accessed August 2010 at: http://mon.gov.ru/work/nti/dok/.

NSF, 2008. Research and Development in Industry: 2004, Division of Science Resources Statistics (SRS), National Science Foundation, 307, accessed March 2010 at: http://www.nsf.gov/statistics/nsf09301/.

Obadia, I. J., 2010. Nuclear Energy RD&D Funding in Brazil, Personal Communication, e-mail correspondence, August 16, 2010.

OECD, 2002. Frascati Manual: Proposed Standard Practice for Surveys on Research and Experimental Development, Organisation for Co-operation and Development, 254, accessed 08/10 at: http://www.lmt.lt/PROJEKTAI/TEKSTAI/Frascati.pdf.
2009. OECD Reviews of Innovation Policy: Mexico, OECD, 1–220, accessed August 2010 at: http://titania.sourceoecd.org.ezp-prod1.hul.harvard.edu/vl=1699107/cl=17/nw=1/rpsv/cgi-bin/fulltextew.pl?prpsv=/ij/oecdthemes/99980134/v2009n16/s1/p1l.idx.
Pemex, 2008. Annual Report 2007/2008, Petróleos Mexicanos, accessed December 2009 at: http://www.ri.pemex.com/index.cfm?action=content&sectionID=135&catID=12320.
Reutov, B. F., 2007. Implementation of Energy Efficiency Projects in Russia within the Framework of Public-Private Partnerships, Eighteenth Steering Committee Meeting on Energy Efficiency – 21 Project, Geneva, accessed February 2010 at: http://www.unece.org/energy/se/pp/ee21_sc/18scMay07/31may/reutov_310507.pdf.
SANERI, 2008. South African Progress on RE Activities: From the Perspective of the SA National Energy Research Institute, SANERI, 21, accessed March 2010 at: http://www.ren21.net/presentations/SouthAfricanprogressonREActivities.pdf.
Siegler, A., 2009. The European Green Cars Initiative, Info Day on the Research PPPs, Brussels, accessed October 10 at: http://ec.europa.eu/research/industrial_technologies/pdf/ppp-infoday-gc-01-a-siegler-the-european-green-cars-initiative.pdf.
Smith, M. A., 2010. The Russian Chronologies: October – December 2009, Defence Academy of the United Kingdom, 128, accessed at: www.da.mod.uk.
Soares, L. R., Melo Junior, A. C. D., et al., 2008. Overview of Brazilian R&D Management: Electricity Regulatory Agency Case, Proceedings of the R&D Management Conference 2008: Emerging and New Approaches to R&D Management, Ottawa, Canada, accessed at: http://www.telfer.uottawa.ca/RnDManagement2008/main.shtml.
State Statistical Bureau, 2009. China Statistical Yearbook on Science and Technology, State Statistical Bureau, State Science and Technology Commission, v., accessed October 10 at: http://hollis.harvard.edu/?itemid=%7Clibrary/m/aleph%7C005059969.
World Bank, 2010. World Development Indicators Online, 2010, online database by The World Bank Group, accessed August 2010 at: http://go.worldbank.org/6HAYAHG8H0.

ACKNOWLEDGEMENTS

This chapter has benefitted from information and feedback by Irina Aksenova, Francisco Aguayo Ayala, Brett Cohen, Gao Changlin, Arnulf Grubler, Isaac Jose Obadia, Balanchandra Patil, Ambuj Sagar, Joisa Saraiva, Zhao Hanwei, and Zhiuan (Charles) Cheng. Data collection took place in 2009, but because of the constant political changes that take place, names and roles of institutions might have changed over the course of writing this chapter. Any omissions or mistakes are the full responsibility of the authors.

23 A Comparative Analysis of Annual Market Investments in Energy Supply and End-Use Technologies

Charlie Wilson and Arnulf Grubler

Authors' Summary

Investments in end-use technologies are integral to energy system objectives including climate change mitigation. Across a wide range of future scenarios, reducing energy intensity by improving the efficiency of end-use technologies and supporting their widespread diffusion is a lower-cost and nearer-term complement to decarbonising the energy supply. Whereas the need to mobilise investment in energy supply technologies is broadly understood, with the current level of investment estimated in the order of \$0.7–0.9 trillion a year, there is a notable absence of analogous investment data for end-use technologies. This exacerbates the difficulty of like-for-like comparisons of supply-side and demand-side (or end-use) investment needs. In response, this case study presents a global, bottom-up estimate of total investments in end-use energy technologies based on volume data and cost estimates for 2005. Total investment in end-use technologies was conservatively found to be in the order of \$0.3–4 trillion depending on the definition of end-use technology used. The magnitude and diversity of end-use investments reinforce the need to improve characterisations of end-use technologies in assessments of future energy system transformation.

Editors' Guide

Relationship of Case Study to Energy Technology Innovation System Framework

Diffusion investments are a financial metric to characterise the *resources mobilised* for the *adoption and use* of energy technologies within the energy technology innovation system. Financial inputs during the *diffusion stage* of energy technologies typically describe the financing requirements for constructing energy supply and distribution infrastructure, from mines and rigs to power plants and pipelines. In this case study, analogous diffusion investments for end-use technologies describe the purchase or installation costs of building energy systems (e.g., heating, lighting), vehicles, and consumer goods. In all cases, diffusion investments are visible as capital stock additions that embody technological *knowledge*. Diffusion investments are also linked to *learning*-related performance improvements associated with commercial experience. The *actors* involved vary widely across technology types, with households and firms playing a major role in the diffusion of end-use technologies, and governments and energy companies dominating the diffusion of energy supply technologies. The financing mechanisms used in each case also vary (e.g., balance sheets, capital markets, disposable household income).

Estimates of future investment requirements generated by scenario modelling studies contribute to the *shared expectations* among the *actors and institutions* that influence and direct energy innovation activities. Yet these *knowledge generation* processes are failing to assess or resolve end-use technologies in a way that allows their investment requirements to be consistently compared with those of energy supply technologies. This biases downwards the perceived importance of end-use diffusion investments within the energy technology innovation system as scenarios relying heavily on (exogenous) efficiency improvements appear less costly than scenarios relying heavily on decarbonisation. Taking a systemic view of energy technology innovation helps reveal this bias.

Insights for Energy Technology Innovation Policy

- Estimating total diffusion investments in end-use technologies relative to those in the energy supply is important to understand the financial needs and magnitudes of energy system transformation. It also provides a common and consistent reference point for international policy making and diplomacy in the context of climate change mitigation, and associated analysis of financing needs and investment flows.
- Diffusion investments are strongly weighted towards end-use technologies if technologies as a whole are considered. This is the most meaningful like-for-like comparison with energy supply technologies: power plants versus vehicles; refineries versus lighting systems. Under a more restrictive definition of end-use technologies, diffusion investments are balanced between end use and supply. In either case, the weighting of resources mobilised for the adoption and use of technologies is a strong indicator of the market-pull drivers of innovation.
- Investments into energy efficiency and energy end use thus emerge as critically important both in overall magnitude and in terms of the necessary incentives and policy inducements to overcome barriers to investment in a sustainable future energy system.

1 Introduction: Investment Needs for Energy System Transformation

Across a wide range of future scenarios, reducing energy intensity by improving the efficiency of end-use technologies is a lower-cost complement to reducing the carbon intensity of the energy supply in the nearer term (see, for example, Ürge-Vorsatz and Metz, 2009). Modelling assessments find energy efficiency improvements in buildings, appliances, vehicles, industrial processes, and so on, to represent the largest and least costly source of potential emission reductions (IEA, 2008a: 40). The ranking of efficiency and conservation as the most important mitigation options is also robust to different scenario and technology assumptions (Riahi, Grubler, and Nakicenovic, 2007). Supply curves of emission reductions similarly highlight the low or negative marginal abatement costs associated with improved efficiency of end-use technologies (Enkvist, Naucler, and Rosander, 2007).

Substantive investments are required to ensure both efficient end-use technologies and low-carbon energy supply technologies diffuse widely throughout the energy system (Nakicenovic and Rogner, 1996). This case study is concerned with the magnitude of these "diffusion" investments in market settings, covering both energy supply and end-use technologies. Energy supply technologies extract, process, transport, and convert primary energy into energy carriers, and distribute them to the point of use (e.g., oil wells, drilling rigs, pipelines, refineries, tankers, gas stations). Energy end-use technologies provide useful energy services to final users (e.g., mobility in the transportation sector, space conditioning in the residential and commercial sectors).

The magnitude of current and future "diffusion" investments in energy supply technologies is relatively well characterised in both reference (or baseline) scenarios and in climate change mitigation scenarios (IEA, 2009), although scenario studies do not typically disclose in detail their underlying base year investment estimates. Analogous estimates for end-use technologies are comparatively weak and patchy. This prevents like-for-like comparisons of capital investment requirements between the supply and demand sides of the energy system.

In this case study, we briefly review current estimates of global diffusion investments in energy technologies and distinguish the different approaches used. We then provide bottom-up, granular estimates of current investments in the principal types of end-use technology worldwide. (By granular, we mean disaggregated to, or resolved at, the level of individual technologies with different unit sizes.) We compare the magnitude and diversity of these end-use investments to analogous estimates for the energy supply. We conclude by arguing for the centrality of end-use technology investments in the analysis of global energy system challenges, particularly climate change mitigation.

2 Energy Technology Investment Estimates

2.1 Current Investments in Energy Technologies

Current levels of investment in the global energy system are estimated by the International Energy Agency (IEA) in the range of \$0.7–0.9 trillion a year (IEA, 2008b, 2009). This is in line with an estimate for the *Global Energy Assessment* of \$0.96 trillion (Riahi et al., 2012). Both calculate energy supply investments only, with more than half the total attributed to electricity plant and transmission infrastructure. (All investment data in this chapter are expressed in 2005\$ using global GDP deflators, unless otherwise noted.) These data broadly compare with earlier estimates of \$0.6 trillion a year in the IEA's World Energy Investment study (IEA, 2003), and annual averages of \$0.7–1.1 trillion and \$1 trillion for the period 1990–2020 taken from, respectively, the Global Energy Perspectives scenario study (Nakicenovic and Rogner, 1996) and the World Energy Council scenario study (WEC, 1993).

2.2 Climate Change Mitigation and Future Investments in Energy Technologies

Estimating future energy technology investment needs is one approach to costing climate change mitigation. Estimates are expressed either in absolute terms as *total investments*, or in relative terms as *incremental investments* needed to move from some reference or baseline scenario to a defined mitigation scenario.

Alternative approaches are to express incremental mitigation costs relative to a reference scenario in terms of economy-wide impacts on GDP (e.g., Edenhofer et al., 2010; Stern, 2006) or in terms of sectorial or economy-wide abatement costs (Levine et al., 2007; Van Vuuren et al., 2009). Ranking technologies by their abatement cost per tCO_2 (y-axis) in combination with their emission reduction potential (x-axis) also generates supply curves for emission reductions (McKinsey, 2009) which are widely used in modelling studies (Cofala et al., 2008). Here, however, our interest is in the investment estimates generated by such studies.

The most widely cited investment figures are published by the International Energy Agency in its authoritative annual report, the *World Energy Outlook*. This details the dominant challenges for the global energy system and projects its development over a decadal timeframe (IEA, 2009). The "Reference Scenario" of the 2009 *World Energy Outlook* identifies energy-related cumulative investments in the period to 2030 equivalent to around $1 trillion a year (IEA, 2009: 104). This *total* investment estimate disaggregates into power generation and transmission (53 percent), upstream oil and gas (23 percent and 20 percent), and coal (3 percent) – all energy supply technologies. *Total* investment requirements are not estimated for end-use technologies.

Incremental investments required in end-use technologies to mitigate climate change are also (partially) estimated. Relative to the costs of the "Reference Scenario" over the period to 2030, the additional costs of a "450 Scenario" are estimated at $0.4 trillion per year (IEA, 2009: 258). The "450 Scenario" describes a stabilisation of atmospheric concentrations of CO_2 equivalents at 450 ppmv. Around 80 percent of the additional investment in the 450 Scenario is in end-use technologies, disaggregated into transport (44 percent), buildings (24 percent), and industry (10.5 percent), with the remainder attributed to energy supply technologies (IEA, 2009: 104). *Total* investments in end-use technologies in the mitigation scenario cannot be estimated as investment subcategory totals and sectoral breakouts are not reported for the reference scenario.

Table 23.1 summarises these and other capital investment estimates, illustrating the asymmetric treatment of energy supply and end-use technologies. *Total* investments are only estimated for energy supply technologies. The corresponding column for end-use technologies in Table 23.1 is blank except for the (IEA, 2008a) study examined later in this chapter. To the extent they are estimated, investments in end-use technologies are expressed in *incremental* terms (i.e., relative to a reference scenario).

The *Global Energy Assessment* scenarios, for example, report cumulative investments to 2050 that combine *total* investments in energy supply technologies and *incremental* investments in end-use efficiency to meet a 2°C stabilisation target (Riahi et al., 2012). *Total* investments in end-use technologies are not reported.

Another recent study used three modelling groups' estimates of the investment requirements to reach a 450ppm CO_2-only stabilisation target (Luderer et al., 2009).

In this case, *total* investment requirements were estimated for both reference and mitigation scenarios, but only for five categories of energy supply technology.

An earlier study based on detailed modelling representations of the residential, commercial, industrial, and transportation end-use sectors did report the *incremental* investment needs for end-use technologies in a 2°C climate stabilisation scenario (Hanson and Laitner, 2006). In this case, however, *total* investments in the reference scenario were not reported.

The lack of total investment estimates for end-use technologies in both reference and mitigation scenarios prevents a meaningful, holistic appreciation of the financing needs of future energy system transformation. Estimating incremental but not total investments creates an additional problem as the magnitude of incremental investments in the mitigation scenario depends on the extent to which total investments are already assumed to characterise the reference scenario (Van Vuuren et al., 2009). This problem is particularly marked for end-use technologies in reference scenarios with strongly falling energy intensity, giving rise to a substantial reduction in apparent end-use investment requirements relative to energy supply investments (Riahi et al., 2007).

2.3 Apples, Oranges, and End-Use Technologies

Certain characteristics of end-use technologies help explain their asymmetrical treatment in assessments of energy system investment needs. Firstly, end-use technologies are not traditionally considered energy sector investments, being rather a scatter of different industrial and consumer goods' investments (Nakicenovic, 1986). Energy conversion – from the final user's perspective – is an often incidental attribute of end-use technologies whose primary purpose is to provide useful services such as lighting (lumens) and mobility (e.g., tonne-kilometres of freight transport). Investments normalised per energy end-use conversion capacity or final or useful energy end-use are therefore less meaningful.

Secondly, to ensure investment estimates for energy supply and end-use technologies are comparable, a common definition of the unit of analysis is needed. Investments in energy supply technologies are quantified at the level of the power plant, refinery, or LNG terminal, usually normalised per unit of energy input or output converted. What is the appropriate scale or system boundary of an end-use technology: engine, car, or road transport system? The one study shown in Table 23.1 that does estimate *total* end-use investments sidesteps this problem by combining *total* investments in cars with *incremental* investments in efficiency improvements in all other end-use technologies (IEA, 2008a). The rationale for defining these system boundaries for estimating energy investments is given (rather unconvincingly) as follows:

> [E]nergy efficiency improvements apply to a wide range of the car's components ... [but] for building improvements, a breakdown of the costs of energy efficiency compared to the fabric or structure of a building would be arbitrary, while including the total construction cost would result in buildings taking up a disproportionate share of investment needs, when their primary role is shelter. (Annex B, IEA, 2008a)

The resulting apples and oranges combination of *total* and *incremental* investments generates an estimate that is hard to interpret and unsurprisingly dominated by cars

Table 23.1. Comparison of future energy supply and end-use technology investment needs globally (annual approximations),

Study	Reported Investment Data (as basis for Annual Estimates)	*Total* Annual Investment Needs (Approximated, 2005$)		*Incremental* Annual Investment Needs (Approximated, 2005$)			Associated Integrated Assessment Models[a]
		Supply	End Use	Supply	End Use	Basis of Incremental Estimates	
Nakicenovic and Rogner (1996)	cumulative investment to 2020 (1990$) in three scenarios	$0.7–1.1 tr	-	-	-	no reference scenario reported	MESSAGE
Hanson and Laitner (2006)	average annual investment to 2060 (2000$)	-	-	-$0.01 tr[b]	$0.07 tr	from reference to mitigation scenario (2°C stabilisation)	AMIGA
IEA (2008a)	cumulative investment to 2050 (2005$)	$0.6 tr	$5.0 tr[c]	$0.1 tr	$0.9 tr	from reference to "BLUE scenario" (50% reduction in CO_2 emissions by 2050)	IEA models
IEA (2009)	cumulative investment to 2030 (2008$)	$1.0 tr	-	$0.1 tr	$0.4 tr	from reference to "450 Scenario" (450 ppmv CO2-eq.)	IEA models
Luderer et al. (2009)	average annual investment to 2030 (2005$)	$0.5–0.6 tr	-	-$0.2 tr[b] – $0.7 tr	-	from reference to mitigation scenario (450 ppmv CO_2 only)	IMACLIM, REMIND, WITCH
Riahi et al. (2012)	cumulative investment to 2050 (2005$) in three scenarios	$1.4–1.8 tr	-	-	$0.3–0.5 tr	no reference scenario reported	MESSAGE
Van Vuuren et al. (2009)	$0.1–1.0 tr abatement costs in 2030 (2000$)	not commensurate with investment data; not disaggregated by energy supply and end use				from reference to mitigation scenario (450 ppmv CO_2-eq.)	AIM, E3MG, ENV-Linkages, IMAGE, MESSAGE, WorldScan

(*continued*)

Table 23.1. (*cont.*)

Study	Reported Investment Data (as basis for Annual Estimates)	*Total* Annual Investment Needs (Approximated, 2005$)		*Incremental* Annual Investment Needs (Approximated, 2005$)			Associated Integrated Assessment Models[a]
		Supply	End Use	Supply	End Use	Basis of Incremental Estimates	
Stern (2006)	-2% to 5% loss of GDP	not commensurate with investment data; not disaggregated by energy supply and end use				from reference to mitigation scenario (500–550 ppmv CO_2-eq.)	various
Edenhofer et al. (2010)	1% to 2.5% loss of GDP	not commensurate with investment data; not disaggregated by energy supply and end use				from reference to mitigation scenario (400 ppmv CO_2-eq.)	E3MG, IMAGE, MERGE, POLES, REMIND

[a] Integrated assessment models vary widely in their structure, treatment of energy supply and demand, resolution of specific energy supply and end-use technologies, technology cost profiles over time, and so on. See corresponding references for details.

[b] Net decrease as lower overall energy supply investments due to demand reductions relative to reference scenario. In Hanson and Laitner (2006), energy supply total also includes systems integration costs.

[c] Reference scenario totals aggregate total investment in cars, and incremental investments in efficiency in other end-use technologies. Consequently, 94 percent of total investments in end-use technologies are in transport. See text for discussion and Chapter 6, IEA (2008a) for details.

Notes: Studies report investment needs differently, typically on a cumulative basis over long timescales (see "Reported Investment Data" column). Here, annual investments are approximated to aid comparability using simple linear assumptions and adjusted to 2005$ using global GDP deflators; these annual estimates are indicative only and should not be attributed to the corresponding studies.

(see Table 23.1). It also points to the difficulties of clearly and consistently identifying what is being invested in. If the primary role of buildings is shelter, is the primary role of cars not mobility? Neither investments in buildings nor investments in cars can be narrowly defined as investments in energy conversion.

Thirdly, investments in (and performance of) end-use technologies are dependent on investments in associated infrastructure such as airports, roads, and buildings. Is it meaningful to quantify investments in a home heating system without quantifying the investment needed for the home's building envelope that influences its heating load?

Although the same system boundary issue exists for energy supply technologies, it is largely addressed by additionally quantifying the investments in associated transmission and distribution infrastructure. The problem for end-use technologies is that the same approach implies a summation of all investments in building structures, roads, railways, ports, airports, industrial machinery, and so on. But limiting the assessment of end-use technology investments to their efficiency improvements or mitigation potentials still leaves the problem of apples and oranges comparisons with energy supply technologies.

3 Method

3.1 Defining End-Use Technologies

Our response to the definitional issues with end-use technologies is to adopt two consistent though still arbitrary definitions of end-use technology investments, and quantify total investments in each category. Our first, broader definition and data set describes end-use technologies as the technological systems purchasable as products by final consumers in order to provide a useful service (Murmann and Frenken, 2006). This implies heating and air conditioning systems not houses, and fridges and ovens not kitchens. Our second, narrower definition and data set describes the specific energy-using components or subsystems of these end-use technologies. This implies engines in cars and light bulbs in lighting systems. Table 23.2 summarises these distinctions for the technologies analysed. In some cases (industrial motors, mobile heating appliances), a distinct energy-using component was not identified and so the data in both cases are the same.

3.2 Bottom-Up Estimation of End-Use Investments

We used volume data (production, delivery, sales, installations) and price estimates to approximate total investment in 2005 in both end-use technologies and their specific energy-using components (see later in this chapter and Table 23.2 for details). We included low and high sensitivities around central estimates, taking account of uncertainties in both volume and cost assumptions.

Our aim is to provide a first order estimate of end-use technology investments to allow a meaningful, like-for-like comparison with estimates of energy supply technology investments. We acknowledge the many approximations and limitations in our data and welcome continued critical revision of our estimates. Full details of our data sources and calculations are available from the authors on request.

Table 23.2. Summary of end-use technologies and their energy-using components included in bottom-up investment estimates.

End-use service	Broad definition: end-use technology	Narrow definition: energy-using component of end-use technology
mobility	commercial jet aircraft	jet engine
mobility	vehicles (cars and commercial)	internal combustion engine
space conditioning	central heating systems (boiler/ furnace, ducts/pipes, radiators, controls, and network connections for new systems)	boiler or furnace
space conditioning	air conditioning systems (AC unit, ducts, controls, and network connections for new systems)	air conditioning unit
space conditioning	mobile heating appliances (e.g., portable convection / fan heaters)	- (same as for end-use technology)
lighting	lighting (light bulb + fixture)	light bulb (or lamp)
food storage, cooking, cleaning	large household appliances (fridges, freezers, clothes washers and dryers, dishwashers, cookers)	compressors, motors, fans, heating elements (depending on appliance)
industrial processes	industrial motors	- (same as for end-use technology)

4 Results

Our first order estimate of *total* global investments in 2005 in end-use technologies is \$1–3.5 trillion, with a central estimate of \$1.7 trillion (see Table 23.3). Our first order estimate of *total* global investments in 2005 in the energy-using components of these end-use technologies is \$0.1–0.6 trillion, with a central estimate of \$0.3 trillion (see Table 23.4).

We emphasise that these *total* annual investment estimates omit many end-use technologies, including propeller-based and noncommercial aircraft, helicopters, all military technologies, mass transit systems, water heaters (residential and other), building envelopes (insulation, windows, doors), information and communication technologies, small appliances, other consumer electronics, and all industrial equipment and process other than motors (e.g., blast furnaces, pulp mills, cement kilns). With the exception of industrial plant and building envelopes, we believe the inclusion of these categories would not substantially increase the narrowly defined investment range (for energy-using components); however, they would substantially increase the broadly defined investment range for end-use technologies.

Various observations can be made from the compiled data. First, the rank contribution of different technologies to total investments is broadly consistent regardless of the breadth of definition of end-use technologies. However, the proportionate cost of energy-using components to their corresponding end-use technology is lowest in vehicles. Second, transport technologies dominate both narrowly defined and broadly defined investments (but to a much lesser extent than in the IEA (2008a) study discussed in Table 23.1, which mixed total and incremental investments). Third, more is invested in retrofitting heating and cooling technologies than in new building installations, reflecting the longevity of buildings and the more

Table 23.3. Estimated annual investments in selected end-use technologies.

"Energy-using components" of end-use technologies in 2005	2005$	Low sensitivity	Central estimate	High sensitivity	Share (central estimate)
Grand Total Investments	**$bn**	**124**	**297**	**713**	**100%**
commercial aircraft – jet engines	$bn	3	7	13	2%
cars – engines	$bn	36	76	159	25%
commercial vehicles – engines	$bn	27	57	119	19%
buildings (retrofits) – central heating units	$bn	13	52	158	18%
buildings (new) – central heating units	$bn	9	20	41	7%
mobile heating units	$bn	2	4	5	1%
buildings (retrofits) – air conditioning units	$bn	5	21	69	7%
buildings (new) – air conditioning units	$bn	4	10	20	3%
lighting	$bn	12	27	59	9%
large household appliances	$bn	11	18	53	6%
industrial motors	$bn	2	6	16	2%

Note: Sources and data available from the authors on request.

Table 23.4. Estimated annual investments in the energy-using components of selected end-use technologies.

End-use technologies in 2005	2005$	Low sensitivity	Central estimate	High sensitivity	Share (central estimate)
Grand Total Investments	**$bn**	**984**	**1,739**	**3,549**	**100%**
commercial jet aircraft	$bn	12	28	50	2%
cars	$bn	540	758	1,194	44%
commercial vehicles	$bn	270	427	672	25%
buildings (retrofits) – central heating systems	$bn	47	250	979	14%
buildings (new) – central heating systems	$bn	33	93	248	5%
mobile heating systems	$bn	2	4	5	0%
buildings (retrofit) – air conditioning systems	$bn	9	42	137	2%
buildings (new) – air conditioning systems	$bn	7	20	41	1%
lighting	$bn	17	38	83	2%
large household appliances	$bn	45	75	124	4%
industrial motors	$bn	2	6	16	0%

Note: Sources and data available from the authors on request.

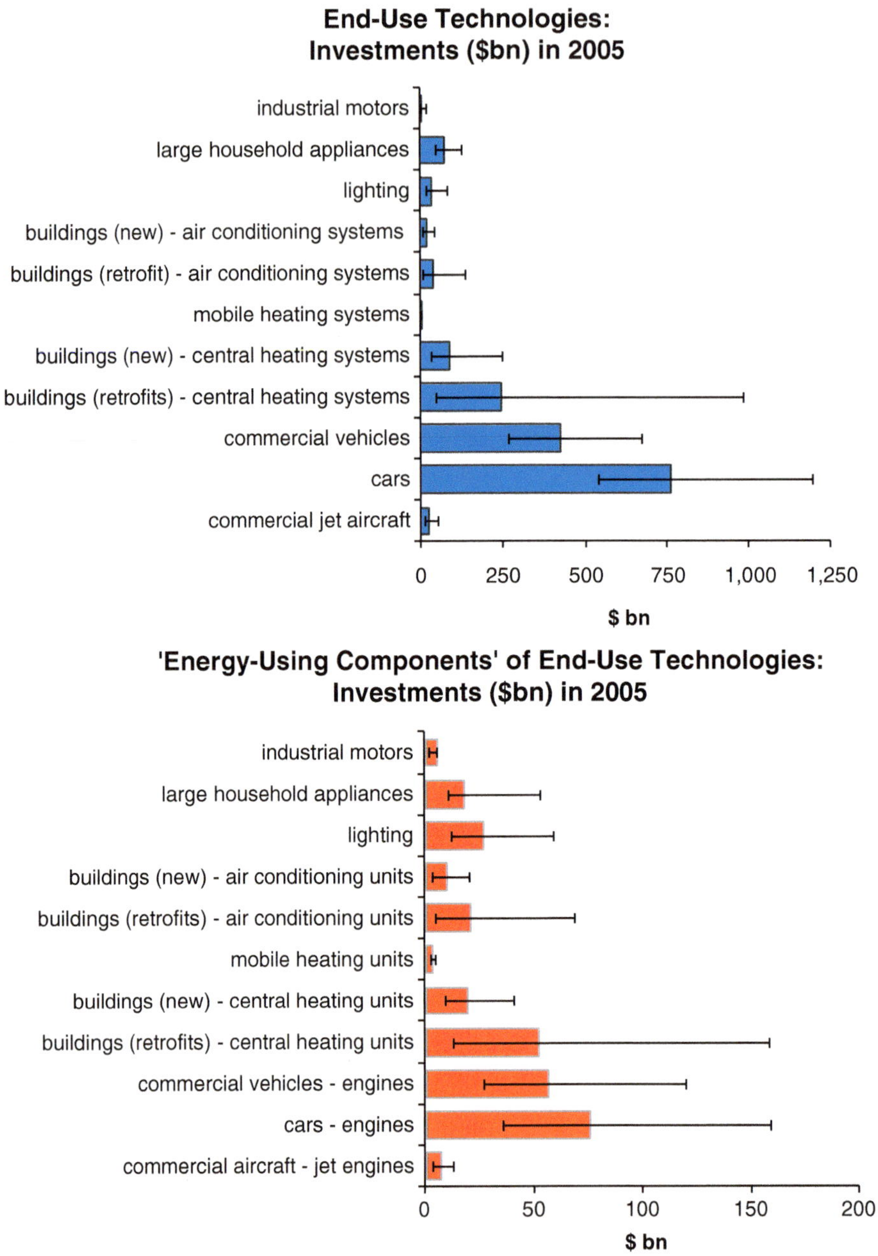

Figure 23.1. Annual investments in end-use technologies (top graph) and their energy-using components (bottom graph). Sources and data available from the authors on request.

pervasive penetration of mechanical space conditioning technology in developed countries (see Figure 23.1). Fourth, technologies providing mobility and space conditioning account for more than four-fifths of total investments, again regardless of whether narrowly or broadly defined (though reaching as high as 92 percent of total investments in the latter case).

Disaggregating the data by region shows that approximately two-thirds of the end-use investments in 2005 are in OECD countries and the former Soviet Union

(FSU), broadly corresponding to IEA countries. The remaining one-third are in developing economies. However, investment data for aircraft and vehicles is by region of manufacture not final use (and purchase), which inflates the developed country shares of these technologies in our estimates.

5 Discussion: The (Relative) Importance of Investments in End-Use Technologies

The range of end-use technology investments is conservatively in the order of $0.3–4.0 trillion, adjusting upwards the range of $0.1–3.5 trillion to take into account the extent of technologies missing from our analysis. This compares with the range of current energy supply investments in the order of $0.7–0.9 trillion. Although the two ranges span the same orders of magnitude, the upper bound of end-use technology investments is some four times greater than its energy supply equivalent. This is in line with the IEA's findings that demand-side investment needs are four times those of the energy supply alone (IEA, 2008a: 227). It is also in line with the one study we found in our review that quantified total investments consistently for different end-use technologies, and estimated "total efficiency investments" in the United States to be three times larger than those in the energy supply (Ehrhardt-Martinez and Laitner, 2008).

Estimating *total* diffusion investments in end-use technologies relative to those in the energy supply is important to understand the financial needs and magnitudes of energy system transformation. It also provides a common and consistent reference point for international policy making and diplomacy in the context of climate change mitigation (Bazilian et al., 2010) and associated analysis of financing needs and investment flows (UNFCCC, 2007). Lastly it helps to put "green growth" policy objectives into perspective. Investments into energy efficiency and energy end-use emerge as critically important both in overall magnitude and in terms of the necessary incentives and policy inducements to overcome barriers to investment in a sustainable future energy system.

To the extent that *total* investment estimates are granular, that is, resolved at the level of individual technologies, a like-for-like comparison of *total* end use and *total* energy supply investments also supports assessments of specific financing implications for different sources of investment (e.g., balance sheets, capital markets, disposable household income) and types of investor (e.g., households, firms, governments). Recognising the importance of investors and sources of capital from outside the traditional energy sector is particularly important. The United Kingdom's Committee on Climate Change, for example, only considers the incremental capital needs of the building stock to be investments if financed via energy companies (UK CCC, 2010: 145). The IEA calls the change in thinking and structure needed to focus on investments in end-use technologies "a paradigm shift" (IEA, 2008a: 223). This is a long-standing view:

> [Traditional definitions of energy investment] do not include investment in end-use technologies, such as furnaces, appliances and vehicles, because they are traditionally counted as durable consumer goods or business investments. However, the fact that the performance of end-use technologies plays such an important role ... is a strong argument in favor of new approaches to evaluating energy sector investment. (Nakicenovic and Rogner, 1996)

The importance of "diffusion" investments in end-use technologies also warrants a more granular treatment in energy system and climate change mitigation analyses. In particular, like-for-like comparisons of financing requirements for both energy supply and end-use technologies avoid the risk that scenarios relying heavily on efficiency improvements and end-use technology investments appear less costly than scenarios relying heavily on decarbonisation via energy supply technology investments. This appearance will be an artefact of the way total investments in energy supply technologies are compared with incremental investments in end-use technologies. This is straightforwardly misleading. In other cases, end-use technology investments are not estimated in any form (see Table 23.1). If *total* investments for both energy supply and end-use technologies cannot be meaningfully compared, then *incremental* investment definitions should be consistently applied (see Luderer et al., 2009 for an example). But even this is unsatisfactory. Informed policy choices and investment frameworks should explicitly consider energy end-use and efficiency options and should account for associated total investment needs in their entirety. Otherwise, they will be structurally biased in favour of the energy supply.

FURTHER READING

Erhardht-Martinez and Laitner (2008) provide one of the few available syntheses of end-use investments although using a different methodology and for the United States only. Nakicenovic and Rogner's (1996) early work establishes the conventional supply emphasis of energy investment analysis.

REFERENCES

Bazilian, M., Nussbaumer, P., Haites, E., Levi, M., Howells, M., & Yumkella, K. K., 2010. Understanding the Scale of Investment for Universal Energy Access. *Geopolitics of Energy*, 32(10–11): 21–42.

Cofala, J., Purohit, P., Rafaj, P., & Klimont, Z., 2008. GHG Mitigation Potentials and Costs from Energy Use and Industrial Sources in Annex I Countries – Methodology. Laxenburg, Austria: International Institute for Applied Systems Analysis (IIASA).

Edenhofer, O., Knopf, B., Barker, T., Baumstark, L., Bellevrat, E., Chateau, B., Criqui, P., Isaac, M., Kitous, A., Kypreos, S., Leimbach, M., Lessmann, K., Magné, B., Scrieciu, S., Turton, H., & Van Vuuren, D., 2010. The Economics of Low Stabilization: Model Comparison of Mitigation Strategies and Costs. *The Energy Journal*, 31(Special Issue: The Economics of Low Stabilization): 11–48.

Ehrhardt-Martinez, K. & Laitner, J. A., 2008. *The Size of the U.S. Energy Efficiency Market: Generating a More Complete Picture*. Washington, DC: American Council for an Energy Efficient Economy (ACEEE).

Enkvist, P.-A., Naucler, T., & Rosander, J., 2007. A Cost Curve for Greenhouse Gas Reduction. *The McKinsey Quarterly*, (1).

Hanson, D. A. & Laitner, J. A. S., 2006. Technology Policy and World Greenhouse Gas Emissions in the AMIGA Modeling System. *Energy Journal*, 27(Special Issue: Multi-Greenhouse Gas and Climate Policy): 355–71.

IEA, 2003. *World Energy Investment Outlook*. Paris: International Energy Agency.

2008a. *Energy Technology Perspectives 2008 – Scenarios and Strategies to 2050*. Paris: Organization for Economic Cooperation & Development.

2008b. *World Energy Outlook*. Paris: International Energy Agency.

2009. *World Energy Outlook*. Paris: International Energy Agency.

Levine, M., Ürge-Vorsatz, D., Blok, K., Geng, L., Harvey, D., Lang, S., Levermore, G., Mongameli Mehlwana, A., Mirasgedis, S., Novikova, A., Rilling, J., & Yoshino, H., 2007. Residential and Commercial Buildings. In B. Metz, O. R., Davidson, P. R., Bosch, R. Dave, & L. A. Meyer (eds.) *Climate Change 2007: Mitigation. Contribution of Working Group III to the Fourth Assessment Report of the Intergovernmental Panel on Climate Change.* Cambridge, UK and New York: Cambridge University Press, 387–446.

Luderer, G., Bosetti, V., Steckel, J., Waisman, H., Bauer, N., Decian, E., Leimbach, M., Sassi, O., & Tavoni, M., 2009. *The Economics of Decarbonization – Results from the RECIPE Model Intercomparison: RECIPE Background Paper*. Potsdam, Germany: Potsdam Institute.

McKinsey, 2009. *Pathways to a Low-Carbon Economy. Version 2 of the Global Greenhouse Gas Abatement Cost Curve*. London: McKinsey and Company.

Murmann, J. P. & Frenken, K., 2006. Toward a Systematic Framework for Research on Dominant Designs, Technological Innovations, and Industrial Change. *Research Policy*, 35(7): 925–52.

Nakicenovic, N., 1986. The Automobile Road to Technological Change: Diffusion of the Automobile as a Process of Technological Substitution. *Technological Forecasting and Social Change*, 29(4): 309–40.

Nakicenovic, N. & Rogner, H. H., 1996. Financing Global Energy Perspectives to 2050. *OPEC Review*, 20(1): 1–23.

Riahi, K., Dentener, F., Gielen, D., Grubler, A., Jewell, J., Klimont, Z., Krey, V., McCollum, D., Pachauri, S., Rao, S., Van Ruijven, B., Van Vuuren, D. P., & Wilson, C., 2012. Chapter 17 – Energy Pathways for Sustainable Development. *Global Energy Assessment – Toward a Sustainable Future*. Cambridge, UK and New York: Cambridge University Press and Laxenburg, Austria: The International Institute for Applied Systems Analysis.

Riahi, K., Grubler, A., & Nakicenovic, N., 2007. Scenarios of Long-Term Socio-Economic and Environmental Development under Climate Stabilization. *Technological Forecasting and Social Change*, 74(7): 887–935.

Stern, N., 2006. *The Stern Review on the Economics of Climate Change*. Cambridge, UK: Cambridge University Press.

UK CCC, 2010. *The Fourth Carbon Budget: Reducing Emissions through the 2020s*. London: UK Committee on Climate Change.

UNFCCC, 2007. *Investment and Financial Flows to Address Climate Change*. Bonn, Germany: UN Framework Convention on Climate Change.

Ürge-Vorsatz, D. & Metz, B., 2009. Energy Efficiency: How Far Does It Get Us in Controlling Climate Change? *Energy Efficiency*, 2(2): 87–94.

Van Vuuren, D., Hoogwijk, M., Barker, T., Riahi, K., Boeters, S., Chateau, J., Scrieciu, S., Van Vliet, J., Masui, T., Blok, K., Blomen, E., & Krama, T., 2009. Comparison of Top-Down and Bottom-Up Estimates of Sectoral and Regional Greenhouse Gas Emission Reduction Potentials. *Energy Policy*, 37: 5125–39.

WEC, 1993. *Energy for Tomorrow's World: The Realities, the Real Options and the Agenda for Achievements*. London: Kogan Page.

PART VII

CONCLUSIONS

24 Lessons Learnt from the Energy Technology Innovation System

Charlie Wilson and Arnulf Grubler

1 Synthesising the Case Studies

The twenty case studies examined in this volume cover a wide range of innovation histories for both energy supply and energy end-use technologies of markedly different characteristics and vintages. The market and policy contexts in which these innovation histories unfold also differ hugely. Yet as the editors' guides in each case study illustrate, each and every one of these innovation histories can be interpreted and understood in terms of the energy technology innovation system that we set out in detail in the introductory chapters (reproduced here as Figure 24.1). Our aim is that this systemic perspective serves as an analytical framework with the necessary generality to apply across all contexts and forms of energy technology innovation.

The remaining two chapters offer a final synthesis to illustrate the energy technology innovation system in practice, and confirm its usefulness in analysing energy-related innovation and informing innovation policy.

In this chapter, examples are drawn from throughout the collection of case studies to validate the various elements of the energy technology innovation system. This is organised using the four main dimensions of the innovation system – knowledge, adoption and use, actors and institutions, and resources – with the interrelationships and dependencies between these dimensions drawn out.

The second and final concluding chapter extracts and synthesises the principal lessons learned for technology and innovation policy. These take the form of principles, guidelines, and approaches – not recommendations for specific policy instruments whose effectiveness and appropriateness will be strongly context dependent. Based on evidence in the case studies, the central argument of this final chapter is that innovation policies and activities that enable or support successfully functioning innovation systems are more likely to produce innovation successes than partial, fragmented, narrowly focussed efforts.

2 Patterns and Linkages in the Energy Technology Innovation System

The first two case studies of the book provide an overall perspective on innovation and technological change in the energy system as a whole from a long-term perspective. The case study on energy transitions reviews patterns, drivers, and dynamics of energy systems, both historically and in future scenarios. A key finding is that

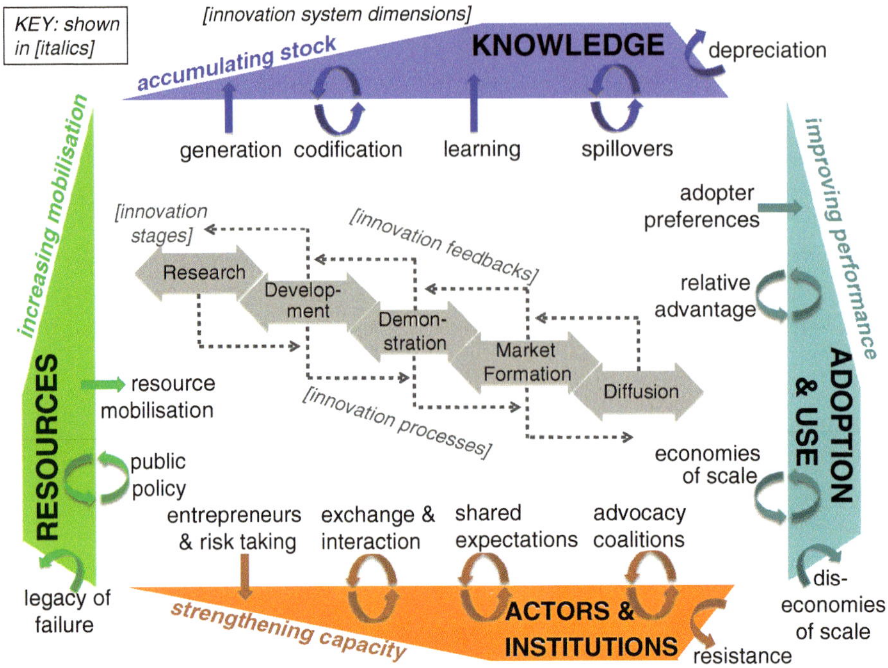

Figure 24.1. The energy technology innovation system.

knowledge spillovers and interdependence between innovations create amplifying effects as well as inertia and path dependence. This implies slow rates of change at a systemic level, with future innovation trajectories shaped and constrained by historical developments. These long time constants of change are further evidenced in the second case study on technology diffusion.

The characteristics of change in the energy system historically result from the interplay between different elements of the innovation system:

- the generation, exchange, and transfer (or spillover) of knowledge between actors and applications;
- the shifting preferences of technology adopters first for the performance advantages and later for the cost advantages of the energy services provided by energy supply and end-use technologies;
- the institutional capacity needed to support standardisation, mass production, up-scaling (to the extent scale economies at the unit level are available), and broad-based industrial growth;
- and the magnitude of resource mobilisation implied by the accumulation of knowledge stocks, the strengthening capacity of actors and institutions for sustained improvements in technological performance driving ever greater adoption and use, and the investments needed to adopt technologies and create an industrial base for their production.

The opening case study on energy transitions thus exemplifies these four dimensions of the energy technology innovation system as well as the important interplay between them.

The case study on technology diffusion focuses more immediately on the adoption and use of technologies. The diffusion patterns observed historically again point to many of the different elements of the innovation systems framework and their interactions. Knowledge is generated through a process of experimentation and learning with many small-scale units during the market formation stage of a new technology. This contributes to improving performance, strengthening the capacity of actors and institutions to support subsequent widespread diffusion, and capturing available economies of scale at the level of individual technologies (unit economies of scale) as well as their industries (industry or manufacturing economies of scale). Knowledge can also spill over from early to later adopting markets, but this is a not a free lunch. Rates of change trade off against extents of change: the faster the diffusion, the less pervasive it is. Time is *not* a substitutable resource input.

The case study on assessment metrics also emphasises the interrelationships between the elements of the innovation system. Although input metrics predominantly describe the financial resources mobilised as investments or expenditures for knowledge generation (particularly at the R&D stage), resource inputs are also integral to the strengthening of actors' capacity to sustain innovation. Although output metrics predominantly describe the widespread adoption and use of technologies, patents and publications serve as proxies for the codification of knowledge embodied in new technologies, and meetings and movements of skilled innovators serve as proxies for the exchange of knowledge among actors. Some metrics explicitly try to link different elements of the innovation system:

- learning curves link deployment experience (and associated learning processes) to performance improvements (typically in terms of lower costs);
- social returns on investments link resources mobilised with the broader economic and environmental outcomes of technology development and adoption;
- modelling studies link technology-push and market-pull drivers (e.g., R&D investments and demand subsidies) to assess the relative influence of each on specific innovation outputs.

But no single metric can cover all the elements of the innovation system, and so all metrics inevitably offer a partial view. The innovation systems perspective warns against the use of reduced form metrics – such as learning curves and derived learning rates – to characterise innovation. This theme is picked up in subsequent case studies and summarised in the section on knowledge later in this chapter. Again the central message is that the different elements of the innovation system interact, and so understanding innovation outcomes means understanding these relationships in greater detail.

The fourth case study on technology portfolios addresses the irreducible uncertainty of innovation outcomes by reviewing the tools and approaches that can be used to insure innovation investments and expectations against the possibility of failure. Analysis of optimal technology portfolios invariably finds that diversity wins out over a reliance on silver bullets (or one-shot, large-scale technological solutions). A diverse portfolio of innovation investments in turn signals the importance of entrepreneurial risk taking, variety, and experimentation in the early formative phase of a new technology, learning processes to sustain performance improvements through commercial deployment, and support for and protection of niche markets by public policy. Just as with the case studies on energy transitions, technology diffusion, and assessment metrics, the case study on technology portfolios brings to life all four dimensions of the innovation system (see Figure 24.1).

Taken together, this first set of case studies substantiates the energy technology innovation system introduced at the beginning of the book. They also bring out some of its signature characteristics. These emerge repeatedly through the case studies and are worth noting from the outset: interdependence, uncertainty, and complexity.

Interdependence means that different elements of the energy technology innovation system mutually influence one another. The strength and direction of these influences may change.

Uncertainty most strongly characterises the outcomes of the innovation process. Such uncertainties are irreducible – technology A is not ensured successful diffusion if recipe B is followed.

Complexity results from the increasing interdependencies in energy systems as well as from the number and variety of variables and mechanisms of energy technology innovation and their shifting forward and backward linkages. Context dependence further adds to this complexity. For this reason the case studies in this book cover a wide range of contexts, technological (end use and supply) as well as geographical and developmental (the United States, Europe, Japan, China, Brazil, Kenya). This addresses a limitation in the traditional innovation systems literature with its principal focus on novel energy supply technologies in a Northern European context.

3 Knowledge in the Energy Technology Innovation System

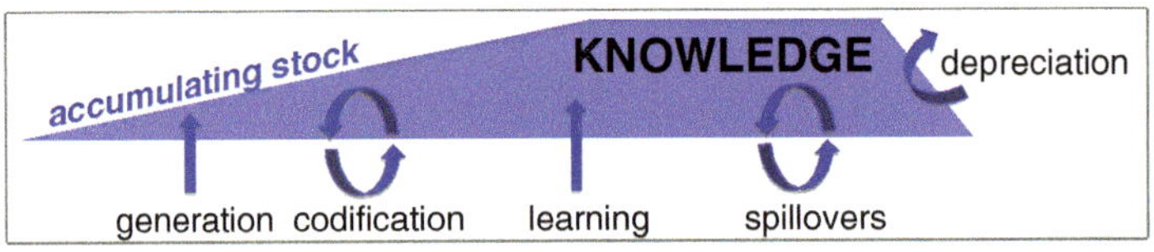

The key processes associated with the *knowledge* dimension of the energy technology innovation system are most clearly illustrated by the case studies summarised in Table 24.1.

3.1 Knowledge Generation, Codification, ... and Depreciation

The case study on solar water heaters illustrates many of the processes associated with the knowledge dimension of the innovation system. Rapid growth in solar water heater installations in the United States in the early 1980s led to learning improvements in the performance of the technology, particularly in terms of its reliability. These improvements and associated R&D or knowledge-generation activities were codified in the form of patents and research papers. This codified knowledge helped sustain some of the accumulated knowledge stock through a turbulent period during which many firms and skilled personnel exited the industry. A similarly important role played by knowledge codification is highlighted in the case study on solar thermal electricity, which further showed how documented cost improvement efforts facilitated the exchange and spillover of knowledge from

Table 24.1. Case studies emphasising knowledge in the energy technology innovation system.

Short name	Summary description	Key points
Solar Water Heaters	Early success and later failure of the solar water heater industry, particularly in the United States.	Lasting legacy of industry failure, including knowledge depreciation. Lack of robust linkages and knowledge exchange between innovation system actors. Importance of codification.
Heat Pumps	Different stages of heat pump diffusion in Sweden and Switzerland, emphasising the role of public policy.	Networks of actors sustaining knowledge exchange and interaction during industry downturn. Interactions between supply of and demand for innovation. Importance of policy consistency and comprehensiveness supporting R&D and market formation.
Knowledge Depreciation	Loss or obsolescence of knowledge, with examples in the context of energy innovation.	Importance of stable, sustained inputs to knowledge generation. High potential rates of knowledge depreciation in energy technologies, particularly related to learning improvements.
Nuclear Power (France)	Review of pressurised water reactor programme in France, including cost escalation.	Interaction between learning effects and market institutions, including standardisation and regulatory stability. Limitations of learning curve model to explain cost reductions.

pioneer firms in the Californian market to later adopters elsewhere in the United States and abroad.

However, discontinuities in market activity can result in the depreciation of knowledge. The case study on solar water heaters shows that the tacit, practical knowledge of the solar water heater installers was particularly vulnerable to knowledge depreciation. The case study on knowledge depreciation explains in detail why and how knowledge stocks can shrink as well as grow. This is important, as innovation studies typically emphasise knowledge generation (R&D) and learning as sources of technological improvement. But the knowledge dimension of the energy technology innovation system is not a one-sided story. Drawing on evidence from the early boom years of the Californian wind industry, the case study highlights how knowledge gained through learning by doing is particularly susceptible to rapid depreciation given its reliance on tacit (non-codified) knowledge acquired and held by skilled workers and managers who can retire, move, or leave, particularly under inconsistent or stop-start public policy support.

3.2 Knowledge Exchange and Spillovers

Knowledge is not a disembodied construct. Most evidently, knowledge becomes embodied in technological artefacts. But it is also inseparable from the actors and institutions involved in the innovation system, and the financial and other resources mobilised to build and sustain this involvement. Relevant actors range from large organisations down to managers or individual employees. Relevant institutions are those that facilitate exchange and interaction within the innovation system, processes that support the maintenance and sharing of knowledge. In the case of solar water heaters in the United States, boom-and-bust markets and depreciated knowledge stocks ultimately led to a swathe of poor-quality installations that left a lasting legacy of failure surrounding the industry. Why? One reason was the lack of robust linkages between different groups of actors – particularly technicians and researchers on one hand and installers operating in the market on the other. The necessary exchange and interaction between the developers and users of the technology was insufficient to maintain a strong base of support as public resources dried up. Some of these lessons on how to sustain accumulated knowledge stocks have been applied in various European countries, as well as China. In these later markets, the installed capacity base of solar water heaters is steadily growing, supported by a suite of policy measures that target different innovation processes together, creating strong actor networks and the buildup and recharge of technological knowledge.

The case study on wind power argues that knowledge spillovers have been important sources of innovation and design improvements both between industries (materials, aerodynamic simulations, and designs) and between manufacturers (facilitated by testing stations). Denmark's success in developing wind power involved not just R&D but also the support for feedback between users and producers of wind turbines (Garud and Karnoe, 2003). This positive feedback process, based upon inter- and intra-firm knowledge spillovers where learning and absorptive capacity production are mutually reinforced, can potentially augment an industry's intensity of knowledge production (Foray, 2004). The case study on solar photovoltaics gives further evidence for the benefits of inter-industry knowledge spillovers.

3.3 Linkages to Actors and Institutions

The case study on heat pumps also highlights the important links between knowledge-related processes and the actors and institutions involved in the innovation system. In the early stages of Swedish and Swiss heat pump developments, the capacity of a diverse range of actors was built and strengthened through conferences, training, and testing activities supported by public policy. This capacity subsequently proved critical for sustaining the innovation system through the period of declining energy prices in the 1980s: the opposite of the story in the U.S. solar water heater market. In the heat pump case, the constellation of developers, installers, and users of the heat pump systems continued to exchange knowledge and were able to respond to early technical difficulties by instigating a testing and certification process to ensure reliability and credibility. Knowledge generated through these evaluations was commonly codified and shared, in turn supporting knowledge spillovers within the industry.

Public policies also played an important role in nurturing the generation and exchange of knowledge among the innovation actors. Sweden used a procurement programme to link manufacturers and end users. A Swiss federal agency used its strategic position to coordinate innovation and deployment activities among both private and public actors. These network-strengthening and coordination efforts helped build up shared expectations around the market potential for the technology, sustaining confidence and facilitating a continued mobilisation of resources as the markets grew.

The case study on nuclear power in France describes a very different technological and market context but shares a very similar central finding: successful innovation outcomes depend on the linkages between knowledge-related processes and the actors and institutions in the innovation system. The successful scale-up of nuclear power in France through the 1970s benefitted from some peculiar characteristics including a small number of key state and industrial actors who were tightly networked, who shared clear expectations around the technology's deployment, and who were able to mobilise substantial resources supported by clear and stable public policy. The special role of ÉDF in acting both as principal and agent in reactor construction also stands out as a unique actor and institutional configuration within an innovation system. Knowledge spillovers from earlier U.S. experiences were also substantial, achieved via licensing of a proven Westinghouse design that was built out in a programme of standardised plant construction and operated with low and stable costs.

But the French nuclear case also illustrates the consequences of foregoing accumulated knowledge in favour of bespoke "Frenchified" designs and ever larger reactor sizes in an industry whose hitherto aligned actors were increasingly fragmenting. In the later years of the French nuclear programme, costs began to rise, construction times lengthened, and diseconomies of scale with this large, capital-intensive, and complex technology began to emerge.

3.4 Learning

The case study on nuclear power in France and the case study on heat pumps in Sweden and Switzerland also shine very different lights on the nature and consequence of learning processes in the innovation system. Various case studies, including the case study on assessment metrics, noted the attractiveness of the learning curve model linking deployment experience with cost reductions, but also its pitfalls. By isolating a single relationship from diffusion to performance improvements, learning curves ignore the many interdependencies between elements of the innovation system and simplify the contingency of observed cost reductions on diffusion contexts.

The prevalence of significant unit-scale economies was also repeatedly emphasised in the case studies, including those on wind power, solar thermal electricity, nuclear reactors, and ethanol distilleries. By using cumulative installed capacity or output, much of the existing learning curve literature not only conflates economies of scale with learning effects, but also misses the locus of knowledge generation: the actual project being built or the concrete process or product being adopted and experimented with. A simple analytical remedy would be to analyse learning curves per cumulative *numbers of units* rather than capacity or output. Applied prospectively to novel energy technologies (particularly renewables), the learning curve model also negates the possibility of negative learning or cost escalation associated with diseconomies of scale, knowledge depreciation, or changing market or technological conditions.

The learning curves from the case studies on nuclear power and heat pumps are shown in Figure 24.2 along with learning curves from other case studies on solar

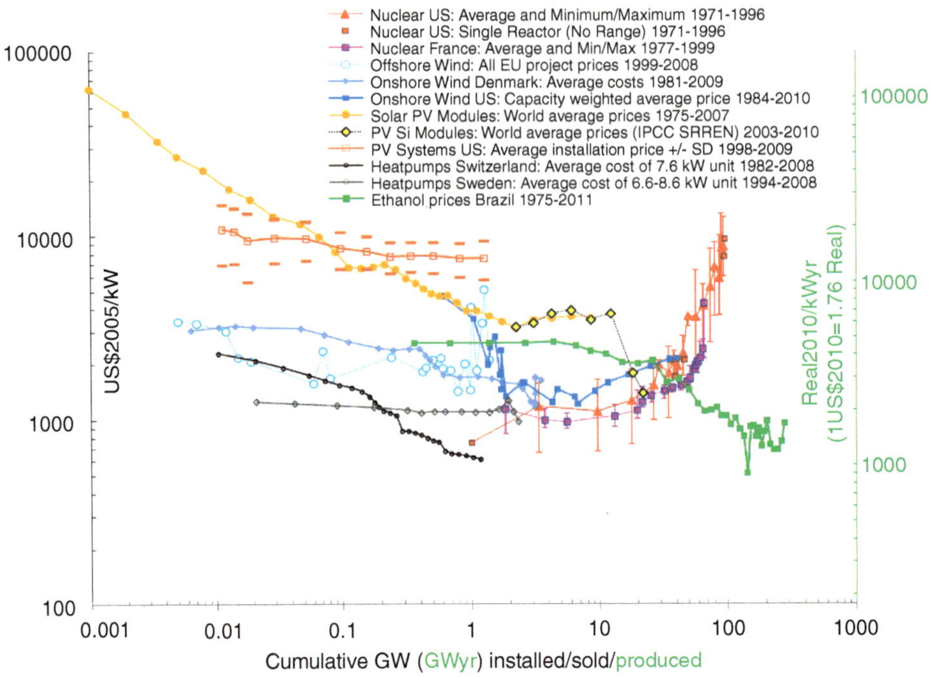

Figure 24.2. Learning curves from five case studies of energy technology innovation.
Notes: Learning curves show US$2005 investment cost per kW installed capacity for nuclear power, on- and offshore wind power, solar photovoltaics and heat pumps (left axis), or production cost per GWyr energy output for ethanol (right axis) versus cumulative GW installed (or GWyr energy output). See respective case studies for details and data sources; wind data from Edenhofer and colleagues (2011), and Junginger, Van Sark, and Faaij (2010). This synthesis figure should not be used for like-for-like economic comparisons as data are not commensurate (e.g., units vs. systems, construction investments vs. operating costs, costs vs. prices, differences in load factors, and so on).

photovoltaics, and ethanol in Brazil using the customary learning curve literature metric of cumulative capacity (or output in the case of ethanol) Additional learning curves for onshore and offshore wind power are also included. A lack of unit capacity data meant cost trends could not be analysed as a function of cumulative numbers of units as suggested previously. (Unit capacities are important additional data that should be reported in the learning curve literature.) Figure 24.2 is a compiled summary figure that illustrates *comparative* cost trends only and is not suitable for like-for-like economic comparisons as data are not commensurate (e.g., units versus systems, construction investments versus operating costs, costs versus prices, differences in load factors, and so on).

The negative learning seen in the later stages of the French nuclear power build out is striking, but not unique (see also the case of U.S. nuclear reactors). Onshore and offshore wind power, as well as solar photovoltaics, show periods of cost inflation or stagnation. In the wind case, this was associated with the (over)ambition of market-pull policies in the face of manufacturing capacity constraints and rising profit margins. These reasons were very different from those characterising the French nuclear case, but had a similar consequence for costs. So although accumulated market deployment experience can *generally* be associated with declining costs (or prices), these learning effects are very context specific and highly variable across technologies and over time.

4 Adoption and Use in the Energy Technology Innovation System

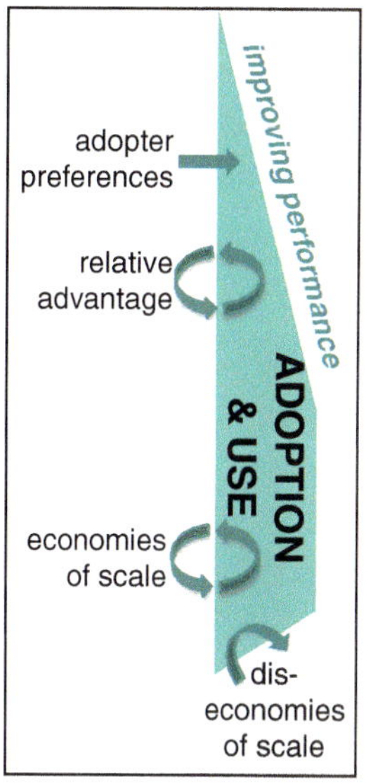

The key processes associated with the *adoption and use* dimension of the energy technology innovation system were most clearly illustrated by the case studies summarised in Table 24.2.

4.1 Relative Advantage and Performance Attributes

It is not coincidental that the group of case studies selected to illustrate the adoption and use dimension of the energy technology innovation system contains two innovation histories relating to cars, one on efficiency improvements, and one on hybrid engines. Of all the energy technologies, cars perhaps best demonstrate that the performance attributes that make up a technology's overall relative advantage are multidimensional. As such, improvements in performance and/or cost can often involve trade-offs. This is seen in the case study on vehicle efficiency, which shows how efficiency gains do not inherently lead to lower fuel consumption, but can rather lead to changes in other technology attributes such as weight or acceleration. The case study on hybrid cars also emphasises the contingency of efficiency improvements on the specific

Table 24.2. Case studies emphasising adoption and use in the energy technology innovation system.

Short Name	Summary Description	Key Points
Solar Thermal Electricity (United States)	Early experience of solar thermal electricity in the United States, and spillovers to later stage production.	R&D and learning interact to support cost reductions. Codification of knowledge supports sharing among firms and spillovers between countries. Large-scale plants carry high investment risks.
Vehicle Efficiency	The "CAFE" standard for vehicle efficiency in the United States and its influence on technological change.	Interaction between policy standards and changing market characteristics, including prices. Variability in policy regimes between countries.
Hybrid Cars	Development of hybrid electric vehicles in Japan, the United States, and China, emphasising role of public policy.	Importance of market demand and end user preferences for driving innovation. Portfolio approaches and adaptive policies to avoid pitfalls of picking innovation winners. Importance of market formation activities.
Solar Photovoltaics	Development of solar photovoltaics in different markets worldwide, focussing on drivers of cost reduction.	Long-term R&D support complemented by market formation activities in multiple niches to stimulate commercial learning.

configuration of the hybrid engine system. Indeed the technology development programme in the United States recognised that one attribute of relative advantage (fuel efficiency) should not substitute for other attributes (performance, size, safety). The later success of the larger, more powerful, second generation of hybrid cars in the U.S. market supports this end-user perspective.

The case study on vehicle efficiency shows how performance improvements are shaped by innovation responses to consumer preferences and market demand, but also by public policy in the form of fuel efficiency standards that enshrine shared expectations as to the direction and rates of required technological change. Policy also acts directly on costs through subsidies and fuel prices (e.g., via taxes), and indirectly on consumer preferences through labelling, disclosure requirements, information campaigns, and differential tax schemes to help form and develop markets for fuel-efficient vehicles.

In the case study on ethanol in Brazil, it was the performance and economic disadvantages faced by the initial wave of ethanol-only vehicle owners in the 1980s that placed the adoption and use of end-use technologies centre stage in the renewed innovation efforts of the 2000s to develop flex-fuel vehicles. The relative advantage of these vehicles to end users was enhanced by purchase incentives to ensure they were cost competitive and by a preferential tax regime that ensured ethanol remained cost attractive as a fuel.

4.2 Adopter Preferences – Heterogeneity and Diversity

The characteristics, heterogeneity, and complexity of adopter preferences prove consistently important for understanding innovation outcomes. The case study on energy transitions shows that although mass market adopters are persuaded by better performance or lower cost, early adopters are attracted to new energy services or distinct performance advantages beyond costs. These relatively price-insensitive early adopters play an important role in helping the technology consolidate its position in niche markets, reducing adoption uncertainty and building market demand among the mass of more risk-averse adopters. This greater clarity and confidence in expected market potential underwrites the necessary resources needed to be mobilised for industrial-scale production and deployment.

This general pattern is identical to the specific case of hybrid cars. The adoption and use of hybrids by early adopters in the United States and Japan helped reduce perceived risks with the technology and seeded the effects of social influence and imitation that played an ever-stronger role as the technology diffused. The case study on hybrid cars also shows how end users were not only responsive to increased fuel prices and fiscal incentives, but also to intrinsic values and beliefs regarding the environment and energy security.

The diversity of adopter preferences and the difficultics with anticipating them have a further implication for technology developers: diversity and choice, rather than singular visions of narrowly specified technologies. This was clear in the case study on hybrid cars as early bets on hydrogen-fuelled and battery electric vehicles as the alternative vehicle of choice foundered, at least in part because of adverse consumer preferences on top of unresolved technological challenges.

The case study on solar photovoltaics draws out the importance of successive market niches from powering satellites, to powering electronic gadgets, to powering off-grid homes, each with its distinctive set of adopter preferences for particular performance (or cost) characteristics. These market niches provided the context for learning processes in commercial settings that helped further improve the technology's relative advantage for a wider and more cost-sensitive set of potential adopters. In this way, the diversity of end-user preferences links to the role of experimentation in multiple market niches during the early deployment of a technology.

4.3 Linkages to Learning

The adoption and use of a technology drives the learning processes that help improve performance and cost and so sharpen the relative advantage of the technology over its competitors. But these learning processes do not occur in isolation. Many of the case studies show how learning and other knowledge-generation activities are complements, not substitutes. The case study on vehicle efficiency shows how diesel engines were rapidly and pervasively adopted under appropriately aligned and supportive regulatory and market conditions. But this concentrated period of improving performance in a market context was founded on lengthy earlier R&D, testing, and market formation activities.

The case study on solar photovoltaics analyses how R&D, economies of scale, learning by doing, and knowledge spillovers from other technologies have all played a role in reducing system costs. Similarly in the case study on solar thermal electricity, R&D activities and learning processes associated with deployment experience interact to produce the substantial decline over time in the operating costs of solar thermal systems. More resources mobilised in the form of R&D investments could not have substituted for the applied experience or learning by doing associated with installing and operating solar thermal electric facilities in market settings. This conclusion further cautions against the use of oversimplified innovation models in which these different forms of learning are treated as substitutable resource inputs into an aggregate production function (as done in so-called two-factor learning curve models). These interactions were also concentrated in the market formation stage during which a combination of investment subsidies, tax credits, and guaranteed returns (through power purchase contracts) provided sufficient market-pull to complement the technology-push effects of the R&D programmes.

4.4 Economies of Scale – Unit Level and Industry Level

Another important element associated with the adoption and use of technologies is scale. The case study on solar photovoltaics provides a good example of how cost reductions associated with manufacturing scale economies and cost reductions driven by learning processes can be conflated. Econometric or bottom-up engineering-type models are needed to disentangle these different influences on cost improvements.

The case study on technology diffusion provides an overview picture of up-scaling patterns observed in various energy technologies. The case studies on wind power, solar thermal electricity, and nuclear power all give more specific examples of how up-scaling can harness available economies of scale at the unit level, or in the case of large, complex technologies like nuclear power operating near the scale frontier, face potential diseconomies of scale.

Distributed, small-scale energy conversion and end-use technologies are more likely to be characterised by manufacturing scale economies. In the case study on solar photovoltaics, economies of scale associated with a growing manufacturing base explain almost half of the substantial cost reductions observed in the United States over the period from 1980 to 2001. In countries like Japan and Germany, the clear and long-term confidence in future market demand that enabled large investments in manufacturing facilities to bring down costs through economies of scale rested in no small part on stable, multiyear, demand-side policy programmes. These policies also yielded important international spillover effects. Chinese PV manufacturers were initially less incentivised by local PV demand, but by perceived (and subsequently realised) export potentials, enabling them to capture significant manufacturing scale economies. Reaping scale economies is therefore contingent on robust shared expectations for future markets, a classic market-pull driver associated with the adoption and use of a technology.

In the case study on hybrid cars, differences between the Japanese and Chinese markets are in part attributable to industry-level economies of scale. The cost premium of hybrids in Japan were supported by the use of a feebate scheme, but costs were already sufficiently competitive, based in part in the large-scale manufacturing effort in response to clear expectations about market demand, both domestically and in the United States. By comparison, the lack of available scale economies in China with its smaller domestic markets, trade restrictions, and lack of component standardisation meant the cost barriers remained prohibitively high from consumers' perspective.

Other examples of manufacturing scale economies are included in the case studies on heat pumps and ethanol in Brazil. In the ethanol case, this was linked to particular characteristics of the actors in the innovation system. A high degree of concentration in the sugar and ethanol market among a few large-scale sugar producers enabled scale economies to be readily captured as production expanded. Capital was available, knowledge could readily spill over, and – as in the solar photovoltaics and hybrid car cases – potential markets were clear.

5 Actors and Institutions in the Energy Technology Innovation System

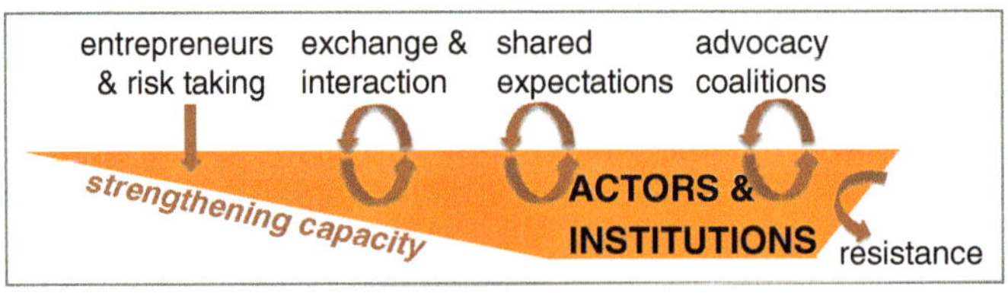

The key processes associated with the *actors and institutions* dimension of the energy technology innovation system are most clearly illustrated by the case studies summarised in Table 24.3.

Table 24.3. Case studies emphasising actors and institutions in the energy technology innovation system.

Short Name	Summary Description	Key Points
Wind Power	Successive innovation strategies in early and subsequent wind power markets worldwide, with differing roles played by public policy.	Need to integrate RD&D support with market formation activities. Importance of interaction and feedback between innovation system actors.
End-Use Efficiency (Japan)	Review and evaluation of the Top Runner Program of energy efficiency standards for consumer products and vehicles in Japan.	Adaptive policies tailored to market conditions creating dynamic incentives within a clear overall strategic direction.
Rural Solar (Kenya)	Market dynamics and product quality issues in the Kenyan market for household solar photovoltaic (PV) systems.	Local institutions need to set and enforce standards for quality control and assurance. Enduring market and regulatory institutions needed for quality assurance.
Synfuels (United States)	History of U.S. government investment in synthetic fuels production as oil substitute, and failure to meet programme targets.	Overexuberant expectations exposed by changing market conditions. Contingency of production incentives on contextual factors. Differing public-private roles in innovation system.

5.1 Actors and Institutions

Denmark's success in developing a world-leading wind industry is one of the clearest recent examples of a systemic approach to energy technology innovation. As the case study on wind power shows, this centred on the generation and application of technological knowledge incentivised by both technology-push and market-pull policies. But the systemic approach to wind power in Denmark also embraced a diverse range of actors including the end users of the technology (developers, farmers, and landowners, as well as manufacturers, utilities, technical experts, quality assurers, and regulators), and a diverse set of institutions that facilitated both technological and market developments (testing stations, wind resource maps, grid connection regulations, municipal planning, and permitting). Networks and relationships among these actors were further strengthened by information-based activities as well as more structured opportunities for exchange and interaction (e.g., at the Risø test station). A similar institutional configuration in the Netherlands was not as successful, as the competitive market setup limited the incentives for mutually productive knowledge sharing and spillovers between competitive actors.

5.2 Shared Expectations, Exchange, and Interaction

Innovation actors' shared expectations of a technology's future potential has already been mentioned as an important ingredient supporting the growth in manufacturing, up-scaling, and resource mobilisation as innovation systems strengthen. The case study on energy efficiency in Japan shows how a stable, long-term policy framework can reinforce these shared expectations. In the Japanese case, the Top Runner standards provided a clear direction for end-use product development aiming at higher energy efficiency and the elimination of low-efficiency variants from the market. Shared expectations among manufacturers, policy makers, and end users helped underwrite innovation investments on the part of technology developers by providing clear and transparent confidence in future market demand.

Shared expectations were further reinforced by the multilateral public policy process through which the standards were set. This process centred on informal networks to enable exchange and interaction among both public sector actors (national and local governments and academics) and private actors (industry, consumer goods manufacturers, and media). The small number of domestic manufacturers, each with high technological competencies, also meant a strong compliance capacity.

Exchange and interaction among actors in the Kenyan PV market has also proven important. As the case study on rural solar in Kenya shows, reliance on foreign product testing and certification weakened potential market feedbacks between suppliers and end users, and contributed to the quality control problems that undermined initial market growth. But opportunities for exchange and interaction between local innovation system actors have strengthened more recently through discussions coordinated by the Kenyan Renewable Energy Association and involving local industry and module importers, as well as nonmarket actors, including academics and researchers. This advocacy coalition that formed behind the need for credible quality assurance ultimately led to the removal of poor-performing modules from the market.

5.3 Quality Assurance, Advocacy Coalitions, and Resistance

In the Kenyan PV market, the recurrent emergence of quality problems emphasises the importance of enduring regulatory and market institutions, as one-time testing efforts or the improvement of individual low-performing brands do not constitute lasting solutions. Product quality was eventually achieved through a combination of public policy, including standards, and market institutions, including quality assurance, testing, certification, and supplier credibility. Standards were matched to adopter preferences not just for maximum rated power output but also durability and longevity, and benefitted by knowledge spillovers by drawing on international precedents.

Quality assurance, such as certification schemes, are an important market institution to build confidence and shared expectations around a technology's potential. As was seen earlier in the case study on solar water heaters, reliability and technical problems with new technologies risk undermining investor and end-user confidence and so creating resistance to further diffusion. The case study on wind power emphasises the importance of the turbine certification scheme and the Risø testing station in the early Danish market. In the U.S. case, the lack of operational testing ultimately led to technical problems that undermined investor confidence.

The case study on synfuels in the United States also shows how technical performance issues can undermine market confidence and breed resistance to a new technology. The experience of the U.S. Synthetic Fuels Corporation in the 1980s makes salient the intertwined influences of shared expectations and advocacy coalitions on successful (and failed) innovation outcomes. The oil crises of the late 1970s established a broad coalition of support for synfuel production, from successive presidents and policy makers (including fiscal conservatives) to technology developers, energy experts, and industrialists. The breadth of this coalition, and the strength of its expectations for the technology, led to an aggressive and well-resourced programme for rapid commercial deployment and up-scaling, managed by a new form of institution at arm's length from government and designed to respond flexibly and competently to business needs. Although these institutions were undoubtedly influential in moving the technology through its demonstration and market formation stages, they could not sustain further developments alone. Technical difficulties plagued the (overambitious) performance and production levels of many of the sponsored projects. And market-pull deployment incentives in the form of production guarantees were undermined by the collapse in oil prices in the mid 1980s. This led to initial shared expectations being recast as overexuberant and unrealistic, undermining political support and investor confidence, and leading ultimately to the termination of the programme.

5.4 Entrepreneurialism and Risk Taking

End-use adopters need to be confident of new technologies' performance and reliability. As noted earlier in the discussion of adoption and use, entrepreneurial early adopters play an important role in demonstrating the functioning of novel technologies to the more risk-averse market. However, entrepreneurial risk taking is generally associated more closely with technology developers. Shared expectations are an important means of reducing uncertainty and stimulating entrepreneurial activity (Borup, Brown et al., 2006; Van Lente and Rip, 1998). For many technologies, particularly "granular," small-scale, modular, or dispersed technologies, the technology development process and market formation is characterised by variety creation, testing, and experimentation, which need to take place before a dominant design emerges around which future innovation efforts coalesce. Shared expectations help guide the search of actors within the innovation system by selecting preferred alternatives from the variety created by knowledge-generation activities. The case study on wind power shows how entrepreneurial experimentation led to a large variety of wind turbine designs from which the three-blade vertical axis turbine was eventually selected as the dominant design. The case study on hybrid cars shows the importance of entrepreneurial experimentation among Japanese car manufacturers for initiating the buildup of an innovation system to which public policy then had to play catch up. The case studies on energy transitions and technology diffusion provide a more general picture that emphasises the importance of experimentation during the formative phases of many different energy technologies.

5.5 End Users

As discussed earlier with reference to the adoption and use of technologies, end users are pivotal actors within the innovation system. The case study on solar thermal electricity shows how interactions between researchers and firms commercialising and using the technologies created feedbacks so that real-world problems could stimulate improvements in design. The case study on rural solar in Kenya shows how interactions between firms and users are important to communicate how users adapt technologies into their daily practices, thereby revealing potential quality and technology problems that in turn form the basis for further design improvements.

In the case study on energy efficiency in Japan, end users were relatively neglected. It remains unclear how the efficiency improvements interacted with other performance attributes of the products subject to Top Runner standards. This stands in contrast to the case studies on vehicle efficiency and hybrid cars that clearly identify trade-offs between fuel efficiency gains and changes in the weight, acceleration, or other attributes of cars. Working in the Top Runner standards' favour, however, was their application to the average of a product range. This meant manufacturers could continue to offer a differentiated range of products to end users as long as their average efficiency complied with the standards. Heterogeneous adopter preferences could thus be satisfied.

6 Resources in the Energy Technology Innovation System

The key processes associated with the *resources* dimension of the energy technology innovation system are most clearly illustrated by the case studies summarised in Table 24.4.

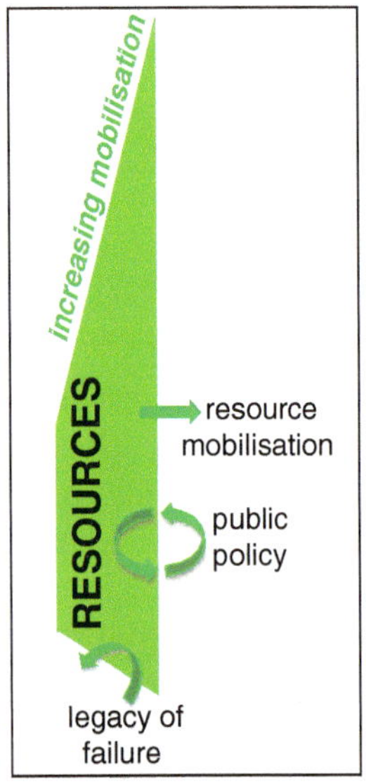

6.1 Resource Inputs

All the dimensions of the innovation system depend on resources, whether financial (investment, expenditure), human (labour, training, skills), or other (time, effort, policy attention, etc.). The importance of the resources dimension has been highlighted already. Knowledge, and its processes of generation, codification, learning, and depreciation, is contingent on resources. As knowledge stocks accumulate in strengthening innovation systems, so too does the magnitude of resources needing to be mobilised. The adoption and use of technologies requires resources in the initial capital investment and associated infrastructures, as well as in the operational phase. The learning processes linking this deployment experience back to knowledge generation activities are mediated by market actors and institutions that also require resources to support and sustain their roles.

The case study on assessment metrics provides an overview of the main indicators used to measure the types and magnitudes of resources mobilised in the innovation system. Labour resources are an important input, and offer a proxy for the tacit knowledge of skilled workers. However, metrics of innovation inputs predominantly describe the financial investments or expenditures that underpin the accumulation of knowledge, the strengthening of actors and institutions, and the improving performance of technologies.

Table 24.4. Case studies emphasising resources in the energy technology innovation system.

Short name	Summary description	Key points
Ethanol (Brazil)	History of ethanol production and developments in automotive technologies in Brazil.	Sustained, continuous mobilisation of resources and public policy support throughout innovation system. Coalitions and shared expectations among innovation system actors.
Global Financial Resources	Comparative assessment of financial resources mobilised globally at innovation, market formation, and diffusion stages for both energy supply and end-use technologies.	Rapidly growing venture capital investments in energy innovation during market formation. Strong energy supply emphasis, particularly nonfossil.
RD&D Investments (Emerging Economies)	Review of public and private RD&D investments in six major emerging economies, disaggregated by technology type.	Increasing importance of emerging economies in global energy technology innovation system. Energy supply technology emphasis. R&D dominated by industry, but with strong government role.
Global End-Use Investments	Bottom-up estimation of current market investments in end-use technologies and their energy-using components.	Dominant position of end-use technologies by share of resources mobilised. Challenges of systemic energy technology innovation analysis.

6.2 Financial Resources – A Global Picture

Financial metrics can focus on early innovation stages (particularly R&D) as data are more readily available and the processes and consequences of knowledge generation activities are clearer. Resulting knowledge stocks can be codified (e.g., as patents or research publications), intangible (e.g., as know-how in engineers or designers), or embodied (e.g., as prototypes or demonstration projects). Public financial inputs for knowledge generation are more widely known in developed country IEA members who regularly collate and report such data. The case study on RD&D in emerging economies synthesises complementary data for the major BRIMCS countries (Brazil, Russia, India, Mexico, China, and South Africa). One of its central findings is that energy RD&D in BRIMCS countries at some $14 billion (in PPP terms) exceeds the entirety of the public energy RD&D budget of all IEA countries combined ($12.7 billion). This highlights the importance of the major developing economies in global innovation efforts, but also reiterates a serious gap in energy RD&D statistics. Private sector energy RD&D data are hardly available, and traditional system boundaries between public and private sectors become increasingly blurred outside the IEA because of the importance of (semi-)state owned energy enterprises in BRIMCS and other developing economies.

The focus of financial metrics on early innovation stages makes it difficult to assess the innovation system as a whole, extending to the adoption and use of technologies in commercial markets. The International Energy Agency as well as some modelling studies do provide estimates of the magnitude of financial resources invested in energy supply infrastructure. The case study on global end-use investments provides novel, analogous, comparative data for energy end-use technologies. All these "diffusion investments" are visible as capital stock additions that embody technological knowledge. Diffusion investments are also linked to learning-related performance improvements associated with commercial experience. The actors involved vary widely across technology types, with households and firms playing a major role in the diffusion of end-use technologies, and governments and energy companies dominating the diffusion of energy supply technologies. Just as the case study on RD&D in emerging economies reveals the importance of the often overlooked non-IEA member countries, so too does the case study on global end-use investments reveal the importance of the always overlooked end-use technology investments. End-use investments plausibly exceed energy supply investments by a factor of around four.

The case studies on RD&D in emerging economies and on global end-use investments fill in missing parts of the picture of financial resources mobilised globally within the energy technology innovation system. The case study on global financial resources paints this picture in its entirety, with the following key findings: (1) the scale of resource mobilisation increases dramatically across successive innovation stages; (2) six major emerging energy economies now account for a significant proportion of global energy innovation investments; (3) the structure of current energy investments is highly asymmetrical, with end-use technologies dominating diffusion investments but energy supply technologies, particularly fossil fuels and nuclear energy, dominating early stage innovation investments. The case studies on assessment metrics and on technology portfolios provide further evidence of the asymmetries (or bias) within current resource flows for energy technology innovation, compounded by the continuing vast amounts of fossil fuel subsidies worldwide. Directed innovation efforts are therefore "pushing" renewable energy-supply technologies to mitigate climate change into a market occupied by heavily subsidised fossil incumbents. Efficient end-use technologies are marginalised throughout (see Figure 24.3).

6.3 Private Actors and Public Policy

Both the magnitude and structure of the US$1–5 trillion resources mobilised annually in the global energy technology innovation system are influenced by market incentives and related public policies. The alignment and consistency of these incentives across a wide range of policies is therefore critical. Many of the case studies draw out the central role of public policy in supporting innovation systems as well as mobilising resources in terms of financial inputs, market institutions, exchange and interaction between innovation system actors, and so on. The role of public policy and guidelines for innovation policy abstracted from the case studies are synthesised in more detail in the next chapter.

Resource mobilisation is strongly linked to the particular institutional configuration of specific innovation systems. The case study of RD&D in emerging economies provides a variety of national examples. In the United States, the federal government and the network of national laboratories are the principal actors, with the more recently established ARPA-E specifically tasked with entrepreneurial risk taking in search of breakthrough technologies. In Russia, the major national energy

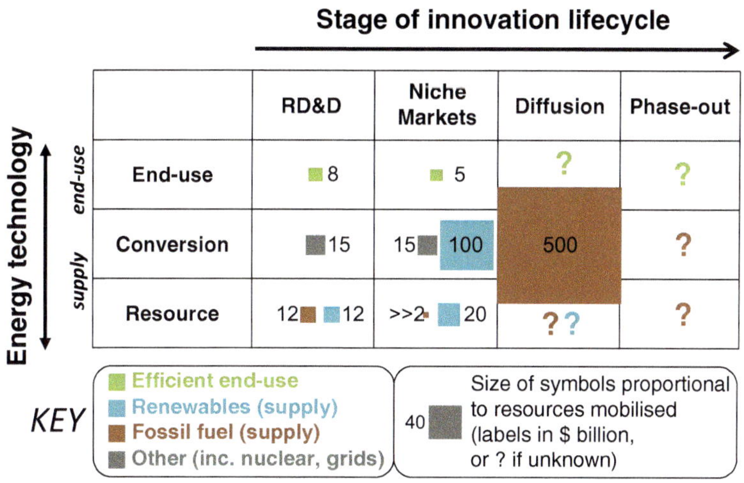

Figure 24.3. Global resource mobilization (billion US$) in 2005 for energy technology innovation by innovation stage and by technology type.
Notes: Energy end use (green); resource extraction – fossil (brown) and renewable (blue); energy conversion – fossil (brown), renewable (blue), nuclear with energy grids and storage (grey). The phase-out stage of the innovation life cycle is included to highlight its importance for capital stock retirement and replacement; however insufficient data exist to provide a quantitative estimate.
Source: Wilson, Grubler, and colleagues (2012).

companies that enjoy significant economies of scale in their respective markets play a lead role in R&D programmes organised in technology streams. In India, an expansive hierarchy of government departments and agencies are involved in organising and directing RD&D investments, in particular to partially state-owned enterprises.

Several case studies assess the changing importance of public and private actors in mobilising resources as innovation systems develop. Private actors (firms, industries, enterprises) play increasingly important roles in the RD&D process, although this depends on the policy and market environments in specific countries and for specific technologies. State-owned enterprises and government mandates for revenue allocation to RD&D play a stronger role in the major developing country RD&D efforts. In the case study on synfuels in the United States, a hybrid approach to public-private roles was ultimately undermined by how mobilised resources were used, given overambitious targets and the technical difficulties of meeting them within an extremely short time span.

Brazil perhaps provides the clearest example of the interrelationship between resource mobilisation, public policy, and the actors and institutions in the innovation system. The central narrative of the case study on ethanol in Brazil is one of sustained resource mobilisation and comprehensive public policy support to grow, maintain, and diversify a major national industry both through periods of relative advantage and through downturns. But the broader institutional configuration of the Brazilian sugar and ethanol markets was integral to this story. Regulatory authority was enshrined in a preexisting market institution that not only managed prices and quantities, but could also arbitrage between sugar and ethanol markets (depending on relative prices). An established cooperative of sugar producers provided a means for exchange and interaction between the main market actors. Knowledge generation associated with R&D investments was managed through a public-private partnership between agricultural researchers and sugar mill / ethanol plant owners. As the market matured, these various institutions extended to coordinate the engagement of all the main industry stakeholders, from growers and processors, to distributors (Petrobras) and end-use (auto) manufacturers. This ensured the interests of potentially disparate actors were aligned behind ethanol innovation efforts. Market concentration among a few large-scale sugar producers enabled economies of scale to be readily captured as production expanded (capital was available, potential market returns were clear, and knowledge could readily spill over in the spatially concentrated sugar industry). Similarly, the concentration of infrastructure requirements (ethanol-only distribution network and filling pumps) in the hands of a monopoly provider (Petrobras) working under clear market incentives tempered the resistance to change characteristic of "locked-in" energy systems.

6.4 Legacy of Failure

As with many of the other technology-specific case studies, ethanol in Brazil has not been without its setbacks. The technical and economic disadvantages faced by the initial wave of ethanol-only vehicle owners led to concerted innovation investments in a new generation of flex-fuel vehicles that removed end-users' exposure to uncertain relative prices of ethanol and gasoline. Purchase incentives as well as a preferential fuel tax regime for ethanol helped overcome the legacy of failure and mistrust associated with ethanol-only vehicles whose market had collapsed once oil prices had fallen.

The legacy of innovation failure can be deep and long-lasting. The case study on synfuels in the United States discusses how the cancellation of the synfuels programme has affected the U.S. government's broader appetite for supporting large-scale demonstration projects, and the R&D effort into thermochemical synfuel production more specifically. The case study on solar water heaters in the United States shows how a combination of poor-quality installations and boom-and-bust market subsidies left a lasting legacy of failure leading to the depreciation of the tacit, practical knowledge of the installers as the market collapsed. The case study on solar thermal electricity in the United States similarly illustrates the risks associated with policy intermittency, which can lead to the exit of key actors and the undermining of market confidence on an enduring basis.

7 Limitations of the Energy Technology Innovation System Framework

The introductory chapters, the editors' guides accompanying each case study, and this concluding chapter drawing together the evidence from the case studies are all designed to substantiate our view of the energy technology innovation system. We have argued that this systemic perspective is useful for understanding and comparing innovation histories, for analysing the reasons why some innovation efforts succeed whereas others fail, and for designing effectively functioning innovation systems oriented towards climate change mitigation or other goals.

But our analytical framework also has its limitations. Most importantly, it cannot be used to optimise or tune innovation processes to produce intended innovation outcomes. Complexity, irreducible uncertainty, context dependence, and the incomplete understanding of key interrelationships make this – in our view – impossible. There are no precise formulations for models or prescriptive policy recipes to determine "if you want to achieve Y, do X."

The dynamic characteristics of innovation systems are also poorly characterised. In the initial stages of the innovation process, only a few actors are involved in developing a new technology. Over time, other actors enter, the knowledge base starts to grow, the legitimacy of the new technology increases, and more financial resources become available. Through this "formative phase," the innovation system around a new technology builds up. At some point the innovation system becomes large and developed enough for technology diffusion to take place during a "growth phase" (Jacobsson and Lauber, 2006). For new technologies that are incremental improvements to existing ones, innovation systems may already be in place. Mature innovation systems may be characterised by large firms with high R&D spending, strong networks between suppliers and consumers of the technology, established markets, and well-aligned institutional infrastructures. In contrast, innovation systems may need building up for radical innovations that strongly deviate from existing technologies and practices (Van De Ven, 1993). Innovation systems emerging around such technologies may be characterised by poorly developed markets, misaligned institutional settings, poorly structured knowledge networks, and small firms with limited resources to develop and market the new technology (Alkemade, Kleinschmidt et al., 2007).

Finally, data on which the energy technology innovation system framework is based are themselves partial, incommensurate, or otherwise limited. This is discussed in greater depth in various case studies, particularly those on global financial resources and RD&D investments in emerging economies. Though rich and detailed in certain areas, the evidence base is weaker in others. Understanding of innovation processes and linkages is incomplete and of insufficient depth. Feedbacks between components of the innovation system are a good example. Studying innovation from a systemic perspective in the energy domain is a relatively youthful endeavour, with an empirical bias towards national, sustainable, and energy supply technologies. Policy experiments and "field experience" are largely still ongoing, particularly in a U.S. and Northern European context which, together with Japan, provide many of the case studies.

Nonetheless, the energy technology innovation system framework outlined in this book offers a promising way forward. It can help to synthesise patterns and lessons from disparate case studies of success and failure in energy technology innovation, as exemplified in the twenty case studies included in this book. And it can guide future analysis and modelling efforts, as well as sorely needed data collection and synthesis.

REFERENCES

Alkemade, F., Kleinschmidt, C., et al., 2007. Analysing Emerging Innovation Systems: A Functions Approach to Foresight. *Int. J. Foresight Innov. Policy*, 3: 139–68.

Borup, M., Brown, N. et al., 2006. The Sociology of Expectations in Science and Technology. *Technology Analysis & Strategic Management*, 18(3–4): 285–98.
Edenhofer, O., Pichs Madruga, R., Sokona, Y., Seyboth, K., Matschoss, P., Kadner, S., Zwickel, T., Eickemeier, P., Hansen, G., Schlömer, S., & Von Stechow, C., 2011. *IPCC Special Report on Renewable Energy Sources and Climate Change Mitigation*. Cambridge, UK and New York, NY: Cambridge University Press.
Foray, D., 2004. *The Economics of Knowledge*. Cambridge, MA: MIT Press.
Garud, R. and Karnoe., P. 2003. Bricolage Versus Breakthrough: Distributed and Embedded Agency in Technology entrepreneurship. *Res. Policy*, 32: 277–300.
Jacobsson, S. and Lauber, V., 2006. The Politics and Policy of Energy System Transformation – Explaining the German Diffusion of Renewable energy technology. *Energy Policy*, 34(3): 256–76.
Junginger, M., Van Sark, W., & Faaij, A. (eds.) 2010. *Technological Learning in the Energy Sector: Lessons for Policy, Industry and Science*. Cheltenham, UK: Edward Elgar.
Van De Ven, H., 1993. The Development of an Infrastructure for Entrepreneurship. *Journal of Business Venturing*, 8(3): 211–30.
Van Lente, H. and Rip, A., 1998. The Rise of Membrane Technology: From Rhetorics to Social Reality. *Social Studies of Science*, 28(2): 221–54.
Wilson, C., Grubler, A. et al., 2012. Marginalization of End-Use Technologies in Energy Innovation for Climate Protection. *Nature Climate Change*, 2(11): 780–88.

25 Policies for Energy Technology Innovation

Arnulf Grubler and Charlie Wilson

1 Guidelines for Innovation Policy

Most of the case studies examined in this book explicitly discuss the role public policy has played in their respective innovation histories. The editors' guides in each case study distil key lessons learnt for policies to support innovation systems. In this final chapter, these lessons are drawn together and summarised under seven broad guidelines for designing effective policy to support energy technology innovation systems. These guidelines argue for policy that is:

- systemic;
- aligned and consistent;
- stable, continuous, but adaptive;
- supportive of experimentation and diversity;
- portfolio based;
- supportive of learning, knowledge feedbacks, and codification; and
- supportive of international spillovers and institutions.

These guidelines are necessarily generic. We have drawn on a diverse range of case studies to develop and then substantiate the energy technology innovation system framework. This demonstrates its broad and general applicability. But it also prevents specific policy recommendations, as policy must be tailored to the technological and market characteristics of an innovation as well as to local capabilities and capacities. Before explaining our policy guidelines in more depth, we provide a brief summary of the rationale and instruments available for public policy in the innovation system for energy technologies.

2 Rationale and Instruments for Energy Technology Innovation Policy

Private firms innovate to meet the perceived needs of energy technology users. Capital investments in the global energy market stand at around US$1 trillion per year in terms of energy supply, and – according to the case study on global end-use investments – conservatively in excess of US$4 trillion per year in terms of energy end-use technologies. The potential private returns to innovation in terms of novel or improved energy technologies and services are therefore vast. But companies tend to invest in innovation likely to bear fruit in the near term; more fundamental, basic

RD&D is less attractive (Rosenberg, 1990). This is especially true during times of economic turmoil or recession. Price volatility in the energy markets also leads to a "lumpy" pattern in which big investments are followed by precipitous declines. This undermines the innovation system's need for sustained and steady inputs (Gallagher, Holdren, and Sagar, 2006; see also the case study on knowledge depreciation).

One of the primary rationales for government involvement in the innovation system is to complement and leverage the private sector's efforts. Why? The overarching reason is that energy services contribute to meeting fundamental human needs, and improvement of those services can advance human welfare. This is seen most distinctively in developing countries where billions lack access to electricity or clean cooking fuels (GEA, 2012). In developed countries, government investment in energy innovation is more commonly justified to make energy supply more reliable and secure, to reduce pollution (including greenhouse gases), and to reduce the negative impacts of energy production and use on water and land resources.

Government policy also helps address market failures and remove barriers to innovation. The tendency for private firms to underinvest in RD&D is one such failure (Jaffe, Newell, and Stavins, 2005), and the role of government funding is perhaps clearest during the early R&D stages of the innovation life cycle. But government policies are also needed to correct for market externalities and ensure a level playing field across technologies (e.g., by removing subsidies or pricing pollution).

Governments are also engines of applied energy RD&D and play an important role in leveraging private sector investment by supporting demonstration and market formation activities. These help shepherd new technologies through the so-called valley of death between successful innovation and successful commercial deployment (Frank et al., 1996; Norberg-Bohm, 2000; Weyant, 2011). As shown in the case study on energy transitions, established energy technologies are supported by institutions and infrastructures with which they co-evolved to their positions of dominance (Unruh, 2000). This "lock-in" creates barriers to entry for new technologies, which are often initially more expensive yet may have market promise.

More general economy-wide policies on education, taxes and subsidies, and market regulations can also exert an important influence on the supply and demand of innovation. This reinforces the need for consistency not just between direct energy innovation policies, but also between the broader regulatory and institutional environments for innovation, and of energy markets in general.

Public policy can thus play important roles in supporting basic RD&D, in helping form markets for new technologies, and for levelling the broader market playing field tilted towards the incumbents. There are a wide range of innovation policy instruments to these ends that can be characterised in one of three ways:

- technology-push policies that support the supply of innovations;
- market-pull policies that support the demand for innovations;
- market formation policies that are a particular type of market-pull policy for creating or protecting market niches for the early deployment of innovations.

Policies supporting the supply of innovation or the development of technologies include investments in R&D, intellectual property protection, laboratory and testing infrastructure, training and skills development, financial incentives (such as tax credits for private investments), university-industry collaborations, formal and informal mechanisms of knowledge exchange, demonstration projects, and technology road maps to guide the direction of innovation efforts and to establish shared expectations. Although R&D activities are the clearest source of knowledge generation in the innovation system, problem-solving activities and incremental improvements to existing technologies are crucially important as well, and can be stimulated and supported by public sector policies that lead to the creation of outreach, extension, and technical support programmes.

Policies supporting the demand for innovations include appropriate market incentives created via changes in relative prices (e.g., environmental taxes), production guarantees, fixed price purchase contracts, production subsidies, production quotas tax incentives, standards, regulations (e.g., to streamline planning and permitting), infrastructure investments, or public procurement (Raven, 2007). Market formation policies can involve similar instruments but targeting specific market niches with identifiable end users. Markets for specific renewable technologies are good examples, created and protected from full price competition by policies including supplier obligations, portfolio standards, or guaranteed price purchase agreements (e.g., feed-in tariffs).

3 Innovation Policy That Is Systemic

The case studies repeatedly emphasise that supply-side "technology-push" policies and demand-side "market-pull" policies are complements, not substitutes. Successful innovation outcomes are typically characterised by comprehensive and consistent "packages" of policies that support many different elements of the innovation system and its effective functioning as a whole.

The case study on wind power shows how countries that initially conceptualised the innovation process as linear, driven by R&D and up-scaling to ever larger wind turbines, did not produce viable commercial applications. Conversely, successful countries not only recognised and supported the development of technologies and infrastructure, but also involved a diverse range of actors and institutions including the adopters and users of the technologies (risk-taking entrepreneurs, developers, farmers, and landowners, as well as utilities), manufacturers and testing stations, wind resource mapping, grid connection regulations, municipal planning and permitting, and information-based activities to support exchange and interaction between these various actors.

In the case study on ethanol in Brazil, both the initial ProAlcool Program from the 1970s and the renewed emphasis on ethanol and flex fuel vehicles in the 2000s combined market-pull measures to provide secure returns on investments in ethanol manufacturing and distribution (e.g., low-interest loans to mills and distilleries, blending mandates, guaranteed prices, preferential tax treatment) with technology-push measures to stimulate innovation (e.g., R&D on ethanol and later flex-fuel engines, research into sugarcane varieties, and yield improvements of crops and distilleries).

The case study on heat pumps similarly finds that policy interventions need to consider the development of the technology as well its emerging market. R&D is a necessary but not sufficient part of a systemic policy approach. Testing and certification processes to support technical quality and to build credibility and legitimacy provide a link between R&D initiatives and stable market formation.

The case study on solar photovoltaics provides rich empirical evidence for the multiple, interacting drivers of cost reduction, including R&D, demand subsidies, economies of scale in manufacturing, learning by doing in localised installation industries, and knowledge spillovers between and within industries. The clear lesson is that there is no one-size-fits-all innovation policy, nor a singular entry point through either technology-push or market-pull approaches. They are needed in combination. This is consistent with Watanabe and colleagues' (2002) evidence of a virtuous cycle between R&D, knowledge spillovers, market growth, and price reduction in the Japanese solar photovoltaic industry. Government programmes stimulated public and private R&D in a broad base of industrial sectors and simultaneously created incentives for niche market deployment. Knowledge stocks increased rapidly both from proprietary R&D and spillover benefits. Learning contributed cost reductions that further induced increases in demand and production, closing a positive feedback loop between knowledge generation and market formation.

The importance of systemic innovation policy can also be illustrated by counterexamples of partial or narrowly focussed policies used in isolation to target particular innovation stages, particular technology components, or particular elements of the innovation system. In the case study on synfuels, a singular emphasis on rapid commercial up-scaling without due regard to market conditions was shown to be unsuccessful. In the case study on solar water heaters, both R&D and market incentives placed excessive emphasis on solar collector units rather than on system integration. The overambitious market-pull policy framework also failed to address continuous improvements, reliability, and cost reductions in the heating systems, which ultimately left the industry vulnerable to policy discontinuity.

The systemic nature of innovation also cautions against the use of single metrics to assess innovation processes. Learning curve models have become widely used in public policy both to assess and to justify the magnitude of expenditure and policy initiatives to subsidise demand for new energy technologies. Yet such models omit key variables, have limited predictive power, and capture a limited set of the causal mechanisms linking dependent and independent variables within the innovation system (Van der Zwaan and Seebregts, 2004). Common misuses of learning rates include assumptions that policies can "buy down" technology costs or that the extent of learning-related cost reductions can be reliably forecast. Cost projections for novel technologies are an inherent element in policy analysis. But as the case study on nuclear power concludes, assumptions on the inevitability of future learning and cost reduction need to be firmly grounded within the historical observational space, reflecting both uncertainty and the possibility for cost escalation (i.e., negative learning). Variations in calculated learning rates as a function of different levels of spatial aggregation, observational period, or the use of price data as a proxy for cost information are examples of important uncertainties, as is the conflation of unit economies of scale with learning effects proper when analysing learning curves as a function of cumulative installed capacity or output. Future prices may not only be reduced as a function of learning but may also increase as a result of increasing input prices, quality improvements of the product, or lack of competition in quasi-monopoly markets especially under rapid demand growth.

The central lesson from the use and misuse of learning rates is that metrics offer a useful window into the functioning of energy technology innovation systems, but multiple metrics are needed for a comprehensive view. As the outcomes of energy innovation are many and varied, policy objectives should be clearly stated so appropriate metrics can be used to evaluate progress (and so contribute to policy learning).

The complexity of the innovation system means some degree of trade-off in a systemic policy framework is likely inevitable. The case study on synfuels illustrates well the tensions between managing expectations and building advocacy coalitions, between consistency and flexibility, between commercial scale projects and diverse portfolios, and so on. The remaining policy guidelines help navigate these tensions.

4 Innovation Policy That Is Aligned and Consistent

To function effectively, incentives within the energy technology innovation system must be aligned and consistent. Alignment means that policy signals to innovation actors stimulate innovation activities towards a shared, desirable outcome (e.g., pollution control, cost reduction). Consistency means that these innovation incentives permeate throughout the entire innovation life cycle from R&D to diffusion in a coherent way, and that policies and incentives put in place for particular technologies such as renewables or efficiency are not contradicted by policies continuing to favour incumbent industries and wasteful practices (e.g., fossil fuel subsidies). Both alignment and consistency are hallmarks of a systemic approach to innovation policy.

The case study on solar thermal electricity provides a good example of aligned and consistent policy, at least initially. Market-pull policies expanded market applications, which in turn enabled up-scaling of the technology and learning by doing, which reduced component failures. Technology-push policies including R&D investments (even with declining budgets) led to technological improvements in efficiency and other performance attributes as well as up-scaling of the technology. Combined, capital costs and operation costs declined. This virtuous innovation development cycle subsequently ended when public policy support was discontinued. (The need for a stable and continuous innovation policy framework is a separate policy guideline discussed further later in this chapter.)

The pervasive adoption of nuclear power in France was also enabled by a clear alignment between government and industry interests under conditions of policy and market stability that combined to help nuclear overcome the relative disadvantage of its high upfront costs. This was further helped by ÉDF's choice – as owner, construction engineer, and operator – to replicate a proven and standardised (imported) technology. Signals on technology reliability and the ability of the industry to manage costs were kept consistent with expectations. Alignment also implies the need for mechanisms for innovation actors to interact and exchange knowledge. In the French nuclear case, this was relatively easy given the close-knit, small network of principal actors.

Aligned and consistent innovation policy is critical for supporting technologies during the common points of failure, between successful innovation (R&D) and commercial testing, and between successful demonstration and commercial deployment (Weyant, 2011). This does not just mean innovation policy targeting particular stages. Consistency implies the need for market incentives to support shared expectations around a technology's future potential. Well-designed dynamic standards are one means of providing clear signals to drive innovation and technological change. The case studies on vehicle efficiency, on hybrid cars, and on end-use efficiency in Japan all illustrate the effectiveness of standards (which are discussed further later in this chapter).

If incentive structures are misaligned or inconsistent, tensions emerge within the innovation system, stifling innovation processes and undermining innovator and investor confidence. A common type of misalignment is between policy efforts to improve energy efficiency (e.g., through R&D investments or performance standards) while simultaneously subsidising the price of retail fuels (Morrow et al., 2010). The case study on hybrid cars shows the importance of aligned technology-push and market-pull policies. In China, the effectiveness of R&D and deployment incentives were weakened by misaligned market conditions including low fuel prices. In the United States, policy effectiveness was similarly undermined by a misalignment with weak and static fuel efficiency standards. In contrast, Japan's feebate scheme to address the upfront cost premium of hybrid electric vehicles aligned with the high fuel prices that underwrote rising market demand.

5 Innovation Policy That Is Stable, Continuous, but Adaptive

Uncertain expectations about future policies undermine the confidence of private investors in innovation. This is particularly acute in the energy system as environmental and knowledge spillover externalities are pervasive, and low-carbon and efficient end-use technologies typically have higher upfront investment costs made profitable by lower life cycle costs. Distant payoffs from often sizeable capital investments tend, therefore, to rely heavily on public policy. As innovation is itself a risky endeavour, private actors only respond to innovation policies that are stable, credible, and long-lasting (Bosetti and Victor, 2011; Nemet, 2010). The lack of credible, lasting legislative frameworks that transcend the short-term political cycle increases policy risks and diminishes the potential profitability of low-carbon investments (Scrieciu, Barker, Ackerman, 2013). As Neuhoff and colleagues (2009) observe: "a significant share of the risk associated with low-carbon investments relates to the stability of the regulatory framework."

This important insight repeats itself throughout the case studies. The case study on heat pumps concludes with the recognition that strategic, long-term, and continuous support for energy technology innovations is critical. Initial attempts to introduce heat pumps in Sweden and Switzerland largely failed, and continuous support was needed to overcome technical and market shortcomings. The case study on wind power describes the long and bumpy history of wind turbine development. Innovation successes, such as in Denmark, have been built on stable and continuous support that has sustained the knowledge generated and learnt along the way. The case study on solar photovoltaics emphasises that the investments needed to grow an industry's manufacturing base to capture available manufacturing scale economies and so bring down costs depend on investors and other market actors sharing clarity and confidence in the long-term future of market demand. R&D support similarly needs a long-term commitment, whether through budgets or grants spanning multiple years, or supporting policies as in the cases of Japan's Sunshine Program or the United States' Project Independence, both policy programmes that elevated PV developments to national priority status. If R&D and market demand is created or shaped by public policy, it is critical that this policy is stable, credible, and visible to allow shared expectations within an industry to coalesce. Subsidies in particular are often unsustainable over successive economic and election cycles. A notable exception is Brazilian ethanol.

The singular lesson of the Brazilian ethanol story is the importance of a continuous, consistent, long-term policy framework. This has endured through four turbulent decades and weathered both endogenous difficulties (e.g., technical problems of ethanol-only vehicles, industry consolidation) and exogenous shocks (e.g., oil price volatility, political regime change). This persistence and longevity was underpinned by the coalition of interested stakeholders built to support the ethanol programme. Stability and continuity did not, however, come without cost. Estimated ethanol production subsidies reveal the magnitude of resource mobilisation made necessary not just by the ambition of a national-scale programme of industrial capacity expansion and end-use market transformation, but also by its being sustained through extended periods of clear relative disadvantage (caused by falling or relatively low oil prices).

An alternative way for policies to ensure stability is by including transparent time horizons from the outset, as long as these are credible. Long-term commitments with loopholes for governments to avoid meeting those commitments should compliance prove more difficult than expected may create flexibility but pay the price in terms of credibility and reduced incentives for investment in innovation.

Patience is therefore a virtue, particularly in innovation systems. Pervasive change takes time; experimentation and market formation take time. Time is an important resource, and a non-substitutable one. The potential for large investments in financial resources to short-circuit or accelerate the innovation processes behind the observed diffusion patterns appears limited. The case study on technology diffusion shows how formative phases during which the technological and institutional capacity necessary to sustain diffusion is built typically extend over decades. An important implication is that market-pull policies to support deployment and associated learning processes to reduce costs will likely have to be sustained over very long periods with the large resource commitments this implies.

Neither continuity nor stability simply mean static policy incentives or mandates. Policy can adapt to changing technological and market developments as the case study on ethanol in Brazil shows. Innovation systems develop and change alongside maturing technologies. The case study on heat pumps shows how initial interventions to support entrepreneurial testing need to develop into credible, stable, and transparent strategies to give industry the confidence to make long-term investments in growing the manufacturing base.

In the case study on solar photovoltaics, the main drivers of cost improvements shift from R&D to market demand and shared expectations and then to learning. And as this sequence continues for the original crystalline silicon technology, it potentially begins again for newer thin film technologies. In other words, the timing and target of innovation policy must be sensitive to differences in the maturity of potentially overlapping technological generations. The success of new technological generations may require renewed R&D support even while markets for existing technologies are expanding.

The case study on hybrid cars provides an interesting example of public policy playing catch up. In Japan and the United States, public investments in the RD&D of advanced-vehicle technologies were initially poorly coordinated with market deployment incentives. But Japanese policy responded more rapidly to the commercial introduction of hybrid electric vehicles, providing incentives for market deployment that were sensitive to consumer preferences and the cost premium that disadvantaged hybrid engines. The first mover success of Japan is attributable in part to this adaptive response of policy to changing market conditions.

Standards are effective at establishing clear expectations around future market demand, but should be designed from the outset to respond dynamically to technological improvements. Static standards that remain in place over long periods (like the CAFE standard in the United States) can act as a disincentive for continual performance improvements. The case study on end-use efficiency in Japan offers a counterexample. Not only were the Top Runner standards transparently designed to be periodically tightened (through regular two to three yearly reviews), but they also applied to manufacturers' product range on average, allowing a degree of product diversity consistent with heterogeneous end-user preferences.

The formulation, implementation, and enforcement of standards (and of public policy more generally) also requires appropriate institutional capacity and the resources associated with building and maintaining this capacity. This is highly salient in the case study on rural solar in Kenya. Policy instruments such as standards, as well as the mechanisms and institutions to enforce them, need continuous support. One-off efforts to establish compliance may not have enduring effects. The case study on solar water heaters similarly argues that the continuous verification of deployed technologies is inexpensive yet essential to sustain legitimacy and market development.

Innovation policy can be dynamically adjusted not just in stringency but also in scope. In the case study on end-use efficiency in Japan, the initial series of nine end-use products covered by the Top Runner Program has been expanded to twenty-three products. The expansion of policy "tested" on narrowly targeted initial markets is facilitated by learning spillovers, particularly if the same market actors are involved.

The importance of stable, continuous policy frameworks can be further illustrated through counterexamples. Discontinuous, boom-and-bust cycles in policy support and resulting market activity are damaging to innovation systems, with the U.S. synfuels programme providing a particularly cautionary tale. As the case study on solar water heaters shows, downturns lead to knowledge depreciation through the loss of experience and tacit knowledge that accompanies firms and skilled workers leaving the industry. This can be mitigated by codifying knowledge to the extent possible and by nourishing diverse networks and relationships between innovation actors.

The case study on knowledge depreciation analyses the particular effect of discontinuous public R&D investments, particularly in contexts that threaten knowledge preservation (e.g., industries with rapid innovating cycles or high staff turnover). The pattern of resource mobilisation over time proves as important, if not more so, than the absolute levels of resources mobilised. Continuous knowledge recharge through sustained R&D investments is needed to counter knowledge depreciation. As a result, stable, gradually rising levels of policy support outperform larger-scale boom-and-bust cycles.

An important cause of discontinuity in policy support are changes in the market conditions in which innovation policies were originally forged. The exemplar is the wave of public investments in alternatives to fossil fuels following the oil price escalation and volatility in the 1970s that were subsequently challenged by the sustained decline in oil prices through the 1980s. Policy support removed from fledgling industries led to severe and long-lasting consequences. In the case of solar thermal electricity, for example, the plants built in the 1980s were technically successful. They steadily improved in performance and cost, are still providing power two decades after they were installed, and provide the basis for newly designed plants in Spain and California. The company that built them failed, however, because of falling energy prices and consequent changes in policy. The legacy of failure resulting from this policy intermittency has endured. In the case study on solar water heaters, a large, high-profile failure in the early stage of the industry has also proven extremely difficult to overcome. Trust in the technology's performance and potential was very slow to rebuild following a sharp downturn in market activity and associated knowledge depreciation. This perceived innovation failure further undermined policy support.

Various of the case studies illustrate how stable and continuous policy can avoid these legacies of failure even in the event that market conditions change. The Brazilian ethanol story describes a comprehensive policy framework continually adapting to changing conditions, being resilient even against oil price downturns. The huge mobilisation of resources that this implied was encouraged by a strong advocacy coalition among the main innovation actors. Strengthening networks of actors also proved important in the heat pump industry following the oil price declines of the 1980s. In the case study on heat pumps, policies that supported exchange and interaction between actors helped sustain the industry through the downturn. Finally, the case study on solar photovoltaics explains how a diverse set of market niches with differing demands for specific performance advantages can also help innovation systems weather storms of changing relative prices. This is discussed further in the next guideline for innovation policy to support technological as well as market diversity.

6 Innovation Policy That Is Supportive of Experimentation and Diversity

The case study on technology diffusion finds a consistent pattern of experimentation with many small units as a precursor to widespread successful diffusion. Experimentation to generate knowledge on a technology's performance, efficiency, reliability, and other service attributes enhances the capacity required to support other innovation processes including learning (Wilson et al., 2012). Governments can support experimentation in numerous ways. They can underwrite small-scale demonstration projects for innovations with less immediate or higher-risk private returns. They can support variety in the early deployment phase by creating and protecting differentiated niches. And they can manage the natural commercial tendency to rapidly confirm a dominant design that confers market advantages and potential cost benefits through scale economies but in the process loses technological variety.

Experimentation is most characteristic of the market formation stage of the innovation life cycle. The case study on wind power shows the importance of early market formation efforts by the Danish government for small-scale wind turbines. Policy support for diversity in both technology development and market formation activities proved a critical ingredient of Denmark's success. The case study on solar thermal electricity illustrates the importance of niche markets for developing the necessary capacity to construct, operate, maintain, and assure quality control of a new technology. Niche markets are protected in some way from full commercial competition. Policy instruments to support niche markets include public procurement, standards, production incentives, or market regulations. In the case study on solar photovoltaics, niche markets are a critical feature of the technology's innovation history. Many of these niche markets were enhanced and protected by public policy, but not bankrolled by it as is the case with feed-in tariffs or other procurement policies. This meant solar PV has had to rely on privately appropriable performance advantages that have proven more enduring, and so has been less exposed to the kind of stop-start policy attention that characterises the innovation histories of solar water heaters and solar thermal electricity (see earlier discussion on stability and continuity).

The need for experimentation in a diverse range of market niches links to the importance of tolerating failure. Failure is an inherent feature of a diverse portfolio of innovation experiments. Venture capitalists build energy technology portfolios with an expected 90 percent failure rate, knowing that the 10 percent that break through will support returns for the portfolio as a whole. Accountability for taxpayer monies and the associated political risks of funding failures makes public innovation policies less tolerant. Yet energy innovation portfolios should not be judged negatively if they include some relative failures so long as they also produce relative successes (Anadón et al., 2011; Scherer and Harhoff, 2000).

Concentrating public resources on single technologies magnifies the consequence of failure, particularly if this requires rapid up-scaling (as the case study on technology diffusion shows). In contrast, building diverse portfolios of modular or smaller-scale technologies helps spread financial losses in the event of failure. Experimentation thus supports granularity: individual eggs in many small baskets. Granularity means smaller-scale, modular, dispersed, plentiful, multifarious technologies or projects with low capital requirements per unit. Compared to large-scale, "lumpy," indivisible, one-off, periodic investments, granularity dramatically reduces the consequences of innovation failure. Granularity mitigates innovation risks and lends itself to the portfolio character of innovation policy.

This in turn suggests a natural preference in innovation portfolios towards end-use technologies. The case study on energy transitions suggests that the capital intensiveness of many energy supply technologies magnifies the consequence of innovation failure (foregone investments) and reduces the pool of actors (particularly investors) who can take risk and drive change. A common pitfall of innovation policy is to push for the early up-scaling of technologies in an effort to bring down costs by capturing scale economies. The case study on knowledge depreciation finds ambitions to up-scale technologies or grow industries by pushing rapid knowledge generation are associated with many innovation risks, of which knowledge depreciation is one. The case studies on solar thermal electricity and on synfuels, both in the United States, illustrate how the consequence of failure with energy supply technologies extends beyond the foregone investment to create a more lasting legacy of failure. The early wind turbine design failures in the United States and Germany tell a similar story.

Nevertheless, public policy can still play an important catalyst role to leverage investment for large-scale projects to the extent that markets inadequately provide financing mechanisms to share or securitise upfront capital demands and associated risks. In the case of solar thermal electricity, firms knew they would need to absorb losses on the first few plants but this was overcome by financial investors' interest in alternative energy in the early 1980s, supported by guaranteed tariffs (market-pull) and capital cost subsidies (technology-push).

7 Innovation Policy That Is Portfolio Based

Innovation outcomes are irreducibly uncertain. Experimentation and learning in diverse market niches help reduce this uncertainty and link to the need for diverse innovation portfolios. Following the other policy design guidelines implies portfolios that include technologies across the whole energy system and that spread investments across many projects and investments. Portfolios should similarly encompass all salient elements of the technology development cycle and different channels of technological knowledge creation, including R&D, demonstration, and niche market deployment (Fuss, Szolgayov, et al., 2012). The case studies offer various cautionary tales to warn governments against trying to picking technology winners. The case study on synfuels shows how rapid, large-scale mobilisation of resources behind a single technology could not solve all technological issues nor be sustained. The case study on hybrid cars tells indirectly of how the initial efforts of governments to pick alternative-vehicle technologies (electric vehicles, hydrogen fuel cells) did not bear fruit. In this case, however, R&D investments were still of benefit to the eventual success of hybrid electric vehicles through knowledge spillovers, particularly at the component level. (Innovation policies to support knowledge spillovers are discussed further later in this chapter.) General purpose technologies with inherent flexibility, versatility, or substitutability in both resource inputs and energy service "outputs" are a central feature of optimal innovation portfolios. Electricity generation has been a historic exemplar and is expected to play an ever more important future role.

Portfolios will inevitably be resource constrained. Their design needs to resolve the inherent tension between maintaining technological diversity and improving economics through standardisation, scale, and concentration. Methods and tools are increasingly available to support the design of innovation portfolios under conditions of uncertainty and with due consideration of multiple policy objectives. Some of these are set out in the case study on technology portfolios (see also McCollum, Krey, and Riahi, 2011). On the one hand, innovation policies need to avoid preempting the outcome of decentralised market-based technology innovation, experimentation, and early market deployment decisions that are key in technology development. On the other hand, innovation policy should legitimately counterbalance private innovation preferences against large-scale, investment-intensive technologies to the extent these are important in addressing broader social and environmental goals. Although resource limitations may require a focus on a few strategic technologies, there is a downside if pressures to concentrate on capital-intensive innovation projects result in the public sector disproportionally shouldering innovation risks. The resulting focus on potential innovation "lemons" – if they fail – leaves a lasting legacy and serves to undermine the legitimacy and credibility of public technology policy intervention.

The design of diversified portfolios is more feasible if they are weighted to favour granular, less capital-intensive technologies, such as end-use innovations and smaller-scale supply options. Single, "lumpy," large-scale investments in technologies at scale (invariably energy supply technologies) exhaust innovation resources and expose resulting portfolios to the risk of failure. Such technologies are more appropriately considered in *global* innovation portfolios and collaborations (as is the case with nuclear fusion) rather than at the national scale. The case study on nuclear power reveals the issues and risks associated with up-scaling large, complex technologies. This was managed initially in the French market by the very small number of interlinked actors with clearly aligned interests. It becomes more problematic in more common deregulated markets characterised by multiple actors, and fragmented design, construction, and operation functions provided by diverse actors.

Various case studies identify a bias in public innovation efforts towards energy supply technologies (see also Wilson and colleagues (2012) and the previous chapter). The case study on technology portfolios finds innovation investments (public R&D) as well as the many institutions comprising the innovation system to be strongly weighted towards energy supply technologies (particularly nuclear fission) at the expense of end-use efficiency. The case studies on energy transitions and on global end-use investments find that forward-looking scenario analysis focuses on changes in the energy supply given the difficulties of developing scenarios and modelling future changes in end-use services. These important knowledge-generation (foresight) activities further skew the emphasis of innovation policy towards the supply side. Yet end-use technologies dominate market investments and opportunities for efficiency improvements, and so determine the size and composition of the energy supply. Technology portfolios need rebalancing accordingly using transparent and clear portfolio design criteria (Wilson et al., 2012).

8 Innovation Policy That Is Supportive of Learning, Knowledge Feedback, and Codification

The energy technology innovation system is founded on knowledge generation and flows. As discussed in the next policy guideline, these are increasingly global, yet adapted and applied to local conditions. The generation of knowledge requires independent and stable institutions to balance the competing needs and interests of the market, of policy makers, and of the research and development community. The technology roadmaps and policy regime that the case study on end-use efficiency in Japan describes are good examples.

Knowledge generated needs to spread through the innovation system. Knowledge flows and feedbacks create and strengthen links between different actors. This can take place formally or informally. Examples range from conferences and exchange programmes to reporting protocols, demonstration projects, and industry consortia. Feedback processes are essential for sustained and successful energy innovation (Lester and Hart, 2012). Public policy can support these feedbacks, but can also hinder or even block information and knowledge flows (Kiss and Neij, 2011; Nemet, 2012). The case study on wind power provides a good example of how a policy-supported institution (certification and testing at the Risø test station) supported knowledge feedback from technology demonstration back to R&D. The case study on heat pumps emphasises the dependency of knowledge feedback and spillovers on public policy support for networking to improve the strategic interactions between actors in the innovation system. For other technologies, early experience in production and use (including operation and maintenance) has been fed back into R&D and design changes. The case studies on wind power, solar photovoltaics, and rural solar in Kenya all illustrate the importance of feedback from commercial experience to redesign and engineering, linking market incentives with R&D efforts.

Codification of knowledge generated, particularly during the early growth years of an industry, can help preserve knowledge stocks during slower or downturn years. Careful codification is rarely a priority during the boom years of an infant industry (with patent applications an exception), yet can be written into policy instruments as a quid pro quo of public investment or support. The case study on solar thermal electricity provides a useful example. The codification of knowledge in the form of publicly available, nonproprietary documentation enabled knowledge to be shared and exchanged as the technology matured, and played an important role in preserving knowledge stocks through subsequent periods of industry turnover and shakeouts. The case study on synfuels reaches similar conclusions.

Next to codification, human capital management to prevent rapid staff turnover is key for minimising volatility-driven knowledge depreciation. Innovation policy can therefore help sustain knowledge stocks indirectly through training, skills development, exchange and interaction activities, and, most of all, through stable and sustained policy frameworks (including funding and incentives) in which market actors have confidence.

9 Innovation Policy That Is Supportive of International Spillovers and Institutions

Energy innovation is international. Many of the innovation histories analysed in the case studies centre on energy technologies born and developed as a result of experience and learning that crossed borders.

International policy creates incentive frameworks for the wider diffusion of energy technologies. This includes treaties (e.g., Kyoto Protocol), norms (e.g., ISO technical standards), and institutions regulating trade, finance, investment, environment, development, security, and health issues (e.g., the IEA, World Trade Organization, and the World Bank). The financial requirements for acquiring hardware, machinery, and equipment are a central aspect of international technology diffusion, especially in capital-intensive, large-scale energy technologies. Currently, however, international financial schemes and institutions reinforce the dominance of incumbent technologies by screening out investment allocations to cleaner energy sources, local R&D efforts, and knowledge infrastructures.

International knowledge spillovers through government-sponsored collaboration efforts are insufficient given the challenge of global sustainable energy transition (GEA, 2012). Non-IEA countries have been progressively investing in and developing capabilities in the earlier stages of development of new technologies. The case study on ethanol in Brazil provides the headline example, but as the case study on RD&D in emerging economies clearly shows, this is far from the only example. BRIMCS countries now account for a significant proportion of the global energy technology innovation system although significant regional imbalances persist. The increasing globalisation of energy innovation points to the need for new mechanisms and institutions to support international technology cooperation and coordination. Existing institutions such as the International Energy Agency (IEA) are limited in their scope and membership. The IEA provides support for numerous international cooperation and collaboration agreements in energy technology R&D, deployment, and information dissemination (IEA, 2012). However, many are not really R&D collaboration projects but simply institutional arrangements for information exchange or standardisation. Moreover, the participation of developing economies in the IEA's Implementing Agreements for energy technology collaboration is low (Grubler et al., 2012).

Energy technology is mostly diffused through private means, routinely flowing through foreign investment, licensing agreements, and international trade, often within multinationals (UNCTAD, 2002). The case studies on wind power and on rural solar in Kenya provide good examples of international knowledge spillovers from early to later adopting markets. In the wind case, this is enshrined in the technology development joint venture between Spanish and pioneer Danish manufacturers that led to the rapid growth of the wind market in Spain through the 2000s. Technology transfer arrangements are similarly now evident in China and India, although both countries have mobilised significant resources in domestic market formation activities including the development of local institutions and manufacturing capabilities. In the case study on rural solar in Kenya, knowledge spillovers to later markets increase the scale of market demand and so indirectly increase the leverage the Kenyan government could exert over manufacturers and suppliers to ensure product quality.

The case study on technology diffusion shows how international knowledge spillovers from early to later adopting regions is associated with an acceleration in the diffusion of energy technologies but with lower penetration rates. This is because the effectiveness of all the different modes of knowledge transfer depends critically on firm strategies and on local assimilation capacities (Cohen and Levinthal, 1989). Local policies are necessary complements of policy incentives for international learning and spillover, as local institutional capacity is necessary to absorb and use technology and knowledge produced abroad. A narrow emphasis on knowledge spillover and appropriation through technology transfers is therefore insufficient to guarantee successful diffusion. Local knowledge is needed to import or manufacture, adapt, install, and above all, *use* a new energy technology effectively. The case study on nuclear power in France illustrates the importance of local technological capability in the successful implementation and subsequent scaling up of imported reactor designs. The case study on rural solar in Kenya shows the problems that can arise if local capacity for testing and evaluating technical characteristics (and defects) of imported PV cells are unavailable.

Table 25.1. Illustrative technology innovation and diffusion policies, matching policy approaches to realistic timescales of outcomes and needs for capital stock additions and/or turnover. Source: adapted from Wilson and Grubler (2011).

Timescale of policy outcome	Examples of policy approaches
short-term (e.g., to 2020) capital stock additions (some)	• create, stimulate, and protect market niches around performance advantages of new technologies • deploy market-ready clean technologies through credible and stable incentive mechanisms • develop long-term technology innovation and market deployment strategies in a consultative process to create shared expectations • reduce/eliminate direct or indirect subsidies for technologies not aligned with long-term technology strategy and portfolios • set "sunset" clauses for planned retirement of depreciated inefficient or polluting capital vintages • strengthen the self-organisation of emerging industrial sectors around new technologies by facilitating network building and alignment of interests • provide cheap capital for emerging industrial sectors to finance the scaling up of production facilities.
medium-term (e.g., to 2050) capital stock additions (all), capital stock turnover (some)	• stably expand public and private R&D investments in diversified portfolios designed to manage risks and corresponding with end-use needs (e.g. with a 50 percent share in resource conservation/ efficiency and a diversified "clean" technology portfolio) • underwrite many, granular, and multifarious technology demonstration and learning cycles • support disclosure, interaction, and feedback between innovation actors • engage in multiple international collaborative projects that further knowledge dissemination and technology spillovers • align innovation and market deployment incentives, for example, in recycling externality price revenues to R&D and market deployment incentives • invest in renewable energy education programmes to train the required engineers of the future
long-term (e.g., to 2100) capital stock additions (all), capital stock turnover (all)	• set long-term targets with appropriate monitoring and enforcement mechanisms to sustain shared technology expectations • maintain portfolio diversity to prevent premature lock-in or standardisation • set technology standards for the gradual phase out of "bridging" technologies
throughout (from now to 2100)	• create and nurture formal and informal institutional settings for technology assessment, evaluation, portfolio design, and knowledge sharing

10 Innovation Policy – Conclusions

Energy technology innovation is key for addressing the multiple challenges for global energy systems ranging from energy access, development, security, and human health to environmental protection. History provides powerful examples of both innovation success and failure and the pervasive effects of these innovations on energy systems, human welfare, the economy, and the environment. Yet innovation policy cannot be complacent. The world today is very different from the world in which the past successes of energy technology innovation unfolded. It is globalised, stretched to its limits, and quickly running out of time in responding to the numerous energy challenges in an adequate and timely fashion. A continuation of "business as usual" fragmented, erratic, and isolated national energy technology innovation policies is clearly not an option.

The most important conclusion emerging from this book is that we need not only more energy technology innovation, but new, *systemic policy frameworks* to better harness the potential of innovation in meeting social, economic, and environmental objectives (the basic definition of sustainable development). More important than just *more* innovation (resulting in *more* technology options) will be an improved innovation "ecosystem" nurtured by policies that embrace the systemic nature of innovation set out in this book. A "next" energy transition towards sustainability is both long overdue and achievable. Illustrative strategies as they would have to unfold over the next century are illustrated in Table 25.1. Key novel concepts in these strategies are: the emphasis on international knowledge flows; the need to nurture collective learning processes among diverse actors; improved institutional capability for innovation policy formulation, implementation, and coordination; a focus on innovation portfolios with larger emphasis on more "granular" innovations, particularly in energy end use and efficiency; and lastly, active policies for retirement of older, polluting technological vintages to leverage the effect of new innovations introduced in the energy system. With such new and improved policy frameworks the much needed next energy transition can be unleashed and sustained.

REFERENCES

Anadón, L. D., Bunn, M., Chan, G., Chan, M., Jones, C., Kempener, R., Lee, A., Logar, N., & Narayanamurti, V., 2011. *Transforming U.S. Energy Innovation*, Cambridge, MA: Belfer Center for Science and International Affairs, Harvard University.

Bosetti, V. & Victor, D. G., 2011. Politics and Economics of Second-Best Regulation of Greenhouse Gases: The Importance of Regulatory Credibility. *The Energy Journal*, 32(1): 1–24.

Cohen, W. M. & Levinthal, D. A., 1989. Innovation and Learning: The Two Faces of R&D. *Economic Journal*, 99(397): 569–96.

Frank, C., Sink, C., Mynatt, L., Rogers, R., & Rappazzo, A., 1996. Surviving the "Valley of Death": A Comparative Analysis. *J. Technol. Transf.*, 21(61–68).

Fuss, S., Szolgayov, J., Khabarov, N. & Obersteiner, M., 2012. Renewables and Climate Change Mitigation: Irreversible Energy Investment under Uncertainty and Portfolio Effects. *Energy Policy*, 40: 59–68.

Gallagher, K. S., Holdren, J. P., & Sagar, A. D., 2006. Energy-Technology Innovation. *Annual Review of Environment and Resources*, 31(1): 193–237.

GEA, 2012. *Global Energy Assessment – Toward a Sustainable Future*, Cambridge, UK and New York: Cambridge University Press and Laxenburg Austria: The International Institute for Applied Systems Analysis.

Grubler, A., Aguayo, F., Gallagher, K., Hekkert, M., Jiang, K., Mytelka, L., Neij, L., Nemet, G., & Wilson, C., 2012. Chapter 24 – Policies for the Energy Technology Innovation System (ETIS). *Global Energy Assessment – Toward a Sustainable Future*. Cambridge, UK and New York: Cambridge University Press and Laxenburg Austria: The International Institute for Applied Systems Analysis.

IEA, 2012. *World Energy Outlook. Paris:* International Energy Agency.

Jaffe, A. B., Newell, R. G., & Stavins, R. N., 2005. A Tale of Two Market Failures: Technology and Environmental Policy. *Ecological Economics*, 54(2–3): 164–74.

Kiss, B. & Neij, L., 2011. The Importance of Learning when Supporting Emergent Technologies for Energy Efficiency – A Case Study on Policy Intervention for Learning for the Development of Energy Efficient Windows in Sweden. *Energy Policy*, 39: 6514–24.

Lester, R. & Hart, D., 2012. *Unlocking Energy Innovation: How America Can Build a Low-Cost, Low-Carbon Energy System*. Cambridge, MA: MIT Press.

McCollum, D., Krey, V., & Riahi, K., 2011. An Integrated Approach to Energy Sustainability. *Nature Climate Change*, 1: 428–29.

Morrow, W. R., Gallagher, K. S., Collantes, G., & Lee, H., 2010. Analysis of Policies to Reduce Oil Consumption and Greenhouse Gas Emissions from the U.S. Transportation Sector. *Energy Policy*, 38(3): 1305–20.

Nemet, G. F., 2010. Robust Incentives and the Design of a Climate Change Governance Regime. *Energy Policy*, 38(11): 7216–25.

2012. Subsidies for New Technologies and Knowledge Spillovers from Learning by Doing. *Journal of Policy Analysis and Management*, 31(3): 601–22.

Neuhoff, K., Fankhauser, S., Guerin, E., Hourcade, J. C., Jackson, H., Rajan, R., & Ward, J., 2009. *Structuring International Financial Support to Support Domestic Climate Change Mitigation in Developing Countries*. Cambridge, UK: Climate Strategies.

Norberg-Bohm, V., 2000. Creating Incentives for Environmentally Enhancing Technological Change: Lessons from 30 Years of U.S. Energy Technology Policy. *Technological Forecasting and Social Change*, 65(2): 125–48.

Raven, R., 2007. Niche Accumulation and Hybridisation Strategies in Transition Processes towards a Sustainable Energy System: An Assessment of Differences and Pitfalls. *Energy Policy*, 35(4): 2390–2400.

Rosenberg, N., 1990. Why Do Firms Do Basic Research with Their Own Money? *Research Policy*, 19(2): 165–74.

Scherer, F. M. & Harhoff, D., 2000. Technology Policy for a World of Skew-Distributed Outcomes. *Research Policy*, 29(4–5): 559–66.

Scrieciu, S. Ş., Barker, T., & Ackerman, F., 2013. Pushing the Boundaries of Climate Economics: Critical Issues to Consider in Climate Policy Analysis. *Ecological Economics*, 85: 155–65.

UNCTAD, 2002. *World Investment Report 2002: Transnational Corporations and Export Competitiveness*. UN Conf. Trade Dev (UNCTAD), Geneva.

Unruh, G. C., 2000. Understanding Carbon Lock-In. *Energy Policy*, 28(12): 817–30.

Van der Zwaan, B. & Seebregts, A., 2004. Endogenous Learning in Climate-Energy-Economic Models – An Inventory of Key Uncertainties. *International Journal of Energy Technology and Policy*, 2(1–2): 130–41.

Watanabe, C., Griffy-Brown, C., Zhu, B., & Nagamatsu, A., 2002. Inter-Firm Technology Spillover and the "Virtuous Cycle" of Photovoltaic Development in Japan. In A Grubler, N. Nakicenovic, & W. Nordhaus (eds.) *Technological Change and the Environment*. Washington, DC: Resources for the Future.

Weyant, J. P., 2011. Accelerating the Development and Diffusion of New Energy Technologies: Beyond the "Valley of Death." *Energy Economics*, 33(4): 674–82.

Wilson, C. & Grubler, A., 2011. Lessons from the History of Technological Change for Clean Energy Scenarios and Policies. *Natural Resources Forum*, 35: 165–84.

Wilson, C., Grubler, A., Sims Gallagher, K., & Nemet, G. F., 2012. Marginalization of End-Use Technologies in Energy Innovation for Climate Protection. *Nature Climate Change*, 2(11): 780–88.